Hertfordshire

Perfumes, Cosmetics and Soaps

VOLUME III Modern Cosmetics

W. A. POUCHER

Revised by

GEORGE M. HOWARD

Seventh Edition

London

CHAPMAN AND HALL

First published 1923
Second edition 1925
Third edition 1928
Fourth edition 1932
Fifth edition 1936
Sixth edition 1941
Seventh edition 1959
Reprinted once
Eighth edition 1974
Reprinted 1976 and 1979

Chapman and Hall Ltd
11 New Fetter Lane London EC4P 4EE
Printed in Great Britain by
Spottiswoode Ballantyne Ltd.,
Colchester and London

ISBN 0 412 10660 4

Distributed in the U.S.A.
by Halsted Press, a Division
of John Wiley & Sons, Inc., New York

Library of Congress Catalog Card Number 74-8882

UNIFORM WITH THIS VOLUME

W. A. Poucher's
Perfumes, Cosmetics and Soaps

Volume I	The Raw Materials of Perfumery
Volume II	The Production, Manufacture and Application of Perfumes

List of Plates

Contents

Preface to the Eighth Edition

W. A. Poucher's original 'Perfumes, Cosmetics and Soaps', first published in 1923, were the first comprehensive text books to deal exclusively with the work of the perfume and cosmetic industry, and received recognition and approval throughout the world. Their success and popularity was due primarily to the fact that they provided a practical guide not only to those engaged in the industry, but to a section of readers interested in the mystique and romance which at that time was associated with the perfume and cosmetic arts.

It was also an outstanding feature that the books were easily readable, subject matter being expressed in a clear and understandable fashion without the frills of pseudo-science or advanced technology.

In revising Volume III and editing Volume I it has been my aim to continue in the pattern and style of their author, and to this end I have not intended to write a new book, but have attempted a true revision, and it is a remarkable fact that many of the original writings made nearly 50 years ago can still apply to an industry which since that time has developed beyond recognition.

Two chapters are omitted from the original work—'Smelling Salts' which are no longer included in the range of perfumed toiletries, and 'Theatrical Requisites'—formerly specialized products but now largely replaced by less greasy modern make-up suitably pigmented to satisfy their particular function. These are replaced by new chapters dealing with modern aspects of depilation, and currently fashionable antiperspirants and deodorants.

An introductory chapter on aerosols is also included, and individual aerosol products appear under the relevant chapter headings. There are also sections dealing with the structure of skin and hair which should be useful to formulators concerned with the preparation of functional cosmetic products.

The book does not contain a collection of miscellaneous formulae—just as Poucher's original work provided practical information and usable formulations, this revised edition contains tried and tested formulae which will produce completed saleable products or as guide lines to the creation of individual specialities. This revision is not intended to contribute to the technical and scientific areas of investigation and research with which the cosmetic chemist is concerned, but as a readable and practical guide book containing something of interest to all those engaged in the perfume and cosmetic industry—the student, the cosmetic and beauty practitioner, and the skilled and practising perfumer and cosmetic chemist.

If I have been successful I will have achieved my objective to follow in the steps of the Master.

G. M. Howard

CHAPTER ONE

Aerosols

An aerosol may be defined as a pressurized product of any type which is self-propelled through a valve to produce any type of spray or foam. Aerosol products can be elegantly packed, are handy to use, and there is no problem of spillage or of contamination because the contents are not exposed to the air.

A type of aerosol was first developed as a system of insect control by the United States Department of Agriculture during World War II. It was discovered that insecticides required for use in the Far East were much more effective when dispersed in a fine dispersion or spray. This resulted in the development of the 'bug bomb' consisting of a solution of insecticide mixed with difluorodichloromethane then in use as a refrigerant, and later to be used as a propellent. The mixture was packed in a heavy welded steel container and this was forced through a valve by the vapour pressure of the low boiling liquid and the contents vaporized as a fine dispersion or spray. The principle was applied commercially about 1947 when propellent liquids of lower vapour pressure were developed, and permitted the aerosol to be packaged in light metal or even glass containers. From this period of time there has been an extremely rapid development of products generally referred to as aerosols, but which are often more correctly defined as pressure packs.

The rapid rise in aerosol production in North America occurred in certain definable product categories from approximately 400 million units in 1957 to over 2500 million units in 1970, and this same pattern appears to be following closely in Western Europe some five or six years later, indicating that a similar production figure of 2500 million units can be expected in Western Europe by 1975.

Hair sprays are the largest single product group with the most rapid rate of growth, followed by toiletries and cosmetics, and household products. The range of products of interest to the cosmetic chemist can be conveniently divided into the following categories:

Liquid preparations:
- perfumes, colognes,
- hair sprays,
- hair setting lotions and colourants,
- antiperspirants and deodorants,
- brilliantine oils, shaving lotions,
- sunscreen oils.

Emulsified products:
- hand creams, and body lotions,
- hair dressings, and conditioners,
- shaving creams,
- sunscreen creams, or lotions,
- make-up preparations.

Toiletries:
- talc, and deodorant powders,
- feminine hygiene preparations,
- hair shampoos,
- toothpaste.

Household products:
- polishes, cleaners,
- insecticides.

The four essential components of an aerosol or pressure pack are:

propellent,
containers,
valve, and actuator button, and
product.

In the simplest two-phase aerosol system the product is in solution, with the liquid propellent forming the liquid phase, and a proportion of the propellent vaporizes to form a gaseous phase which exerts excess pressure on the product and on the walls of the container. When the valve is opened by depressing the actuator button, the propellent forces a mixture of propellent and product up the dip tube and on contact with ambient air causes rapid evaporation of the propellent to form a fine dispersion or spray of product. Thus the type of spray pattern of all aerosols depends upon: the type of product to be dispensed, the proportion and type of propellent, and the type of valve and actuator button.

Propellents

The original propellent—difluorodichloromethane—as prepared by Dupont was known as Freon 12, and several chloro- and fluoro-derivatives of methane and ethane were subsequently developed including the following:

monofluorodichloromethane ($CHCl_2F_2$)
difluoromonochloromethane ($CHClF_2$)
difluoromonochloroethane ($CH_3C\ ClF_2$)
trifluorotrichloroethane ($CCl_2F-CClF_2$)

Propellents were eventually manufactured by several companies and these are now available in various parts of the world under different trade names. Some of these are as follows:

Freon	du Pont de Nemours E. I. & Co. Inc., U.S.A.
Arcton	Imperial Chemical Industries Ltd., England.
Isceon	Imperial Smelting Company, Ltd., England.
Frigen	Farbwerke Hoechst A. G., West Germany.
Genetron	Allied Chemical Corporation, New Jersey, U.S.A.

The products are now marketed under identical code numbers which relate to their composition. For example, propellents with two figure numbers indicate methane derivatives and three figure numbers indicate ethane derivatives. In the following text propellents are referred to by the recognized code numbers and particular commercial brands are not given.

These substances differ in their chemical and physical properties, particularly in regard to their boiling points and vapour pressures, factors of prime importance when considering the conditions required for packaging a particular aerosol product. It was eventually found that suitable pressure systems, adaptable for most cosmetic applications, could be obtained by using a more restricted range of propellents using these either on their own, or more frequently, as mixtures.

The three propellents now mainly used in this way are as follows:

Propellent 11	Monofluorotrichloromethane CCl_3F
	Boiling point + 23·8°C
	Vapour pressure at 25°C–0·7 p.s.i.g. (pounds per square inch, gauge)
Propellent 12	Difluorodichloromethane CCl_2F_2
	Boiling point –29·8°C
	Vapour pressure at 25°C 79·8 p.s.i.g.
Propellent 14	Tetrafluorodichloroethane $CClF_2$–$CClF_2$
	Boiling point + 3·6°C
	Vapour pressure at 25°C 16·6 p.s.i.g.

Suitable pressure systems are obtained by mixing the low boiling propellent 12 with the higher boiling materials. A mixture of propellents 11 and 12 in a 50:50 ratio gives a vapour pressure of 43·5 p.s.i.g. at 25°C. Mixtures of propellents 12 and 14 in a 10:90 ratio gives a vapour pressure of 24·6 p.s.i.g. at 25°C, in a 40:60 ratio gives a vapour pressure of 46·3 p.s.i.g. at 25°C, in a 60:40 ratio gives a vapour pressure of 58·7 p.s.i.g. at 25°C.

The choice of propellent depends upon the type and composition of the product and stability of the product in contact with the propellent. Propellent 11, containing a higher proportion of chlorine, hydrolyses more readily in the presence of water to form hydrochloric acid and cannot be used with products containing water as this results in corrosion of the valve and container. In such cases alternative propellents must be used.

n-Butane, *iso*-butane, and propane are used as propellents for household products, and the destenched grades now available are also considered satisfactory for use with many cosmetic preparations. As alternative propellents they are cheaper than fluorinated and chlorinated hydrocarbons, are stable in water-based systems but have the disadvantage of being flammable.

Compressed gases, nitrogen, nitrous oxide, and carbon dioxide, are used as propellents for specific applications. Nitrogen is used as a propellent for toothpaste and foodstuffs such as cake toppings and sauces. Carbon dioxide and nitrous oxide can be used by means of a saturation tower, and a suitable liquid is sprayed onto the gas which is absorbed as it passes down the tower, the rate of absorption depending upon the design of the tower. This pressurized liquid is then injected into an aerosol can through the valve, using a normal pressure filling head.

Containers

A variety of containers are available for aerosol products, the most commonly used being a three-piece tinplate container with a soldered or welded side seam and the top and bottom seamed on to the body. Two piece containers usually consist of a seamless aluminium body with either an aluminium or tinplate bottom. A two-piece container has recently been introduced where the bottom and main body are in one piece and the top is seamed on. These cans are internally lacquered after the can is formed, unlike the three-piece tinplate container when the lacquer is applied to the sheet metal and subject to stress cracking during pressing. All these containers are suitable for medium pressure propellent mixtures and safe to use up to pressures of about 44 p.s.i.g. The aluminium monobloc container spun in one piece is internally lacquered after formation, and withstands high pressure propellent systems. High density polyethene containers made by a blow moulding technique are more expensive than metal containers, and also able to withstand medium to high pressure systems. Polypropylene containers are an attractive alternative method of packaging, but are only suitable for use with low pressure systems.

Valves and actuators

The standard type of valve used for a large proportion of popular aerosol products consists of a standard actuator button with a simple orifice, a valve stem, a stainless steel spring held in position by a valve housing and gasket, and a dip tube. Variations of this standard actuator are the mechanical break-up foam buttons. The component parts of the valve are varied to suit any particular characteristic of a product or prevent attack by any reactive materials which may be present. Antiperspirant aerosols require specially designed valves when active material may react with the spring or gasket, and powder products, where the valve is designed to prevent blockage. A metering valve is designed to dispense a measured dosage of product each time the valve is opened, and these are used for example to avoid wastage of expensive perfumes, or for a product such as a mouth freshener, where controlled dosage is desirable. Most valves are available to use with containers having 15 mm, 20 mm, or 1 inch necks.

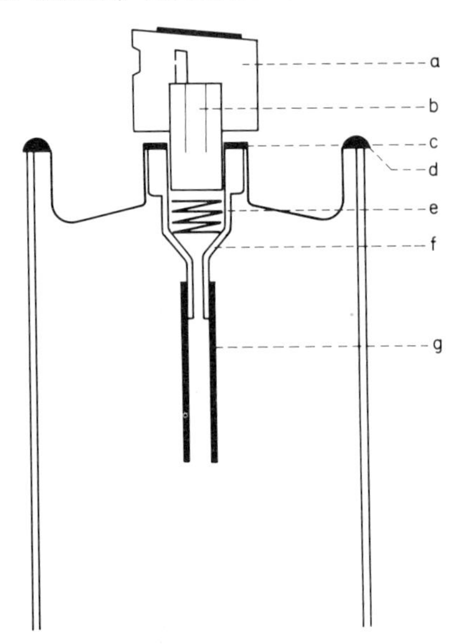

Fig. 1.1 Diagrammatic illustration of Aerosol valve system showing Actuator button (a), valve stem (b), gasket (c), lining (d), spring (e), valve housing, (f), and dip-tube (g).

Filling and testing

Aerosol containers are filled either by freeze filling or pressure filling methods. For freeze filling the propellent is chilled as it leaves the container and a measured quantity is added to the product in the open can which is also chilled to a similar temperature. The container is then sealed.

Pressure filling is carried out using the compressed liquid propellent in a completely closed system. The active ingredient is filled into the container which is then sealed (with the valve) and this is pressed against the special head of the filling apparatus, which opens the valve and allows a measured quantity of the propellent to pass under pressure through the valve opening into the container. When the filled container is removed from the filling machine head, the valve closes automatically. Automatic pressure filling machines are now mainly used for large scale filling of aerosol containers and this is usually handled by contract fillers specializing in this work. More simple types of filling equipment are available suitable for laboratory experimental work and small scale production runs.

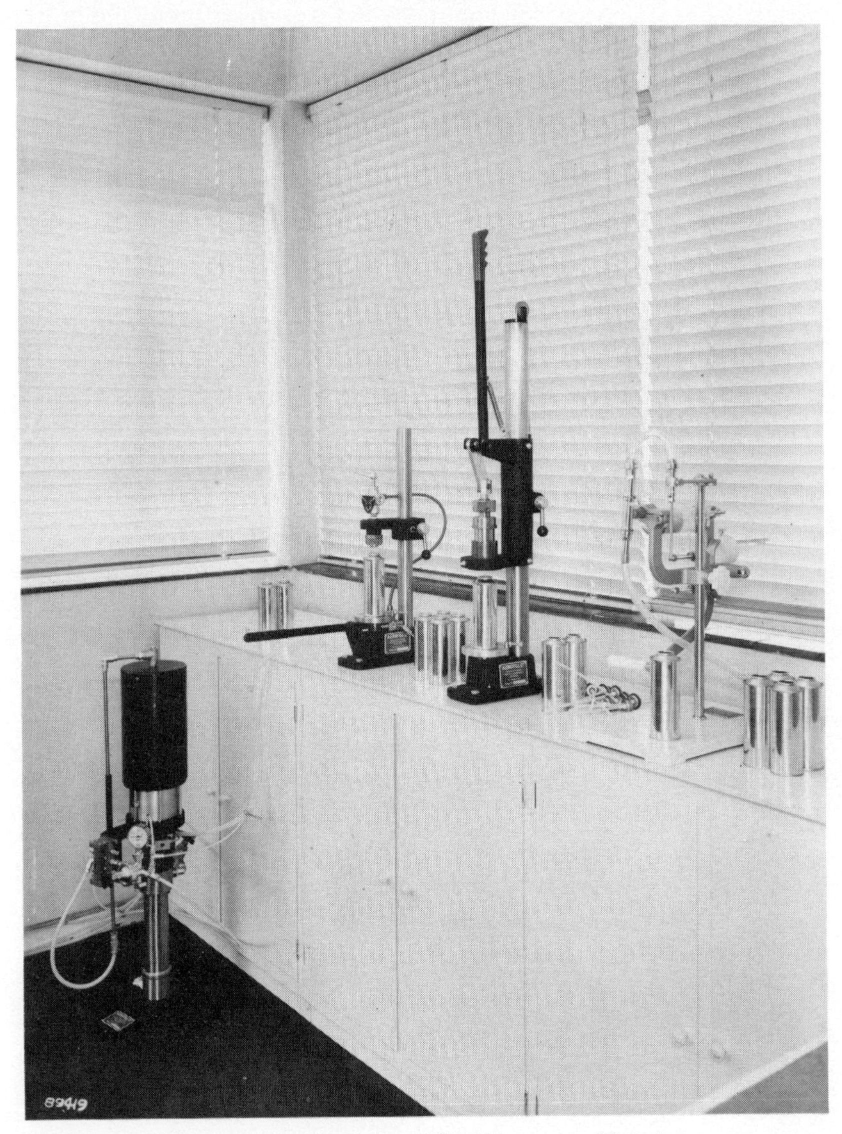

Plate 1 Pilot Pak aerosol laboratory filling line–Aerofill Ltd.

After the aerosols have been filled they are immersed in a water bath at a temperature of about 50°C for a short period of time. This test reveals any leaks due to faulty crimping or incorrect valve assembly. Tests are also made after fitting the actuator button to check whether the valves are functioning correctly. Where product development work is concerned the filled aerosol is checked for performance of the product and the effectiveness of the valve and propellent systems. If these are satisfactory further tests are required to determine the stability of perfume and container under storage conditions. It is a good plan to test a range of perfumes using various types of containers incorporating different internal surfaces. Ideally storage tests are carried out for 12 months at ambient temperatures checking the contents for stability at 3-monthly intervals. Accelerated tests are made by incubating at 40–50°C and checking for stability at monthly intervals. At suitable intervals empty a can and cut in half to examine the conditions of the valve and inner surface of the container.

Plate 2 Automatic Star Pak filling line showing rotary unscrambling table, automatic filling machine and conveyor drive unit—Aerofill Ltd.

Plate 3 View of automatic filling line used for the production of aerosols—The Crown Cork Company Ltd.

CHAPTER TWO

Antiperspirants and Deodorants

Although antiperspirant and deodorant preparations are invariably grouped together, their actions are quite different. Antiperspirants are astringents aimed to reduce the amount of both eccrine and apocrine sweat secretions. They are considered to have a co-agulating effect on skin protein and thereby block the openings of the sweat ducts on the skin surface. Their exact mode of action is still a matter of dispute but it is generally acknowledged that perspiration rate can be checked by using an astringent.

The natural function of the sweat gland is to help regulate body temperature. They are also more active as a result of emotional disturbances. In hot weather, or if the body temperature is raised by exercise, the blood capillaries of the skin dilate so that more blood circulates near the surface. At the same time the sweat glands are active and as the water of the sweat evaporates it takes heat from the blood capillaries and exerts a cooling effect. In cold conditions, when the body temperature falls the skin contracts and the sweat glands are not so active, thus conserving body heat. It can be argued therefore, that artificial contraction of the skin caused by an astringent reduces the activity of the sweat glands.

Deodorants

Deodorants are not designed to check the flow of perspiration. They are based on bactericides or antiseptics which either destroy bacteria or prevent their activity. Fresh perspiration from both the eccrine and apocrine glands is practically odourless. It is bacterial decomposition of sweat which is responsible for development of odour. Most antiperspirants generally act as deodorants but a deodorant need not necessarily have an antiperspirant action. An effective antiperspirant is therefore designed to regulate or control the flow of sweat and prevent development of odour. The preparation should not irritate the skin and not cause deterioration of clothing. A simple way of dealing with excessive perspiration is to apply an absorbent powder to the affected parts of the body. Antiseptic materials can be included in the powder.

No. 2001

Boric acid (powder)	50
Light magnesium carbonate	50
Talc	900
	1000
Perfume	0·3–0·5 per cent

This type of product is also used as a foot powder, in which case stronger antiseptic materials, such as salicylic acid or oxiquinolene sulphate are used.

No. 2002

Salicylic acid	30
Boric acid (powder)	50
Starch powder	500
Talc	420
	1000
Perfume	0·3–0·5 per cent

In cases of hyperhidrosis from 2·5 to 5·0 per cent of an oxidizing substance such as sodium perborate can be included. Hexamine (or hexamethylenetetramine) is also used at concentrations up to 5 per cent. This compound liberates formaldehyde on hydrolysis.

Astringents and antiperspirants

Several materials have astringent properties and have been considered for use as antiperspirants, but salts of aluminium and zinc are mostly used at a concentration of between 12 and 20 per cent. At one time many preparations were based on aluminium chloride or sulphate but in solution these salts have a low pH value (pH 2·0–3·0 at the concentration used) and can be irritating to the skin. The compounds also hydrolyse to acid and can damage fabrics. High temperature hydrolysis and damage to fabric also occurs if these aluminium salts are present when they are ironed. These effects can be prevented to some extent by adding suitable buffering substances such as urea or borax which raise the pH value to about 4·0. Aluminium compounds which have been used as astringents include the acetotartrate, phenolsulphonate, salicylate and borate. Probably the most commonly used material today is aluminium chlorhydrate. In solution this compound has a pH value between 4·0 and 4·5 which approximates closely to the lower limit of the normal pH value of the skin surface. The compound is a good astringent and does not have any harmful effects either on skin or clothing. In common with most other astringent materials, there is a rask of damage to fabrics, if the chlorhydrate is not washed away before ironing, becasue of the hydrolysis which can occur due to the effect of heat. The addition of urea to formulations based on aluminium chlorhydrate reduces the risk of damage to fabric even further, since the heat of ironing liberates ammonia and neutralizes any acid formed through hydrolysis. Aluminium chlorhydrate ($Al_2(OH)_5Cl \cdot xH_2O$) can now be bought from the manufacturer in the form of an odourless and colourless powder or as a 50 per cent solution in water. The dry powder is difficult to handle unless solutions are made by adding the powder gradually to water, mixing well after each addition of water. If water is added to the powder this results in formation of a gelatinous mass which is difficult to disperse. Intending manufacturers are recommended to purchase the material in solution form. This is miscible with alcohol, propylene glycol and glycerin and is thus ready for use in most antiperspirant preparations.

The simplest type of antiperspirant is a clear solution, containing a small proportion of a suitable non-irritant perfume.

No. 2003

Aluminium chlorhydrate (50 per cent solution)	200
Glycerin	50
Alcohol	430

No. 2003 (*continued*)

Water (softened or distilled)	300
Borax	20
	1000
Perfume	0·3–0·5 per cent

Procedure: Dissolve the borax and any colouring material in the water, and the perfume in the mixture of alcohol and glycerin. Mix the two solutions and add to the solution of aluminium chlorhydroxide.

The perfume used in this formula should be soluble in a low concentration of alcohol or alternatively a water-soluble type should be used. Glycerin acts as an emollient facilitating the application and spreading properties of the lotion. Up to 10 per cent of glycerin or propylene glycol can be included. Borax is a useful additive to counteract irritation which may occur if the product is applied to delicate skin. The viscosity can be increased if required by including a small quantity of methyl cellulose.

A useful quick-drying lotion can be prepared as follows:

No. 2004

Aluminium chlorhydrate (50 per cent solution)	300
Propylene glycol	50
Alcohol	450
Water (softened or distilled)	200
	1000
Perfume	0·3–0·5 per cent

A soluble bactericide can also be included in this product. Suitable materials include cetrimide, chlorhexidine diacetate (or dihydrochloride), or a quaternary ammonium compound. Hexachlorophane has also been used successfully for many years as a surface bactericide. In February 1972 the U.K. Secretary of State for Social Services advised that hexachlorophane may be a potential health hazard, particularly when it is applied to raw areas, or when it is applied extensively and not subsequently washed off.

Although there is to date no evidence of actual harm to humans under normal conditions of use, it is now recommended that preparations containing hexachlorophane should be rinsed off after use. It is also recommended that hospitals, doctors, and those concerned with administering hexachlorophane, are advised of the potential hazard; that a cautionary notice should appear on the wrapper of soaps containing hexachlorophane; and particular caution should be observed in the use of products containing hexachlorophane in infancy.

In the case of antiperspirants, bactericides of a phenolic type can cause discolouration of the end-product due to the formation of colour complexes with iron. Aluminium chlorhydrate contains about 0·02 per cent of iron and, although this is not generally sufficient to cause discolouration, it is advisable to include a sequestering agent such as EDTA to prevent reaction with iron, or other trace metals which may be present in any of the constituents of the formulation.

No. 2005

Ingredient	Amount
Aluminium chlorhydrate (50 per cent solution)	300
Propylene glycol	50
Glycerin	20
Alcohol	500
Water (softened or distilled)	125
Chlorhexidine diacetate	5
	1000
Perfume	0·5–0·5 per cent

Procedure: Dissolve the chlorhexidine diacetate in the propylene glycol with the aid of gentle heat and allow the solution to cool. Mix the remainder of the materials together and finally add the chlorhexidine solution. Adjust the alcohol level if required. The quantity given limits the choice of perfume due to the solubility factor.

A surface-active material can be included as a wetting agent and perfume solubilizer, although the amount used should be quite small or the lotion will foam.

No. 2006

Ingredient	Amount
Chlorhexidine diacetate	5
Aluminium chlorhydrate (50 per cent solution)	20
Alcohol	750
Triethanolamine lauryl sulphate	10
Glycerin	20
Water (softened or distilled)	195
	1000
Perfume	*q.s.*

Procedure: Dissolve the chlorhexidine in the alcohol and add the triethanolamine lauryl sulphate. Add this solution to a mixture of the remaining ingredients.

Lotion-type antiperspirants are frequently packaged in plastic squeeze bottles. As an alternative the lotion is applied from a wide mouth bottle fitted with a plastic sponge, in which case thickening agents are not required.

Antiperspirant creams and lotions

Antiperspirant creams and lotions are emulsified preparations containing low proportions of oils and fats so that the products do not grease or soil clothing.

Astringent salts are characterized by acid hydrolysis and acid stable emulsifiers must be used for true antiperspirants. Deodorant creams containing a germicidal material only are not often marketed but if required a conventional vanishing cream base can be used.

An antiperspirant lotion is made as follows:

No. 2007

A	Glyceryl monostearate (acid stable)	150
	Mineral oil (cosmetic quality)	75
	Petroleum jelly	25
	Spermaceti	50
B	Water (softened or distilled)	250
	Glycerin	50
	Aluminium chlorhydrate (50 per cent solution)	400
		1000

Perfume	0·3 per cent
Methyl parahydroxybenzoate	0·15 per cent
Propyl parahydroxybenzoate	0·05 per cent

Procedure: Heat A and B independently to 75°C. Add B to A slowly with stirring. Stir until cool, adding the perfume at about 35°C.

Formulae for creams are as follows:

No. 2008

A	Glyceryl monostearate (acid stable)	150
	Non-ionic emulsifier[1]	10
	Diethylene glycol monostearate	15
	Mineral oil (cosmetic quality)	75

	Petroleum jelly	50
	Spermaceti	50
B	Glycerin	50
	Water (softened or distilled)	200
	Aluminium chlorhydrate (50 per cent solution)	400
		1000

Perfume	0·3 per cent
Methyl parahydroxybenzoate	0·15 per cent
Propyl parahydroxybenzoate	0·05 per cent

[1] Abracol LDS-type–Bush Boake Allen

Procedure: Heat A and B independently to 75°C. Add B to A slowly with stirring. Stir until cool, adding perfume at about 35°C.

No. 2009

A	Cetyl alcohol	20
	Mineral oil (cosmetic quality)	50
	Diethylene glycol monostearate	45
	Non-ionic emulsifier[1]	30
B	Glycerin	30
	Water (softened or distilled)	425
C	Aluminium chlorhydrate (50 per cent solution)	400
		1000

Perfume	0·3 per cent
Methyl parahydroxybenzoate	0·15 per cent
Propyl parahydroxybenzoate	0·05 per cent

[1] Abracol LDS-type–Bush Boake Allen

If additional deodorant properties are required about 0·2 per cent of chlorhexidine diacetate can be added to the above formula. In such cases include 0·1 per cent of EDTA as indicated previously.

Antiperspirants based on aluminium salts have a rough or dry feel due to their astringent effect. With creams in particular, this encourages 'rolling' and glycerin or propylene glycol is included in the formulation to prevent this effect. From 5·0–8·0 per cent is recommended, the lower amount being preferred if the composition also contains mineral oil and petroleum jelly.

Roll-on antiperspirants

Roll-on antiperspirants consist of viscous lotions or thin milks. They are packed in a special applicator fitted with a ball at the neck which disperses a film of liquid as it is rolled over the skin.

The liquids used in the applicator must flow easily but be of sufficient viscosity to prevent leakage from the ball head. Glycerin or propylene glycol in this type of product prevents crystallization of the aluminium salt which could prevent movement of the ball. For the same reason only small proportions of alcohol are used. Hydrophillic colloids such as methyl cellulose or magnesium aluminium silicate* are suitable thickening agents.

No. 2010

Aluminium chlorhydrate (50 per cent solution)	400
Propylene glycol	75
Methyl cellulose	25
Water (softened or distilled)	500
	1000

Perfume	0·3 per cent
Methyl parahydroxybenzoate	0·15 per cent
Propyl parahydroxybenzoate	0·05 per cent

* Veegum—R. T. Vanderbilt Co., Ltd.

This lotion does not contain any alcohol and a water-soluble perfume must be used. The viscosity is controlled by adjusting the quantity of methyl cellulose. An emulsified product is prepared as follows:

No. 2011

A	Ethylene oxide condensate of cetyl/oleyl alcohol[1]	50
	Propylene glycol monostearate (S.E.)	40
	Mineral oil (cosmetic quality)	100
B	Water (softened or distilled)	410
C	Aluminium chlorhydrate (50 per cent solution)	400
		1000

Perfume	0·3 per cent
Methyl parahydroxybenzoate	0·15 per cent
Propyl parahydroxybenzoate	0·05 per cent

[1] Empilan KL10 type—Albright & Wilson Ltd.

Procedure: Heat A and B independently to 75°C. Add B to A slowly with stirring. Stir until cool, adding C at about 40°C. and the perfume at about 35°C.

Antiperspirants and deodorant sticks

Antiperspirants and deodorant sticks are prepared in a similar manner to cologne sticks being based on the gelling of alcohol with sodium stearate. Chlorhexidine diacetate or bithional are effective non-irritant bactericides which retain their activity in the presence of the stearate soap. A suitable formula is given:

No. 2012

Sodium stearate	70
Cetyl alcohol	15
Propylene glycol	50
Alcohol	830
Chlorhexidine diacetate	5
Water (softened or distilled)	30
	1000
Perfume	0·5 per cent

Procedure: Heat all the materials together under total reflux until all the soap is dissolved. Colour solution if required can be added. The temperature is allowed to fall to 65°C and perfume added to the cooling base. Mix well and transfer to suitable moulds.

To obtain a bright and clear near-transparent product it is essential to use a high quality grade of sodium stearate. The proportion of stearate used can be varied from 5·0 to 8·5 per cent but when the lower levels are used a suitable proportion of cetyl alcohol may be necessary to obtain the required rigidity. Use a small proportion of iso-propyl myristate–up to 5 per cent improves the feel of the product during and after application and prevents the residual film of sodium stearate from drying out to a white powder and flaking off.

A variation of this product which is particularly suitable for use in hot climates when storage conditions affect the stability and keeping properties is the non-alcoholic deodorant stick prepared as follows:

No. 2013

Sodium stearate	100·0
Urea	2·5
Water	240·0
Propylene glycol	500·0

No. 2013 (*continued*)

Chlorhexidine diacetate	5·0
Benzyl alcohol	50·0
	897·5

Perfume	0·5 per cent
Methyl parahydroxybenzoate	0·15 per cent

Procedure: Dissolve the mix in the water and the hexachlorophane and preservative in the propylene glycol. Mix the solution and add the sodium stearate and warm gently to affect solution. Add the benzyl alcohol and mix well. Add the perfume during cooling. Mix and transfer to suitable moulds.

Aluminium chlorhydrate is incompatible with sodium stearate and other soaps and is consequently not suitable to use for the preparation of antiperspirant sticks. To overcome this sodium aluminium chlorhydroxylactate is used prepared by complexing aluminium chlorhydrate with sodium lactate. Ionization of the aluminium is thereby suppressed and the pH value of the solution is raised so that it does not react with the soap.

Sodium aluminium chlorhydroxylactate is available as a clear, almost colourless syrupy aqueous solution containing 40 per cent w/w solids. Since it is not such an effective astringent as aluminium chlorhydrate it is often used together with a bactericide to prepare a stick with combined antiperspirant and deodorant properties.

No. 2014

Sodium aluminium chlorhydroxylactate (40 per cent sour)	400
Propylene glycol	30
Sodium stearate	100
iso-Propyl myristate	25
Alcohol	445
	1000

Perfume	0·5 per cent
Methyl parahydroxybenzoate	0·15 per cent

Procedure: Heat the sodium stearate in the alcohol under total reflux. Warm the chlorhydroxylactate solution and add to the alcohol solution. Dissolve the preservative in the propylene glycol and add to mixture. Add the *iso*-Propyl myristate. Mix well and allow to cool adding the perfume to the cooling base. Transfer into moulds.

Perfumes used for both antiperspirant and deodorant products should be carefully checked for stability of odour in contact with an aluminium salt and also, particularly, for stability to colour change.

Aerosol deodorants

Aerosol deodorants are based on an alcoholic solution of a bactericide or they may include an alcohol-soluble aluminium salt or other astringent material, and consequently have antiperspirant properties. In some cases a product sold as an aerosol deodorant and body spray is based solely on an alcoholic solution of a perfume compound. A typical formula for a true aerosol deodorant is as follows:

No. 2015

Deodorant base:	
Diethyl phthalate	15
iso-Propyl myristate	30
Chlorhexidine diacetate	5
Alcohol (99 per cent v/v)	950
	1000
Perfume	0·5–1·0 per cent

Procedure: Dissolve the chlorhexidine diacetate in the alcohol. Add the remaining materials and perfume and mix.

Container charge: (1)	
Deodorant base	40
Propellent 11/12 (50 : 50)	60
	100

Suggested containers: internally lacquered aluminium monobloc on tin-plate.

Container charge: (2)	
Deodorant base	34
Propellent 12/114 (10 : 90)	66
	100

Suggested containers: plastic coated glass.

Dry spray deodorants are based on a suitably perfumed talc powder with or without a bactericide. Prepare as follows:

No. 2016

	Deodorant talc base:	
A	Micron Talc No. 1 LO[1]	900
	Magnesium myristate[2]	60
	Chlorhexidine diacetate	10
B	*iso*-Propyl myristate	30
		1000
	Perfume	0·5–1·0 per cent

[1] Whittaker Clark and Daniels Inc.
[2] Satinex—Bush Boake Allen

Procedure: Mix the talc and magnesium myristate A and perfume and *iso*-Propyl myristate B separately. Weigh parts A and B directly into the aerosol container.

Container Charge:	
Powder Base	13
Propellent 12	87
	100

Suggested containers and valve. Container: plain tin-plate, plain aluminium 2-piece or monobloc. Valve: powder.

Feminine hygiene sprays or vaginal deodorants are a further development of the dry spray. They may consist of a suitable non-irritant perfume in propellent only, a perfume and bactericide or may contain a small proportion of talc. As the active material use chlorhexidine dihydrochloride or diacetate. Formula as follows:

No. 2017

(Feminine hygiene base)

A	Micron talc No. 1 LO[1]	940
	Chlorhexidine dihydrochloride	20
B	*iso*-Propyl myristate or *iso*-Propyl palmitate	40
		1000
	Perfume	0·2 per cent

[1] Whittaker, Clark & Daniels Inc.

Procedure: Mix Part A and mix the perfume with Part B independently. Weigh A and B directly into the aerosol container.

An aerosol provides a convenient method for applying an antiperspirant preparation, but difficulties arise with formulation due to the reaction of aluminium compounds causing corrosion of conventional tin-plate and aluminium containers and valve blockage as a result of reaction with metal parts of the valve. Products based on aqueous solutions of aluminium chlorhydrate can, therefore, only be specially packaged in expensive glass or polypropylene containers, and charged with a low pressure propellent mixture. A special valve without metal parts is also required. The base is prepared as a three-phase system containing an aqueous-alcoholic mixture and a surface-active agent which is used at sufficient concentration to prepare a temporary emulsion when the product is shaken before use. A suitable product can be prepared using the following formula:

No. 2018

(Water-based type)

Antiperspirant base.	
Propylene glycol	50
Aluminium chlorhydrate (50 per cent solution)	100
Alcohol (96 % v/v)	550
Water (softened or distilled)	295
Polyoxyethylene oleyl ether[1]	5
	1000
Perfume	0·5–1·0 per cent

[1] Volpo N. 3 type—Croda Ltd.

Procedure: Mix the perfume and oleyl ether with the alcohol. Add the propylene glycol and water, and mix. Finally add the aluminium chlorhydrate—with continuous stirring.

Container charge:	
Antiperspirant base	50
Propellent—12/114 (10/90)	50
	100

The proportion of alcohol included in this formula is sufficient to allow the addition of 0·25 per cent of a bactericide if an antiperspirant with deodorant properties is required. A low pressure propellent mixture is required as indicated to give a maximum internal pressure of 23 to 25 p.s.i.

Suggested container: plastic coated glass or polypropylene.
Valve: special tilt-action without internal metal parts.

A suitable mechanical break-up button is also advisable to prevent clogging.

A more suitable material for use in antiperspirant aerosols is available as a derivative of aluminium chlorhydrate and propylene glycol which is soluble in alcohol and compatible with halogenated hydrocarbon propellents. A two phase system is obtained with the following formula and consequently the end product does not require shaking before use.

No. 2019

(Alcohol type)

Antiperspirant base:	
Aluminium chlorhydrate propylene glycol complex[1]	100
Propylene glycol	20
Alcohol (99·5% v/v)	880
	1000
Perfume	0·5–1·0 per cent

[1] Rehydrol A.S.C.—Reheis Chemical Co.—Division of Armour Pharmaceutical Company.

Procedure: Weigh out the alcohol and propylene glycol into a reactor fitted with a fast stirrer. Add the aluminium complex with fast stirring to ensure even dispersion of the powder.

Container charge:	
Antiperspirant base	46
Propellent—12/114 (10/90)	54
	100

Chlorhexidine dihydrochloride can be included in this formula if required. Some difficulty may be found in selecting suitable perfumes for this product due to their reaction with the aluminium chlorhydrate complex under low pressure conditions. This reaction can cause a precipitate which clogs the valve.

Suggested containers: plastic coated glass—polypropylene or special internally coated tin-plate. With the latter propellent 12/114 can be used at a ratio of 50:50 or 60:40 which removes problems of precipitation caused by low pressure systems.

Valve: for plastic coated glass and polypropylene containers: special tilt-action without internal metal parts—fitted with mechanical break-up button for tin-plate containers with specially coated cup.

Aerosol antiperspirants are now mainly based on a powdered form of aluminium chlorhydrate. This is available as a white micronized free flowing powder known commercially as micro-dry ultrafine. Formulations are possible with this material without the use of alcohol or water and the products can therefore be packaged safely in lacquered aluminium cans with little fear of corrosion. A suitable product can be prepared as follows:

No. 2020

(Dry spray type)

Antiperspirant base:	
Aluminium chlorhydrate micro-dry ultrafine[1]	460
Silicon dioxide[2]	40
iso-Propyl myristate	500
	1000
Perfume	1·0–1·5 per cent

[1] Reheis Chemical Co.—Division of Armour Pharmaceutical Company.
[2] Aerosil R.972—Degussa.

Procedure: Add the perfume to the *iso*-Propyl myristate and mix. Mix the two powders together and slowly stir in the perfume/*iso*-Propyl myristate mixture. When a slurry has been obtained pass this through a colloid mill to thoroughly disperse the particles.

Container charge:	
Antiperspirant base	10
Propellent—11/12 (50 : 50)	90
	100

Bactericides such as chlorhexidine diacetate or hydrochloride, or trichloro-hydroxydiphenylether[1] can be included in the formula if required. Silicon dioxide is used as a suspending agent to form a thixotropic paste with the remaining ingredients. The paste shows very little separation on standing and can be conveniently handled for filling into the aerosol cans. Other types of suspending agents, including alternative types of silicon dioxide, can be used with satisfactory results but these can affect the

[1] Irgasan DP300—Geigy.

stability of the perfume and each formulation requires adequate storage testing to ensure perfume stability of the particular system used.

Suggested containers: internally lacquered tinplate containers. Valve: special powder valve.

CHAPTER THREE

Bath Preparations

There are several types of preparations which can be used in the bath. Their primary function is to assist in softening hard water and also to impart a fresh and fragrant bouquet to the bath water. The type of bouquet used for this purpose varies considerably and can be of a pine or slightly medicated note, or a sophisticated, expensive, French bouquet. The various kinds of bath preparations will be discussed under their separate headings.

Bath salts

Bath salts or crystals are probably one of the earliest products used to perfume the bath water and give it a pleasant odour. They are still very popular and being available in cheap form they are sold either in bulk or in a cheap container, or they are sold in elegant and expensive packages at a high price. The basic constituents of the preparation are the same, but their final selling price is governed by the cost of the package, presentation and, above all, the cost of the perfume. From the manufacturer's point of view, it is necessary to pay particular attention to the keeping properties, stability, melting point and solubility of the final product. The raw materials used in formulating these products, together with their properties, are enumerated below.

Soda crystals

Soda crystals are probably one of the oldest and most common of the raw materials used to prepare bath crystals. Chemically they are known as sodium carbonate decahydrate and have the composition $Na_2CO_3 . 10H_2O$. They occur as colourless, transparent crystals and are obtainable in various sizes, those which are used for the preparation of bath salts being commonly referred to as 'pea' crystals. They are readily soluble in two parts of cold water or 0·25 parts of boiling water. Although they are useful as a means of softening the water when the hardness is due to the presence of lime salts, they cause precipitation of calcium carbonate which has the disadvantage of making the bath water cloudy, and is often responsible for forming a ring of scum around the bath.

Soda crystals tend to lose their water of crystallization on exposure to air and should, therefore, be packed or stored to prevent this taking place. This efflorescence which takes place eventually reduces the crystals to powder although this can, to some extent, be rectified by coating the crystals with a hygroscopic material such as glycerine or glycol. Very often, the glycol is used in this way as a solvent for the dye-stuff used for colouring the crystals. Another disadvantage of soda crystals arises as a result of their low melting point, 34°C. As a result the crystals sometimes melt when they are packed in a jar and exposed to strong sunlight, either in a shop window or bathroom. The effect of these partially melted crystals is most inelegant. In spite of these disadvantages, soda crystals, suitably perfumed and coloured, still have a certain amount of popularity, particularly in the cheaper range of preparations for use in the bath

Sodium carbonate

Sodium carbonate monohydrate occurs as a white, small crystal or in the form of a crystalline powder, and is available in several grades of crystal agglomerates, the grades being selected according to their particle size. The monohydrate contains 85·47 per cent of anhydrous sodium carbonate which is equivalent to 230·75 per cent of soda crystals or sodium carbonate decahydrate. Since the water content of the monohydrate is only 14·53 per cent, the compound remains stable at ordinary temperatures and atmospheric conditions. One part of the monohydrate is soluble in three parts of water and is thus somewhat less soluble than the decahydrate. If the material is obtained in the form of a small sized agglomerate then the question of solubility is overcome and on account of its stability it is to be preferred as a raw material for the preparation of bath salts. Furthermore, in the form of small crystals it is easy to colour and perfume. This compound can be used for both medium and high priced products.

Sodium sesquicarbonate

Sodium sesquicarbonate is prepared from sodium carbonate and sodium bicarbonate, has the following formula: $Na_2CO_3NaHCO_3 \cdot 2H_2O$. It occurs in the form of fine, needle-shaped crystals and remains stable when exposed to air. It is readily soluble in water and is easily coloured and perfumed. Owing to its stability it does not require any special precautions in packaging, so that it has become one of the most popular raw materials used in the preparation of bath salts. In solution it is only mildly alkaline due to the content of bicarbonate, and although it readily softens hard water it does not have a harsh effect on the skin.

Sodium perborate

Sodium perborate $NaBO_3 \cdot 4H_2O$, occurs as white, odourless crystals which are stable in cool and dry air but decompose with the liberation of oxygen when exposed to warm and moist conditions. In water it is decomposed with the liberation firstly of hydrogen peroxide and then oxygen, and thus it has some antiseptic properties. In bath salts it is sometimes used in proportions up to 10 per cent with sodium sesquicarbonate to give an oxygenated preparation. Although the sesquicarbonate helps to maintain the perborate in a stable condition it is, however, necessary to pack bath preparations containing the perborate in well-closed containers.

Sodium phosphate

Sodium phosphate is the dodecahydrate containing 12 molecules of water and has the following composition: $Na_2HPO_4 \cdot 12H_2O$. It occurs as colourless, translucent crystals or white granules. It is readily soluble in three parts of water, the solution being mildly alkaline with a pH value of about 9. When exposed to air at ordinary temperatures it effloresces and loses 5 molecules of water forming the hepta-hydrate, which is stable in air but is only soluble 1 in 4 parts of water.

Sodium Phosphate also occurs in the tri-basic form when it is generally known as trisodium phosphate and has the following composition: $Na_3PO_4 \cdot 12H_2O$. This material occurs as colourless or white crystals and is more stable than the dibasic salt. In solution, however, it is strongly alkaline and gives a pH value of about 11.

Of the sodium phosphates the trisodium salt is probably the most favoured owing to its stability, but the high alkalinity of the solution is a disadvantage and for this reason it is not advisable to use this material on its own. Trisodium phosphate also has the disadvantage of being difficult to colour, partly due to the fact that the colour used must be resistant to the high pH value. On the other hand, this material is easily compressed and is, therefore, sometimes included as a constituent of bath cubes.

Sodium chloride

Sodium chloride of composition NaCl is the same material as common table salt. It occurs as colourless cubic white crystals, granules or powder, and is soluble 1 part in 3 parts of water, the aqueous solution being neutral. It is anhydrous and only slightly hygroscopic and can readily be tinted and perfumed to present an attractive bath salt. In the bath, sodium chloride has a mild toning and stimulating effect. It does, however, have the disadvantage that it depresses the lathering effect of soap and tends to form a scum ring around the bath. Since it also has no water-softening properties it is seldom used as the only component of bath crystals. Bay salt is a form of sodium chloride, sometimes known as rock salt or sea salt, is produced either by mining or by the evaporation of sea water. It occurs in large, irregular shaped crystals and is a pale, amber colour. It is only used occasionally in bath preparations owing to the difficulty of obtaining crystals of regular shape, and its colour makes it difficult to tint.

Sodium borate

Borax or sodium borate occurs as hard, colourless, and odourless crystals, as white granules or as a crystalline powder. It is becoming increasingly popular as a component of bath crystals, probably because it is obtainable in well-graded and regular crystals. It is also known as sodium borate or sodium tetraborate, and has the following composition: $Na_2B_4O_7 \cdot 10H_2O$. The water of crystallization represents approximately 47 per cent of its weight, which is considerably less than that of either sodium carbonate or sodium phosphate. Although it effloresces in dry air, it shows good stability in the presence of moisture. The solubility of borax is, however, a disadvantage since one part of borax is only soluble in about twenty parts of water. On the other hand, it is only mildly alkaline, the solution having a pH value of 9·5 and does not therefore have a harsh or drying effect on the skin. It has a mild, bacteriostatic action and is very slightly astringent when applied externally. These good points make it acceptable as one of the best of the raw materials available.

Sodium bicarbonate

Sodium bicarbonate occurs in the form of white, crystalline powder or granules, and has the composition: $NaHCO_3$. It is readily soluble in ten parts of water giving a slightly alkaline pH value of 8·3. It is mainly used, together with an acid to prepare effervescent bath salts, but can also be used as an ingredient of bath crystals. In solution it is useful to relieve irritation of the skin, particularly where itching is present in conditions such as urticaria and eczema.

Simple formulae for bath salts are as follows:

Sodium carbonate crystals	100					
Sodium carbonate monohydrate		80				
Sodium sesquicarbonate			100	90	85	
Sodium perborate					5	
Sodium phosphate		20				50
Sodium chloride						50
Borax				10	10	
Perfume, colour	Sufficient					

Colouring and perfumes

Bath salts are generally tinted by spraying an aqueous solution of dyestuff on to the crystals, followed by a mixing process to distribute the colour evenly. This method can also be modified by firstly placing the crystals in a tumbler-type mixer and spraying the colour on whilst the crystals are rotating. This is simple to operate and gives evenly coloured crystals after only a few minutes mixing time. Another method which is sometimes used is to immerse the crystals in a tank containing the solution of dyestuff. This latter operation is sometimes preferred when larger quantities of crystals are being treated. With this method the question of drying becomes an important factor, and special arrangements for removing the excess water must be considered.

The colour solution used with either process can be prepared using either water or alcohol, or a mixture of both as the solvent. Obviously, alcohol is the preferred solvent because it evaporates quickly and does not damage the crystals. Used on its own, however, it is more suitable for the dipping process when losses will be proportionately small. In the spraying process the loss of alcohol is considerable and adds to the cost of the final product. Aqueous solutions are, however, equally effective in giving even dispersion of colour, provided the quantity is carefully controlled to avoid the crystals becoming too damp. The ideal vehicle to use for tinting is probably a mixture of about 90 parts of water with 10 parts of alcohol to help the rate of drying. After the colouring process the coloured bath salts are spread evenly on trays and allowed to dry before packing. An alternative method of tinting is to add a dispersed pigment to the crystals during the mixing process. Dispersed pigments are often stable to the perfume but may cause a coloured 'ring' in the bath.

Any dyestuff used must be resistant to alkaline conditions. Another factor which must be checked is their colour fastness to the perfume. Many of the raw materials included in the preparation of perfumery compounds can noticeably react with a dyestuff to give coloured reaction products. It is probably true to say that the effect of the perfume on any colouring material is the most important factor to consider.

All perfumes should be diluted before they are added to the crystals. A rapid mix in a tumbler-type mixer ensures even distribution of the perfume, and the crystals should then be collected and spread over trays and given a short period of time to allow the perfume solvent to evaporate. The crystals should then be packed immediately to avoid loss of perfume. It should be borne in mind that the solvent is ultimately lost by evaporation and this adds to the cost of the product. If alcohol is used as the perfume solvent, the cheaper types of industrial methylated spirits or *iso*-Propyl alcohol are suitable.

A water-soluble type of perfume can also be used as an

alternative method of adding perfume. Such perfumes generally consist of solutions or clear emulsions of the concentrated perfume compounds, in a mixture of surface-active materials, compounded so that the solution of perfume can be diluted with water. If this type of perfume is used then the concentration of perfume present in the water-soluble form should be obtained from the manufacturer to ensure that the correct quantity is added. When using a water-soluble perfume the solution of colouring material can be prepared together with the perfume and the colouring and perfuming of the bath salts can be carried out in one operation. The quantity of water should be kept at a minimum for obvious reasons, and the end product should be packed immediately after the salts have dried.

Oxygenated salts are used either in the bath or for bathing the feet. A typical formula is given:

No. 2021

Sodium chloride	15
Sodium perborate	50
Borax	100
Sodium bicarbonate	250
Sodium carbonate monohydrate	585
	1000
Perfume, colour	*q.s.*

In the bath they are considered to be of therapeutic value by helping to clean the skin and dissolve impurities. These preparations are also claimed to have a reducing effect. Magnesium sulphate, sodium thiosulphate and magnesium chloride are also used as shown in the following:

No. 2022

Potassium iodide	2
Potassium bromide	3
Potassium chloride	15
Magnesium sulphate	300
Sodium sulphate	180
Sodium chloride	500
	1000

No. 2023

Magnesium sulphate	600
Sodium sulphate	150
Sodium chloride	250
	1000

No. 2024

Magnesium chloride	5
Sodium thiosulphate	450
Magnesium sulphate	545
	1000

On its own magnesium sulphate is used in the bath because it is considered to be of value for treatment of sprains, bruises or other inflamed conditions. It is also used as a treatment for rheumatism, using about 1 pound of the dried form of magnesium sulphate or 2 or 3 pounds of crystals in the bath for each treatment. A useful saline bath taken from the French Pharmacopoeia can be prepared to the following formula:

No. 2025

Magnesium sulphate (exsiccated)	500
Sodium sulphate (exsiccated)	500
	1000

About 500 grammes of this mixture is added to the bath water.

As with other powdered products for use in the bath, care should always be taken to ensure that raw materials are thoroughly dried before mixing.

Bath powders

Bath powders at one time were quite popular but have now largely been replaced by liquid preparations. The object of the product is to supply sufficient material, which may either be in the form of fine crystals or powder, for one bath. They are generally packed in cellophane or plastic containers, and if delicately coloured and presented in assorted shades they can be used to prepare an attractive and expensive looking pack. Individually packed powders are often used as a means of giving the consumer a high concentration of perfume, so that when they are used in the bath their main value is to impart a residual perfume-odour to the body rather than act as an efficient water softener. To achieve this effect the perfumes used should contain a high proportion of fixatives.

The powders can be prepared either from dry sodium carbonate or sodium sesquicarbonate crystals. A small proportion of borax can also be added. They are manufactured in a similar manner to that used for the preparation of bath crystals, the perfume and colour being sprayed on, and mixed until uniform. If they are based on sesquicarbonate they can be mixed in a rotary type mixer, but if true powders they are best prepared by using a machine of the sifter/mixer type. Powders are suitably perfumed and coloured using a trituration process. To do this, the perfume and colour are separately mixed with a small portion of the powder to prepare a concentrated mix. This is then added to the bulk of the material and mixed until both the perfume and the colour are uniformly distributed. Examples of powder products are:

No. 2026

Sodium sesquicarbonate	800
Borax (powdered)	200
	1000

No. 2027

Sodium carbonate (dried)	900
Sodium alkyl sulphate (powder)	100
	1000

Bath cubes

Bath cubes have become increasingly popular, probably due to the fact that they can be elegantly wrapped and the amount of perfume per cube can be controlled. By controlling the amount of perfume the manufacturer can be sure that sufficient perfume is present in each bath cube to give the consumer the desired effect in the bath.

In the main they are prepared by compressing either sodium carbonate monohydrate or sodium sesquicarbonate. Sometimes these two materials are used together and a small amount of borax can also be added. When tablets are prepared mainly from sodium sesquicarbonate they can be fed directly into the compression machine and do not necessarily require the addition of a lubricant. In addition, tablets made from sesquicarbonate disintegrate readily in bath water. Tablets made from sodium carbonate are much more firm and if the degree of compression is not carefully controlled they remain undissolved in the bath water for some considerable time. To help disintegration of tablets prepared from the carbonate, a small proportion of starch powder of the order of 2 to 5 per cent should be mixed with the basic raw material before compression. When starch is used in this way as a lubricator, it is mixed with the bath crystals after they have been coloured and perfumed. To prepare the tablets, the crystals are coloured and perfumed in the same way as described for the colouring of bath crystals. It is, however, important that they should be comparatively dry before being fed to the compressing machine so that the crystals do not stick during the process. As previously mentioned, it is advisable to dilute the perfume with alcohol before adding to the mix.

Bath oils and essences

Liquid bath preparations can either be in the form of bath oils which are immiscible with water, or as essences which disperse or dissolve in water. Bath oils are sometimes referred to as floating bath oils because they float on the surface of the bath water as a very thin layer of oil. The fragrance permeates the bathroom as a result of the temperature effect of the water. A disadvantage of the product is that oil which remains in the bath contributes to formation of scum ring.

Many materials have been suggested as perfume carriers for these preparations including light mineral oil, oleyl alcohol, castor oil, or other vegetable oils, or one of the fatty acid esters. Of the latter *iso*-Propyl myristate and *iso*-Propyl palmitate can be used since they are of the right order of viscosity, are free of odour and do not become rancid during storage The concentration of perfume can vary from 5 per cent to as high as 35 per cent. Tests should always be made to confirm that a particular perfume is soluble in the vehicle used as a carrier. It may be found that a particular perfume is not entirely soluble in mineral oil or one of the other vehicles mentioned above. In such cases a co-solvent system can be used, solution of the perfume being obtained by using a mixture of the oils mentioned above.

Bath oils are also packed in capsulated form. This gives an attractive product and is often considered to be a more convenient way of adding a fixed quantity of perfume than pouring a variable quantity from a bottle. A small proportion of a soluble lanolin oil can be added to the basic carrier. Using such a product it seems likely that some of the oils from the surface of the water may adhere to the body and so give a perfumed effect after bathing. It is usual to include colour in the product and the concentration of perfume can be varied according to the cost in relation to the selling price of the finished product. Typical formulae are given:

No. 2028

iso-Propyl myristate	520
Oleyl alcohol	480
	1000
Perfume	1·0–10 per cent
Colour	*q.s.*

Use of diethyl phthalate as a co-solvent is illustrated in the following formula:

No. 2029

iso-Propyl myristate	300
Diethyl phthalate	100
Perfume	250
Mineral oil (cosmetic quality)	350
	1000
Colour	*q.s.*

It is claimed that a bath oil containing mineral oil, *iso*-Propyl myristate, *iso*-Propyl sebacate and polyethylene glycol 400 di-oleate relieves itching and retards evaporation of skin moisture. (Dome Chemicals).

Bath essences are soluble or water dispersable preparations and may contain a lower proportion of perfume than a bath oil.

Suitable products can be made simply by using sulphonated castor oil or triethanolamine oleate to disperse the perfume.

No. 2030

Triethanolamine oleate	10
Perfume	10
Water (softened or distilled)	980
	1000

Solution of formaldehyde (40 per cent) is a suitable preservative for bath essences. Use 0·1 to 0·2 per cent.

An increase in the popularity of bath essences is to some extent due to the availability of modern surface-active agents, which besides being used to disperse the perfume are good additives for use in hard water. Surface-active agents are used in this way to prepare what are sometimes referred to as transparent emulsions. An additional foam builder can also be added if required. Triethanolamine alkyl sulphate can be used to dissolve the perfume and the mixture diluted with water according to the price level required for the finished product. Sodium lauryl ether sulphate is also useful as a perfume solvent and is used in low or high concentrations according to the solubility factor of the perfume in the dilution. It is compatible with soap and also has good foaming properties. From 25 to 50 parts sodium lauryl ether sulphate can be used as the perfume diluent making the solution up to 100 parts with water. The perfume is first mixed with the surface-active agent before the addition of water. In addition to preparing the final product in this fashion, the solubility of a new perfume which is being examined for use in bath essence can be determined by experiments on similar lines.

The same solubility effect is obtained by using an alcoholic solution of the perfume compound. When a perfume is solubilized in a transparent emulsion system, however, the rate of perfume release is much less than that which is obtained by an alcoholic solution, and thus a more lasting effect is obtained in the bath. A typical formula is given below:

No. 2031

Perfume	50
Polyoxethylene sorbitan monolaurate[1]	400

Water (softened or distilled)	550
	1000

[1] Tween 20—Atlas Chemical Company.

This is prepared by mixing the perfume essence with the solubilizing agent and adding the water. It should be noted, however, that according to the composition of the perfume essence the amount of solubilizer will vary within fairly narrow limits, and it might be necessary to determine the amount of solubilizer required by experiment. Further, it is a good plan when these experiments have been carried out, to allow the finished product to stand for several days to ensure that no separation of oil takes place. It is, of course, not necessary to use as high a percentage of perfume as that given above, but a minimum of 5 per cent is suggested since the odour value is the main attribute of these products.

As an example of bath essence which contains a foam builder in addition to surface-active agents, the following formula is given:

No. 2032

Triethanolamine ammonium alkyl sulphate[1]	50
Triethanolamine lauryl sulphate[2]	360
Alkylolamide	30
Perfume	50
Water (softened or distilled)	510
	1000

[1] Empicol TC 34 type—Albright & Wilson Ltd
[2] Empicol TL 40 type—Albright & Wilson Ltd

Tinting of bath essences with fluoresceine or similar water soluble dyestuff adds to their attractiveness. These products add fragrance during bathing and the use of surface-active materials not only helps to soften the water but contributes towards the absence of scum film on the sides of the bath. To give a refreshing, exhilarating and relaxing effect it is a good plan to use at least 5 per cent of perfume. Solubilization of the perfume is, therefore, a matter of prime importance and the amount used and choice of solubilizer will depend to a large extent upon the composition of the perfume, since those perfumes which contain a high proportion of solids or resins will require a higher proportion of surface-active materials, not only to effect solubility but also to prevent precipitation of solids taking place during storage. The following formulae illustrate the proportions of surface-active agent to water which can be used to prepare clear solutions of almost all perfumes normally used.

No. 2033

Triethanolamine lauryl sulphate	500	to	600
Perfume	50		
Water (softened or distilled)	350	to	450
	1000		1000

No. 2034

Triethanolamine lauryl sulphate	360
Alkylolamide	30
Perfume	50
Water (softened or distilled)	560
	1000

A formula for a complete pine bath essence is given below in which triethanolamine alkyl sulphate and ethylene glycol are used as the solubilizers of the perfume.

No. 2035

Triethanolamine lauryl sulphate	400
Ethylene glycol	45
Colour solution	5
Perfume	150
Water (softened or distilled)	400
	1000

A pine perfume suitable for use in the latter formula can be prepared as follows:

No. 2036

Terpineol	50
Terebene	50
Bornyl acetate	20
Geranium bourbon	20
Patchouli	10
	150

Procedure: Mix the perfume ingredients and add the solubilizers. Mix and add the colour and finally add the water with gentle stirring to avoid incorporation of air.

In the following formula a surface-active agent is used to disperse a mixture of a perfume compound and mineral oil. This is referred to as a dispersible bath oil.

No. 2037

Perfume	50
Lauryl alcohol ethoxylate[1]	150
Mineral oil	800
	1000

[1] Empilan KB3 type—Albright & Wilson Ltd.

Include a small proportion of a suitable oil-soluble dyestuff.

Procedure: Mix the perfume with the ethoxylate lauryl alcohol and add the mineral oil and mix. Finally, add any colour.

A viscous product with similar dispersion properties in water is made as follows:

No. 2038

Glycerin	250
Propylene glycol	250
Ethoxylated cholesterol[1]	250
Triethanolamine lauryl sulphate	250
	1000

[1] Solulan C24 type—American Cholesterol Products, Inc.

Perfume, use up to 5·0 per cent, and include a suitable water soluble dyestuff.

Procedure: Warm the ethoxylated cholesterol to disperse in part of the water and add the remaining ingredients and mix.

Bubble baths

Bubble baths, or foam baths, as their name implies, are intended to fill the bath with a light, frothy lather, and are now probably one of the most popular of bath preparations. As a general rule they are liquid products based on surface-active materials as the foaming agents and sometimes containing an additional foam stabilizer. In a similar manner to bath essences these products do not leave any scum ring on the bath.

Although there are many references to the preparation of products with good foaming properties, the main problem when formulating is concerned with the addition of the perfume, particularly if a good quality product containing a relatively high proportion of perfume is required. It is often necessary to experiment before a formula can be finalized because even when a satisfactory formula has been devised, it does not necessarily follow that the same formula will be satisfactory if it is used with a different perfume. The formulae which follow indicate the use of several types of surface active agents. Use 0·1 to 0·2 per cent of solution of formaldehyde (40 per cent) as a preservative. Up to 10 per cent of most perfumes can be used in the following formula:

No. 2039

Triethanolamine alkyl sulphate[1]	400
Nonyl phenol condensate[2]	100
Water (softened or distilled)	500
	1000

Perfume use up to 10 per cent

Procedure: Mix the perfume with the nonyl phenol condensate and dissolve with the aid of gentle heat if necessary. Add the triethanolamine alkyl sulphate followed by the water and mix.

In some cases either alcohol or hexylene glycol can be used as a mutual solvent or diluent to obtain a clear bright product, as in the following formula:

No. 2040

Sulphated lauryl alcohol ether[3]	325
Nonyl phenol condensate[4]	150
Hexylene glycol	25
Water (softened or distilled)	500
	1000

Perfume use up to 10 per cent

[1] Empicol TLP type
[2] Empilan NP20 type
[3] Empicol ESB 30 type
[4] Empilan NP 20 type
} Albright & Wilson Ltd., Marchon Division

Procedure: Mix the perfume with the nonyl phenol condensate and dissolve with the aid of gentle heat if necessary. Add the sulphated lauryl alcohol ether followed by the water and mix. Finally add the hexylene glycol.

If an alkylolamide is added to the formula the perfume solubility is affected and the product becomes cloudy. In the next formula, ethoxylated lauric *iso*-Propanolamide is used to increase the amount and stability of the foam.

No. 2041

Sulphated lauryl alcohol ether	200
Hexylene glycol	100
Lauric *iso*-Propanolamide condensate[1]	200
Water (softened or distilled)	500
	1000

Perfume use up to 10 per cent

[1] Empilan 303 type—Albright & Wilson Ltd., Marchon Division.

Procedure: Mix the perfume with the lauric *iso*-Propanolamide condensate and dissolve with the aid of gentle heat if necessary. Add the sulphated lauryl alcohol ether followed by the water. Finally add the hexylene glycol and mix.

Bubble baths with good foam stability can be prepared with amphoteric surface-active materials. Because these substances have some substantive properties, it follows that they will give a pleasant perfumed effect to the skin after bathing. The choice of materials used in the formula is again decided by their ability to solubilize the perfume.

In the following formula a solubilizing material is used in addition to the main foaming agent.

No. 2042

N. Lauryl, myristyl β-amino proprionic acid[1]	400
Sodium salt of N. Lauryl β-iminodiproprionic acid[2]	100

No. 2042 (*continued*)

Water (softened or distilled)	450
Perfume	50
	1000

[1] Deriphat 170 C (50 per cent paste) General Mills, Chemical Division, Kankakee, Illinois.
[2] Deriphat 160 C (30 per cent paste) General Mills, Chemical Division, Kankakee, Illinois.

Procedure: Warm the ingredients together until homogeneous.

In certain formulae a synthetic water-soluble gum such as sodium carboxy methyl cellulose can be used to increase the viscosity and improve the foam stability. Choice of perfumes for these products was originally towards a lavender or pine type, but today a wide variety of perfumes are used including fancy floral bouquets and sophisticated aldehydic types. It is an essential feature of the product that a fragrance should be used which will linger for the entire period of the bath.

Bath jellies or pastes are used as additives to the bath water. Based on comparatively high concentrations of surface-active agents they can also be applied to the wetted body in the bath or during a shower when they act as cleansers in the manner of a shampoo. The products are usually packed in a tube and suitable formulae are as follows:

No. 2043

Sodium lauryl ether sulphate	800
Lauric diethanolamide	200
	1000

Perfume use up to 3·0 per cent
Solution of formaldehyde (40 per cent) 0·1–0·2 per cent

Include a water-soluble dyestuff.

Procedure: Mix the lauric diethanolamide with part of the ether sulphate and any colouring solution and warm gently. Add the perfume followed by the balance of the ether sulphate with slow stirring to avoid entrapment of air.

No. 2044

Sodium lauryl ether sulphate	740
Coconut diethanolamide	100
Water	160
	1000

Perfume	up to 3·0 per cent
Solution of formaldehyde (40 per cent)	0·1 to 0·2 per cent
Water soluble dyestuff	*q.s.*

Procedure: Gently warm the coconut diethanolamide with part of the ether sulphate and any colour solution. Add the perfume followed by the balance of the ether sulphate with slow stirring to avoid entainment of air. Prepare a 25 per cent solution of sodium chloride in water and add this with gentle stirring to the mixture to adjust to the required viscosity. The solution of sodium chloride and water used corresponds to 160 parts as given in the formula.

An interesting bath product in the form of a firm gel can be made as follows:

No. 2045

Gelatin	25
Sodium lauryl ether sulphate	300
Glycerin	55
Water	620
	1000

Perfume	use up to 5·0 per cent

Procedure: Soak the gelatin thoroughly in water and dissolve in the glycerin, previously heated. Mix the ether sulphate and perfume and add the remaining water. Warm gently if necessary and mix. Pack in jars.

The preparation of bubble baths in powdered form is limited by the availability of stable and solid forms of surface-active materials. Sodium alkyl sulphonates and sodium alkyl aryl sulphonates are two suitable materials available as free flowing powders. They can be obtained commercially together with a built-in foam stabilizer. A bubble bath in powder form can be prepared as follows:

No. 2046

Sodium dodecyl benzene sulphonate (containing foam stabilizer)[1]	250
Sodium alkyl sulphonate	250
Carboxy methyl cellulose	20
Sodium sesquicarbonate	480
	1000
Perfume	2·0–3·0 per cent

[1] Nansa UC (H)—Albright & Wilson Ltd.

Use of sodium alginate as a foam stabilizer is illustrated in a formula of the American Society of Hospital Pharmacists.

No. 2047

Sodium alkyl sulphate	350
Sodium sesquicarbonate	630
Sodium alginate	20
	1000

Water softeners

Water softeners can be defined as materials which promote the formation of lather or foam. Their prime object is to remove the permanent hardness of water caused by sulphates of calcium and magnesium and chlorides of magnesium and sodium. This can be done either with chemical substances or by using surface-active agents.

The simplest form of water softener is probably washing soda crystals $Na_2CO_3 . 10H_2O$, or the dried form of sodium carbonate. These materials cause precipitation as carbonates those metallic ions which are responsible for the hardness of water.

The water softening properties of the sodium metaphosphates are due to their sequestering action, a term used to describe any instance when an ion is prevented from exhibiting its usual properties due to the formation of a co-ordination complex with an added material. Thus calcium and magnesium soap precipitates are not produced from hard water treated with these compounds. They are, therefore, useful materials for use in bath preparations or other products where soap is used as the cleansing agent.

Equally effective when used with synthetic detergents they are also usefully employed in preparations for domestic cleaning, home laundering, and in industries concerned with washing, dyeing and textile processing.

Sodium metaphosphate, $NaPO_3$ exists in a number of forms and varieties. In addition, certain mixtures are known by this name which leads to confusion in associating names and compositions with properties. Generally referred to as glassy sodium metaphosphates the compounds are members of a series of polyphosphates with varying ratios of Na_2O to P_2O_5. They are available either as glassy plates or in powder form, and are completely soluble in water but insoluble in organic solvents. 'Calgon' (Albright & Wilson (Mfg.) Ltd.) is the registered trademark for a sodium phosphate glass commonly called sodium hexametaphosphate. This material has a molecular ratio of 1·1 Na_2O: 1 P_2O_5 and a typical analysis of the compound is given as follows:

P_2O_5	67%	Fe_2O_3	200 ppm
As_2O_3	1 ppm	SO_4	100 ppm
Pb	2 ppm		

Sodium hexametaphosphate is a suitable sequestering agent which prevents precipitation of calcium soaps in hard water. Tetra sodium pyrophosphate and sodium tripolyphosphate can also be used for their sequestering effects. These materials can be used either on their own or in combination. A useful general purpose formula is as follows:

No. 2048

Sodium tripolyphosphate	32–50
Tetra sodium pyrophosphate	20–30
Sodium hexametaphosphate	to 100

This formula combines the sequestering action of the hexametaphosphate for calcium salts and the sequestering action of the pyrophosphate for magnesium salts. Sodium tripolyphosphate is a very stable material during storage and a suitable basic builder for the mixture.

Surface-active agents

Surface-active agents added to water cause a change in the surface tension so that they lower the surface energy. This effect is known as surface-activity, and materials which show these properties have become known as surface-active agents. It should be noted, however, that the term surface-active agent includes not only wetting agents and dispersing agents but also emulsifying, foaming, and penetrating agents. These are in fact materials which modify the properties of any liquid either at the surface or at an interface. Probably the oldest and best known of surface-active agents is soap. At this point it is of interest to define a surface or interface as a line which divides the two phases. Since a single line can only exist between two phases it follows that emulsification, wetting, spreading or foaming or any combination of these effects is governed by what happens at the interface of the two phases. It should be remembered however, that the term surface-active agent is very often loosely used, but in the main can be considered as a soluble substance which when added to a vehicle will alter the properties of that vehicle. When suitable surface-active agents are mixed with oils and the mixture added to water a fine dispersion of the oil is obtained giving the effect of solubilization. The properties of the aqueous vehicle have been modified to cause formation of a clear emulsion system of the oil and water. Mixtures of surface-active agents are used in this way for the preparation of so-called water soluble perfumes.

When considering surface-active agents and their various applications in widely divergent products, it is useful to remember that their different properties depend upon two main factors governing their constitution. These are firstly, the number and type of hydrophilic or water-loving groups in the molecule, and secondly the manner in which the hydrophilic groups are connected with the hydrophobic or oil-loving groups of the molecule. In order to visualize the classification of a surface-active agent according to the nature of the hydrophilic and hydrophobic groupings, consider the case of sodium stearate which ionizes in solution in the following way:

$$C_{17}H_{35}COONa \rightleftharpoons C_{17}H_{35}COO^{-}Na^{+}$$

In this example the long chain negatively charged stearate ion forms the hydrophobic or oil soluble portion of the molecule, and the positively charged sodium ion forms the hydrophilic or water soluble portion of the molecule.

A solubilizer is generally regarded as a substance which enables insoluble oil to be made water-soluble or water-miscible. Best results are obtained by using a mixture of surface-active agents rather than single substances. The mechanism of the system depends upon the formation of micelles and complete dispersion or solubility depends upon the concentration of micelles.

Miscellaneous bath products

Miscellaneous products for use in the bath include antiseptic and germicidal liquid preparations. These are often considered as pharmaceutical products but concern the cosmetic chemist because they are often used for toilet purposes and it is also becoming popular to include a perfume. These preparations generally contain a germicide, such as one of the chlorxylenols. Either parachlormetaxylenol (P.C.M.X.) or dichlormetaxylenol (D.C.M.X.) can be used, the latter being more active than the para salt although it has a stronger cresol-type of odour. When the strong antiseptic odour of the chlorxylenol is not required to be predominant a strong type of perfume must be used in order to cover the basic odour of the antiseptic. Pine type perfumes based on terpineol, terpinolene or dipentene are popular. Floral perfumes are also used. Castor oil soap is generally used as the solvent for the germicide and is made as follows:

No. 2049

Castor oil	500
Caustic soda (NaOH)	65
Water (softened or distilled) sufficient to prepare	1000

Basic formulation for the complete product are given:

No. 2050

D.C.M.X.	20
Terpineol	20
Terpinolene	20
Castor oil soap (50%)	120
Water (softened or distilled) sufficient to prepare	1000

No. 2051

D.C.M.X.	50
Terpineol	50
Dipentene	50
Castor oil soap (50%)	500
iso-Propyl alcohol	100
Water (softened or distilled) sufficient to prepare	1000

If a floral perfume is used, strong odours such as violet, sweet pea, or hyacinth are the most effective to cover the basic odour of chloroxylenol. Terpineol and dipentene should then be omitted from the formula and solubility of the perfume oil and germicide effected by adjustment of the proportions of alcohol and soap. When a higher concentration of germicide is used the amount of soap solution present may not be sufficient to give a clear bright product. Solubility and clarity can be obtained by the addition of alcohol.

Several products can now be prepared in aerosol form either for use in the bath or as shower aids. Such a product is an aerosol body shampoo which is sprayed onto the wetted body, and massaged into the skin to remove surface soil. The foam is washed off in the bath or under the shower. A basic formula is as follows:

No. 2052

Shampoo base:

Triethanolamine/diethanolamine lauryl sulphate[1]	200
Polyoxyethylene lanolin derivative[2]	10
Water (softened or distilled)	790
	1000

Perfume 1·0–2·0 per cent
Preservative: Solution of formaldehyde (40 per cent) 0·1–0·2 per cent

[1] Empilan TDL.75 type—Albright & Wilson Ltd.
[2] Solulan 98 type—American Cholesterol Products Inc.

Procedure: Gently warm the triethanolamine/diethanolamine lauryl sulphate in part of the water. When dissolved add the perfume and polyoxyethylene lanolin derivative. Finally add the remaining water, and mix well.

Container charge

Shampoo base	80
Propellent—Arcton 12/114 (50:50)	20
	100

Suggested containers and valves. Containers: internally lacquered aluminium monobloc. Valve: tilt action with lacquered cup fitted with a micromist button. Use of this valve and button causes the liquid to foam as it impinges on the surface of the body and reduces residual foam on the button to a minimum.

A floating type bath oil is prepared as follows:

No. 2053

Bath oil base:

Mineral oil	750
iso-Propyl myristate	200
Silicone oil[1]	50
	1000
Perfume	2–5 per cent

[1] Type MS 200/10 (oil soluble)—Midland Silicones Ltd. or type ESP.2565 (alcohol soluble)—Midland Silicones Ltd.

An oil soluble colour can be included if required.

Procedure: Mix the perfume with the *iso*-Propyl myristate and stir in the silicone fluid. Finally add the mineral oil and mix.

Container charge:

Bath oil base	40
Propellent—11/12 (50 : 50)	60
	100

Suggested containers and valves. Containers: plain tin plate or plain aluminium monobloc. Valve: standard.

The following formula gives an instant bubble bath foam:

No. 2054

Foam bath base:

Ammonium triethanolamine lauryl sulphate[1]	500
Coconut diethanolamide[2]	5

No. 2054 (*continued*)

Ethoxylated lanolin alcohol[3]	25
Water	470
	1000

Perfume 2·0–5·0 per cent
Preservative: solution of formaldehyde (40 per cent) 0·1–0·2 per cent

[1] Empicol TC34 type—Albright & Wilson Ltd., Marchon Division.
[2] Empilan CDE type—Albright & Wilson Ltd., Marchon Division.
[3] Solulan 16 type—American Cholesterol Products Inc.

Procedure: Gently warm the lanolin derivative and perfume with the lauryl sulphate and part of the water. Stir until dissolved and add the remainder of the water with gentle stirring to prevent frothing.

Container charge:

Foam bath base	80
Propellent—12/114 (50 : 50)	20
	100

The spray is directed on to the bath water until the required amount of foam is obtained. If more foam is required the propellent ratio can be altered to 60:40. Some purfumes, particularly when used at high concentration may give a cloudy solution but this can be cleared by increasing the proportion of lauryl sulphate.

Suggested containers and valves. Containers: internally lacquered tinplate or internally lacquered aluminium. Valve: standard with a lacquered cup. Use a button with a simple orifice to give a jet of product on to the bath water.

A moisturizing cream used after bathing prevents dryness and flaking of the skin and can be conveniently applied in aerosol form. A suitable product is prepared as follows:

No. 2055

Cream Base:

A	Mineral oil	7·5	
	Silicone oil[1]	2·5	
	Lanolin	10	
	Emulsene 1220	50	(split)
	iso-Propyl myristate	20	
B	Propylene glycol	20	
	Glycerin	40	
	Water	850	
		1000	

Perfume	0·3–0·5 per cent

[1] Silicone MS.200/10 type–Midland Silicones Ltd.

Preservative: methyl parahydroxybenzoate 0·15 per cent

Procedure: Heat part (A) and part (B) independently to 65°C. Add (B) to (A) with stirring. Cool with stirring, adding the perfume at 35°C. Continue stirring until cold.

Container charge:	
Cream base	91
Propellent–12 or 12/114 (50 : 50)	9
	100

Suggested containers and valves. Containers: internally lacquered tinplate or internally lacquered aluminium. Plastic coated glass or polypropylene are also suitable but require a low pressure propellent (12/114 10:90). Valve: for metal containers use lacquered cup fitted with a foam button. For plastic coated glass or polypropylene use a standard valve fitted with a foam button.

Another type of product used to lubricate the body after bathing is the after bath body spray. This is based on an alcoholic solution of a suitable oily material.

No. 2056

Spray base:	
Alcohol (90 per cent V/V)	800
iso-Propyl myristate or *iso*-Propyl palmitate	100
Silicone oil[1]	100
	1000

Perfume	0·5–1·0 per cent

[1] Silicone ESP.2565 type–Midland Silicones Ltd.

Procedure: Thoroughly mix all the raw materials. Add the perfume and mix.

Container charge:	
Spray base	60
Propellent–12	40
	100

Suggested containers and valves: Container: standard internally plain tinplate or aluminium. Valve: standard fitted with a micromist button.

CHAPTER FOUR

Dental Preparations

Dental preparations concerned with the cleansing of teeth are considered to help the mechanical cleaning action of the toothbrush. The primary object of cleaning is to remove food debris from pits or crevices on the outer surface of the tooth or between adjacent teeth. Food debris which is allowed to accumulate on the teeth is responsible for the metabolism of bacteria, and it is generally believed that this bacterial action causes formation of acid products and leads to decalcification of tooth enamel and the formation of cavities. These eventually extend to the dentine and pulp of the tooth. It is of interest to examine briefly the anatomical structure of teeth.

The teeth are set in the alveoli or sockets which occur in the jaw bones, and the gums form the covering membrane. The bulk of the tooth is made of a bone-like substance known as dentine which is normally covered by the hard calcified tissue called enamel. The enamel forms the crown of the tooth and covers that portion of the dentine which projects through the gums into the mouth. The remainder of the tooth, which forms the root, is covered with a special calcified connective tissue known as the cementum. The part of the tooth between the enamel of the crown and the cementum of the root is known as the neck, and the visible line at the junction of these structures is known as the cervical line. The dentine structure of the tooth covers the space in the tooth known as the pulp cavity—an area of connective tissue which is well supplied with nerve fibres and blood vessels. This sensitive pulp is

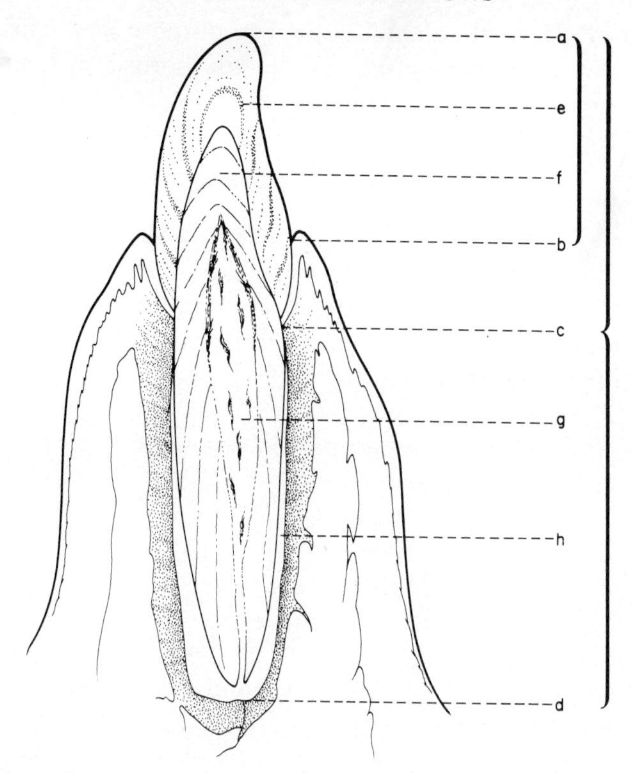

Fig. 4.1 Diagrammatic representation of tooth crown a–c, and dental tissue c–d, showing crown (a), enamel (e), dentine (f), pulp (g), cementum (h), gingiva (b), and cementoenamel junction (c).

often called the 'nerve' of the tooth. After a tooth has erupted, a portion of the gum tissue extends on to the crown of each tooth and forms the gingiva, and at this stage the tooth is considered to exhibit the area known as the clinical crown which may or may not be identical with the anatomical crown already described. As the gingiva recedes, however, it becomes attached to the tooth at the cervical line and as a result of any further recession which often occurs particularly in older people, the gingiva may become located against the cementum. The surface of teeth is coated with a thin film of mucin, food debris, and micro-organisms, this film being known as the dental plaque. Bacterial metabolism within the dental plaque causes damage to the tooth enamel and results in formation of cavities. The cavities develop progressively and eventually the sensitive dental pulp is affected. When this occurs surgery is required. The life of a tooth, therefore, is dependent upon the health of the dental pulp. The purpose of a dentifrice is to assist the mechanical effect of the toothbrush to remove the dental plaque.

Toothpastes

Toothpaste is the most popular form of dentifrice although powders, liquids and solid blocks are still used. It is important that the extent of abrasive action of all cleaning preparations should not be so harsh as to cause damage either to the gums, or to the soft dentine structure which may be exposed near to the cervical line.

The ingredients used in the preparation of toothpaste can be classified as follows:

Cleansing and polishing materials.
Detergent and foaming materials.
Humectants.
Binding agents.
Sweetening and flavouring materials.
Miscellaneous ingredients (therapeutic ingredients, whitening agents, preservatives).

Cleansing and polishing materials

Cleansing and polishing materials are used as a means of removing stains on the teeth and usually contribute about half the total weight of a toothpaste. Materials considered for use are as follows:

Calcium pyrophosphate $Ca_2P_2O_7$ occurs as a white powder which is insoluble in water. It is mildly abrasive and can be used as the sole ingredient for cleansing and polishing. It is also used as the abrasive of fluoride toothpastes because it is compatible with stannous fluoride and does not inactivate it.

Dicalcium phosphate (calcium phosphate dibasic or dicalcium orthophosphate) $CaHPO_4 . 2H_2O$ and $CaHPO_4$ occurs as an odourless white tasteless powder which is very slightly soluble in water. It is probably one of the most widely used of raw materials as a cleansing and polishing agent, and special grades are available for use in dental preparations.

Tricalcium phosphate (calcium phosphate, calcium orthophosphate, or calcium phosphate tribasic) consists chiefly of a mixture of calcium hydrogen orthosphosphate ($CaHPO_4$) and tricalcium orthosphosphate ($Ca_3(PO_4)_2$) and is available as a white colourless and tasteless amorphous powder. It is insoluble in water, and used as a mild abrasive and polishing material.

Calcium carbonate (precipitated calcium carbonate) $CaCO_3$ is a white odourless and tasteless microcrystalline powder which is stable in air and practically insoluble in water. It is available commercially in several grades of different densities described as light, medium, dense, and extra dense. The light and medium varieties find most acceptance and are often used as the sole cleansing and polishing material.

Chalk (creta, or prepared chalk) a native form of calcium carbonate, $CaCO_3$, is available as a white or greyish-white amorphous friable mass or as a powder. It is stable in air and insoluble in water but is not generally used in high proportion owing to its off-white colour. It is more often used as an ingredient of tooth cleaning powders. Magnesium trisilicate is a hydrated magnesium silicate corresponding approximately to the formula $2MgO.3SiO_2$ containing water of crystallization. It is a fine, white, odourless and tasteless powder, free from grittiness and insoluble in water. It is sometimes used with other cleansing and polishing agents as a mild antacid. As an antacid it forms magnesium salts, but such action is only exerted slowly, and it is doubtful whether this can be of any beneficial effect as an ingredient of dentifrice.

Aluminium hydroxide (aluminium hydrate) $Al_2O_3.3H_2O$ or $Al(OH)_3$ is used with other substances as a cleansing and polishing agent. The material is commercially available in two forms, as a white microcrystalline powder, and as a gel suspension. The former is used for its abrasive and cleansing properties, whereas the latter is a slow acting but effective antacid and is useful to replace calcium compounds when these are incompatible with other ingredients. The gel form of aluminium hydroxide has certain thixotropic properties depending upon the method of preparation and these should be considered in relation to the other ingredients of a formula since they affect the storage characteristics of a paste.

Ammonium phosphate dibasic (diammonium phosphate) $(NH_4)_2HPO_4$ occurs as white crystals or powder and is soluble in water giving a slightly alkaline reaction. This material is sometimes used as an ingredient of ammoniated dentifrices.

Detergent and foaming materials

Detergent and foaming materials are used in toothpastes and their cleansing action is achieved by lowering the surface tension. This promotes penetration of the paste thereby assisting removal of deposits and debris. They also cause emulsification and subsequent removal of mucus.

Soap is still used in toothpaste, although it has the disadvantage of being strongly alkaline, and specially prepared dental soap powder is now available. It is white or creamy-white in colour, of neutral pH value, and comparatively tasteless. Although specially prepared soap of this type is to be preferred, powdered castile soap, cocoa-butter soap, and soft soap can be used. All samples of soap should be examined for their odour and taste before use since some are unpleasant and slightly bitter.

Synthetic detergents have now replaced soap as the foaming agent in many modern toothpastes. These have better keeping properties, are equally effective in either acid or alkaline media, and do not form precipitates of calcium salts with hard water or saliva. Detergents for use in toothpastes are as follows:

1. Sodium lauryl sulphate.
2. Magnesium lauryl sulphate.
3. Sodium lauryl sarcosinate.
4. Sodium lauryl sulphoacetate.
5. Di-octyl sodium sulphosuccinate.
6. Monoglycerides, sulphates, and sulphonates.

These materials can be classified according to their odour and taste, their foaming, emulsifying and cleansing properties, their stability to acids and alkalis, and their compatibility with other ingredients of the paste. The most popular are sodium lauryl sulphate and special grades of this material are now available manufactured from medium and narrow cut lauryl alcohol containing only small percentages of unsulphated alcohol. Purification includes a solvent extraction process which removes bitter flavouring notes and consequently improves the taste. These pure grades are available in the form of needles and as spray-dried powders.[1]

[1] Empicol LZ or LZV: purified sodium lauryl sulphate.
Empicol LZH: high active purified sodium lauryl sulphate.
– Albright & Wilson Ltd., Marchon Division.

Humectants

Humectants are used to prevent drying out and at the same time give some degree of plasticity to the paste. The amount required in a particular formula depends to a large extent upon the specific gravity of the powders used in the product. If the powder ingredients are of low specific gravity, then less humectant is required. The proportion of humectant can vary from 5 or 10 per cent to as high as 30 per cent or more. As a general rule low proportions of humectant are not recommended. Generally only three humectants are considered for use in toothpaste. These are glycerin, propylene glycol, and sorbitol. Of these glycerin is the most widely used since it produces a glossy product and its properties and behaviour in contact with the other ingredients of the paste are well known. In addition, it is useful as a sweetening agent. Propylene glycol tends to give pastes of a softer consistency and may be preferred for use with powders of high specific gravity. It does, however, have a slightly bitter taste. Sorbitol syrup is often used to replace glycerin and is sometimes considered to have certain advantages over glycerin. The viscosity of sorbitol syrup is greater than that of glycerin and when used in the same proportions gives firmer pastes with good plasticity. It is also less sticky than glycerin and effective in preventing separation of water.

Binding agents

Binding agents are used as the excipient in toothpastes to improve and maintain their consistency. Several hydrophilic colloids are available which provide viscous aqueous systems or mucilages to maintain the colloidal balance and prevent separation of the pastes under extremes of temperature.

The amount of colloid required to give a stable paste is generally of the order of from 1–2·5 per cent depending upon the gel strength of the particular colloid used and its behaviour in combination with the remaining ingredients of the product. Glycerin of starch, liquid glucose, and simple syrup were at one time used as excipients and these were followed by the mucilages prepared from karaya, gum arabic and gum tragacanth. Of these natural products only gum tragacanth is still used in modern toothpaste formulations. Known in commerce as Persian tragacanth it is obtained by incision from various species of

Astralagus and is available either in ribbon-like flakes or ground as a fine white powder. It is obtainable in several qualities but only finely pulverized material of high quality is recommended for use in toothpastes. The proportion required depends upon the viscosity of the aqueous solution or gel prepared from the gum. To ensure consistency of the final product from batch to batch it is essential to determine the gel strength or viscosity of the grade selected and maintain the same quality of material for subsequent manufacture. To facilitate the preparation of mucilages of tragacanth it is convenient to wet the powdered gum with a sufficient quantity of either glycerin or alcohol before any water is added. This avoids the formation of lumps in the mass which become difficult to handle. Toothpastes prepared with tragacanth normally maintain their consistency and do not alter during storage. They are also stable under heat-cold conditions and are resistant to breakdown at varying pH levels.

Aqueous mucilage prepared from Chondrus or Irish moss is also used as a binding agent. Known also as carrageen, Irish Moss consists of the seaweed Chondrus crispus which after collection is partly bleached, by watering and exposing to the sun, and then dried. The dried material becomes gelatinous and glutinous when soaked in water and a 3 per cent solution in boiling water forms a thick jelly on cooling. Extracts are available in powder form in several varieties of colour, viscosity and gel strength but only those of pale yellow-green colour should be selected. Derivatives of Irish moss are also available. Aqueous mucilage and gels prepared from these materials are stable to normal temperature fluctuations but initially they are of a softer nature than gels prepared from tragacanth and more thixotropic. When formulating with this type of substance therefore, a period of time should be allowed for the paste to mature before assessing the final consistency of the product. Pastes made with Irish moss extracts have an attractive soft texture which tend to spread on the toothbrush rather than remain in a firm ribbon.

Several other materials have been suggested for their binding qualities. Of these methyl cellulose and sodium carboxymethyl-cellulose (CMC) have been considered. These synthetic materials disperse or swell in water to give clear to opalescent viscous colloidal solutions. Methyl cellulose is obtained as a greyish-white fibrous powder and the sodium derivative is available as a free flowing powder and as granules. Both materials are stable, non-toxic, odourless and tasteless. Depending upon the degree of etherification of the cellulose, giving colloidal solutions of low, medium, or high viscosity.

Purified sodium alginate also has good hydrophilic colloidal properties. The material is extracted from brown seaweed or Kelps and is available in powder form in both technical and refined grades. The technical grades are dark brown in colour and only refined grades which have been bleached and purified should be used for toothpaste. The viscosity of alginate mucilages is increased by the addition of calcium ions. These ions are responsible for the precipitation of calcium alginate and this effect must be considered when using materials in the formula which yield calcium ions.

Carbopol 934 (Honeywill & Stein Ltd.) supplied as a white fluffy powder is a carboxy vinyl polymer. It has useful thickening and suspending properties and pastes prepared with this material have a high gloss, are smooth in texture, and show good viscosity stability on storage. The pastes appear to be firm and of high viscosity but when slight pressure is applied they flow readily from a tube pack. Carbopol 934 is supplied in free acid form and the gel is prepared by neutralizing the acid solution. To prepare a paste it is recommended that the acid form is mixed with the liquid and soluble ingredients of the formula, and the resulting acid solution neutralized before adding the cleansing or polishing materials.

Bentonite (Berk Ltd.) and Veegum (R. T. Vanderbilt Co.) have also been suggested as materials suitable for use as binding agents. Bentonite is a colloidal clay composed principally of aluminium and silicates and this swells by absorption of water. Gels prepared from Bentonite are remarkably stable but dark in colour. Veegum is a magnesium aluminium silicate which absorbs water to give gels of a good colour. Their viscosity is influenced by the presence of electrolytes.

Sweetening and flavouring materials

Sweetening and flavouring materials are most important ingredients since they have a great influence on the acceptance of the final product.

Saccharin sodium (soluble saccharin) is generally used as the sweetening material at concentrations of from 0·05 to 0·3 per cent. The concentration required in a particular formula should be determined in relation to the sweetening effect of any glycerin being used. Chloroform is also useful for sweetening and flavouring and is particularly useful to cover the chalky taste and dry feeling in the mouth. It gives a sharp fresh sweetness which differs from the metallic type of sweetness associated with saccharin. It is also a useful ingredient due to its preservative effect.

The flavour is obtained by blending suitable oils which together with the sweetening agent is required to give a smooth but distinctive taste to the product, and leave a pleasant, clean and refreshing feeling in the mouth after use. It is probably true to say that the most popular toothpaste flavours are based on peppermint oils or blends of peppermint and spearmint oils often with the spearmint oil predominating. Only the finest quality peppermint oil should be used. Oils obtained from the flowers of *Mentha piperita* have a characteristic delicate aroma and should not be confused with Arvenis type oils of Brazilian or Chinese origin. Arvenis oils are sometimes used to adulterate the Piperita varieties and have a harsh aroma and a particularly fatty and straw-like flavour. For use in toothpastes, blends of American type oils are to be preferred; Italo-Mitcham oils are also used. Other flavouring agents used in small quantities to modify peppermint-spearmint flavours are: anthol, eugenol, eucalyptol, cinnamon, cassia, caraway and wintergreen. Cinnamon, cassia, and cloves are used in very small quantities owing to the intensity of their flavour, but as the others are pleasanter, larger proportions may be employed. It is always advisable to allow blends of oils to mature together before use. Thymol is another useful material which is sometimes used as a bactericide and used in small proportions has a characteristic flavouring effect. Similarly, small proportions of oil of thyme are used to obtain acceptable variations of the main flavour theme. Menthol is often added to flavouring compounds particularly those of a peppermint-spearmint type to enhance the refreshing and cooling effect in the mouth. The best effects are obtained with natural menthol. Some synthetic menthols compare favourably, whereas others are slightly bitter to the taste and have a less cooling effect.

The proportion of the flavour blend normally used varies from 1·0 to 3·0 per cent according to the combined effect which is obtained with the sweetening agents, and whether a soap or synthetic detergent is used as the foaming material. The final flavour of the product therefore depends upon the blending of suitable oils, and their taste in relation to almost all the other ingredients. When revaluating a blend of oils several trials should always be carried out to examine the different effects obtained with varying proportions of sweetening agents. Trial batches should be assessed both when they are freshly prepared and also after a period of ageing. This allows for alterations which occur by the absorption effects of the solids on the flavouring oils. Although mentioned under the heading miscellaneous materials it is important to include a preservative to prevent growth of moulds and bacteria. Some preserving action is obtained by the flavouring oils and any chloroform present, but it is advisable to add a mixture of 0·15 per cent of methyl parahydroxybenzoate together with 0·02 per cent of propyl parahydroxybenzoate. These materials are best incorporated by first dissolving them in the humectant, and will be found effective in most formulations. Sodium benzoate is not recommended for use as a preservative in toothpaste.

Liquid paraffin is sometimes used as a lubricant particularly in pastes containing soap. Even if the paste stiffens noticeably on storage a small proportion (about 1·0 to 2·0 per cent) of liquid paraffin is sufficient to allow the paste to come out of the tube without undue pressure. Among other materials suggested for use in toothpaste, 2·0 per cent of pancreatin has been used for the softening and aid to the removal of tartar without damaging the enamel in any way. Sodium perborate, magnesium peroxide, and stabilized forms of hydrogen peroxide are sometimes used ostensibly as stain removers and to whiten the teeth. Alkalis or urea compounds are used as acid neutralizers to prevent decomposition of carbohydrates and development of acidity in the saliva. Silicones are recommended for use in non-foaming toothpastes (British Patent 686429) in proportions from as low as 0·5 per cent to as high as 45·0 per cent. These materials spread easily leaving a thin water repellant film on the teeth which is considered to prevent adhesion of food particles. Since they are not miscible with water it is necessary to include an emulsifying agent in the formula in order to obtain a stable product.

Chlorophyll is included in pastes for its deodorizing effect in halitosis, but there is no clear evidence that it does have this effect. Chlorophyll has been used for the treatment of wounds when it tends to reduce odour and give them a healthy granulating appearance. In these cases a high concentration of chlorophyll is kept in constant contact with the wound. In toothpaste which generally contains from 0·1 to 0·5 per cent of chlorophyll, it is clear that the material does not remain in contact with the mouth long enough for it to have any marked deodorant effect.

It is accepted that in Great Britain the general condition of teeth, particularly those of children is poor and selected materials are constantly being examined for treatment of periodontal conditions to promote oral hygiene and cleanliness of teeth. Desirable effects include, for example, reducing the formation of bacteria or increasing resistance of the tooth enamel.

Sodium lauroyl sarcosinate [Sarkosyl NL 30 (30 per cent solution) Geigy Industrial Chemicals. *Arch. Biochem. Biophys.*, 55, 356] has already been referred to as a foaming agent. In addition the compound functions as an anti-enzyme and bacteriostat and is considered to have anti-cariegenic activities (German Patent 675827). In concentrations as low as 0·03 per cent it inhibits hexokinase and at a concentration of 0·25 per cent inhibits the bacterial flora of human saliva. The compound has comparatively long activity because it is only adsorbed slowly onto the dental plaque. It is recommended for use in toothpastes and related products at a concentration of from 1 to 3 per cent.

Use of strontium chloride applied in a paste has been investigated as a treatment of hypersensitivity (*J. Period.* 22, 49-53, 1961). Strontium is present microscopically in teeth and bone and it is suggested that the capacity of these substances for further storage is quite high. A strengthening effect is claimed by a process known as strontium calcification which helps to overcome porosity and improve retention of calcium.

In certain areas of North America where the drinking water contained 1 part per million of fluoride it was observed that children in these areas suffered less from tooth decay than did the children of other areas where fluoride was not present in the drinking water. This observation led to investigation of the effects of fluorides on tooth enamel and eventually to a proposal for large scale fluoridation of public drinking water. This proposal, however, does not meet with full approval of the experts, mainly on the grounds that the effects of long term daily ingestion of fluoride have still to be determined. Daily supplements of fluoride to the diet are most effective during the actual formation of the teeth and children who drink fluorinated water from birth can be expected to experience a reduction in tooth decay. These benefits can be expected to last into adult life. The fluoride is taken up by the enamel of the tooth during its formation in the gum, but when the crown of the tooth is fully formed, the tooth enamel does not have any further access to the ingested fluoride. It is, therefore, argued that the adult cannot be expected to receive any appreciable benefit by continuous ingestion of fluoride.

It is believed that under certain brushing conditions stannous fluoride combines with tooth enamel to form an insoluble film of tin oxide, tin phosphates and calcium fluoride which inhibits acid penetration. Many inorganic fluorides including stannous fluoride are hydrolysed and become inactive by most of the materials normally used for cleansing and polishing. This difficulty is overcome by formulating a paste at an acid pH value and using calcium pyrophosphate as the main polishing agent. For this reason, topical application of fluoride by continuous use of toothpaste is considered by some authorities to be preferable to fluoridation of water. It is the sodium compound which is normally added to water supplies but this does not respond in a paste. No less than 200 compounds were screened for their ability to reduce the solubility of tooth enamel before stannous fluoride was selected as the most effective materials for topical application (*Proc. Soc. Exp. Bio. & Med.* 1956, 92, 849). Chemical trials eventually confirmed that a paste based on a mixture of calcium pyrophosphate and stannous fluoride was effective in reducing the incidence of dental caries and the American Dental Association has endorsed publication of the value of stannous fluoride as being an effective anti-caries agent in toothpaste when it is used regularly.

It is generally accepted that the condition of teeth, particularly those of children, is poor, and the successes claimed by the use of fluoride represents a significant attempt to rectify this. Recent reports indicate that molybdenum also has some effect on oral hygiene and it seems likely that the amount of molybdenum taken in the diet has some influence on the incidence of dental caries.

Formulation of a toothpaste is indicated by first referring to the following basic formula showing approximate percentages of the main ingredients:

Cleansing and polishing materials	40 –60
Detergent and foaming materials	1·5 – 3·0
Humectants	5·0 –45·0
Binding agents	0·5 – 2·5
Sweetening and flavouring materials	0·5 – 3·0
Miscellaneous ingredients preservatives	0·15– 0·2
Water (softened or distilled)	to 100

A formula in which dicalcium phosphate is used as the cleansing and polishing material follows:

No. 2057

Dicalcium phosphate	600
Sodium lauryl sulphate	25
Glycerin	75
Propylene glycol	175
Saccharin sodium	5
Gum tragacanth	10
Water (softened or distilled)	110
	1000

The above can be modified on the following lines:

No. 2058

Dicalcium phosphate	600
Sorbitol 70	150
Gum tragacanth	10
Sodium lauryl sulphate	10
Saccharin Sodium	1
Water (softened of distilled)	229
	1000
Flavour, oils, chloroform, preservative	*q.s.*

Tricalcium phosphate is used in the following:

No. 2059

Tricalcium phosphate	500
Glycerin	75
Propylene glycol	150
Sodium alginate[1]	5
Saccharin sodium	0·5

Mineral oil	10
Sodium lauryl sulphate	15
Water (softened or distilled)	245
	1000·5
Flavour oils, chloroform, preservative	*q.s.*

[1] Manucol SA/KP.

Pastes made with calcium carbonate follow:

No. 2060

Calcium carbonate	400
Glycerin	300
Propylene glycol	50
Gum tragacanth	12
Saccharin sodium	0·5
Mineral oil	10
Sodium lauryl sulphate	12
Water (softened or distilled)	216
	1000·5
Flavour oils, chloroform, preservative	*q.s.*

No. 2061

Dicalcium phosphate	350
Calcium carbonate	140
Glycerin	200
Gum tragacanth	12
Saccharin sodium	0·5
Sodium lauryl sulphate	100
Water (softened or distilled)	198
	1000·5

In the following formula suggested by Rovesti and Ricciardi sorbitol 70 is used as the humectant:

No. 2062

Dicalcium phosphate	600
Sorbitol 70	150
Gum tragacanth	10
Sodium lauryl sulphate	10
Flavour oils	10
Water (softened or distilled)	220
	1000
Preservative	*q.s.*

Use of calcium pyrophosphate as a polishing agent in a paste containing stannous fluoride is now patented. The paste is made to the following formula:

No. 2063

Stannous fluoride	4·0
Calcium pyrophosphate	390·0
Glycerin	300·0
Stannous pyrophosphate	10·0
Miscellaneous ingredients	46·3
Water (softened or distilled)	249·7
	1000·0

The following formula for a paste used as a treatment for hypersensitivity has been published:[1]

No. 2064

Strontium chloride	100·0
Polyols	240·0
Colloidal substance	18·0
Detergent materials	22·0
Diatomaceous silicon	230·0
Water (softened or distilled)	378·0
Sweetening and flavour oils	3·5
Colour and preservative	8·5
	1000·0

Manufacture

Manufacture is carried out in stainless steel or tin lined mixing vessels. A kneading machine is suitable or a dough mixer fitted with slowly rotating blades. To prepare a mix an aqueous gel or mucilage is first made with hydrophilic colloid. In the case of gum tragacanth it is convenient to disperse the gum with a suitable quantity of the humectant before adding any water. Chloroform or a small amount of alcohol is also suitable for use as a dispersing agent. Other colloids are dispersed by adding the material to water and gently stirring the mix. The solutions are usually allowed to stand overnight or until required for use. A preservative is necessary if the solution is likely to be stored for a long period. Sodium carboxymethyl cellulose should first be wetted with sufficient hot water, whereas a methyl cellulose gel is best

[1] Sensodyne.

prepared by first wetting the material with cold water. The remaining water is then added to the concentrate and the mix allowed to stand overnight. If the humectant is not used to prepare the gel a proportion of this can be added at this stage.

The powder ingredients (not including the detergent) are sifted together and added gradually to the mucilaginous mixture with continuous gentle stirring. Additions of powders and humectant are alternated to maintain the mix in a creamy consistency. Some manufacturers add all the humectant and the detergent before adding the mixed powders, but this procedure tends to affect the thixotropic characteristics of the colloidal gel. As the mixing process is continued the thixotropic or liquid/solid transformation nature of the mix can be observed.

It will be seen that the consistency varies with narrow limits during this part of the mixing process and when each separate addition has been thoroughly mixed the paste should always be of a reasonably firm consistency. At no time during the mixing procedure should the paste become either too stiff or too soft. At this stage flavouring agents are added and finally the soap or other surface-active material. These materials both have a marked effect on the consistency of the paste, and as mixing is continued it becomes much softer. If at this stage it becomes too soft it means that the formula is unbalanced and the paste will not have the correct thixotropic properties to assume a gel-like consistency. The final texture of the paste is not obtained until the effects of the surface-active agent and flavouring oils have been added, and some manufacturers add the flavouring oils last because of their marked effect on the consistency. It is important to avoid aeration or foaming occurring during the mixing process. This can be avoided by using vacuum mixing, although if the formula of the paste is well balanced and the correct type of mechanical agitation is used aeration and foaming does not occur.

Mixing is continued until even distribution of all the constituents has been attained, and the product is then sieved or milled. The finished paste is allowed to stand for up to 24 hours, and during this time it becomes firmer but more elastic as the final consistency of the gel is obtained.

Filling of the paste into tubes requires some attention to detail to standardize the amount packed in each tube, to prevent either over or underfilling and to avoid air spaces in the sealed tube.

Hand operated machines are obtainable for small scale production together with special devices for hand sealing and crimping. For large scale production fully automatic line filling machines are used.

Aerosol toothpastes

To prepare an aerosol toothpaste a fairly thick paste is prepared containing a reasonably high proportion of humectant which contributes to a stable viscosity. The paste is dispensed through a long spout type actuator and the humectant also prevents the residual product in the nozzle from drying out to cause blockage of the valve mechanism. The following formulae illustrate the use of two thickening agents:

No. 2065

Ingredient	Amount
Dicalcium phosphate[1]	410
Glycerin	320
Sodium carboxy methyl cellulose	9
Saccharin sodium	1
Sodium lauryl sulphate (needles)	13·6
Water (softened or distilled)	246·4
	1000·0

Flavouring 1·0–1·5 per cent
Methyl parahydroxybenzoate 0·2 per cent
Chloroform is also included as a flavouring material and acts as a preservative as previously indicated.

[1] Albright & Wilson Ltd.

Procedure: Dissolve the saccharin sodium and sodium lauryl sulphate in the water. Add the sodium carboxymethyl cellulose slowly with gentle stirring and leave the mix to stand overnight. Add the glycerine and preservative and mix, add the dicalcium phosphate a little at a time, mixing thoroughly after each addition. Finally add the flavouring and mill.

No. 2066

Ingredient	Amount
Dicalcium phosphate	370
Glycerin	250
Gum tragacanth	8
Saccharin sodium	1
Sodium lauryl sulphate (needles)	10
Liquid paraffin	10
Water (softened or distilled)	351
	1000

Flavouring	1·0–1·5 per cent
Methyl parahydroxybenzoate	0·2 per cent
Chloroform—a sufficient quantity.	

Procedure: Dissolve the saccharin sodium and sodium lauryl sulphate in the water. Dissolve the preservative in the glycerine and use this to prepare a paste free from lumps, with the gum tragacanth. Add the aqueous solution slowly with stirring to the tragacanth paste. Add the dicalcium phosphate a little at a time mixing thoroughly after each addition.

Container charge: fill the paste to two-thirds total container capacity and pressurize to 85 p.s.i. with nitrogen. Container: internally lacquered tin plate or aluminium. Valve: special as supplied for toothpaste for use with nitrogen.

Tooth powders

Tooth powders are still in demand although they are less popular than toothpaste. They were the original form of dentifrice as indicated by the old pharmaceutical preparation known as camphorated chalk.

No. 2067

Camphor	100
Calcium carbonate	900
	1000

The camphor is first dissolved in a sufficient quantity of warm alcohol and calcium carbonate is gradually added. The powder is then passed through a fine sieve.

Continuous use of camphor is now considered to cause cracking of the tooth enamel.

Carbolic tooth powders were popular at one time and are still sold.

No. 2068

Carbolic acid (phenol B.P.)	25
Kieselguhr	575
Heavy calcium carbonate	400
	1000
Colour	*q.s.*

First prepare a triturate of the acid and a small proportion of the kieselguhr. This is sieved and mixed with the bulk.

It is interesting to note that these simple preparations were used for a long time as effective cleansers. With the probable exception of fluoride it would, therefore, seem that for many years the advances made in dental cleansing preparations have been more in the manner of presentation than by improvement in the efficacy of the products. The composition of modern tooth powders is basically that of a toothpaste without the humectants, water, and binding agents. Any of the substances listed as cleansing and polishing materials can be used as the basic constituents although light precipitated calcium carbonate is still largely used mainly because it is cheap and provides suitable bulk.

Specially prepared dental soap powder or any other suitable detergent may be used as the foaming agent. Dental soap powder prepared from selected fatty acids is white or creamy white in colour, and tasteless. Soaps made from unsuitable fats have a strong odour and a bitter soapy taste. Flavouring is similar to that of toothpaste although materials with strong flavour notes such as phenol, and the oils of wintergreen, cinnamon, and clove are accepted more readily in powders. A suitable amount of saccharin should also be included. The method of manufacture is simple, but the necessity for mechanical mixing and sifting is essential in all large-scale production. The flavouring is incorporated with a part of the base which is then re-sifted two or three times, and if any colouring material is being used this is also added in solution at this stage. The concentrated trituration is then added to the bulk and mixed in a sifter-mixing machine using a sixty-mesh sieve. Formulae for tooth powders are as follows:

No. 2069

Calcium carbonate	938
Dental soap powder	50
Flavouring oils	10
Sodium saccharin	2
	1000

Smokers tooth powders are intended to remove tartar and tobacco stains from the teeth. Abrasives, alkalis, and oxidizing agents are used but the formula should be designed so that the enamel is not damaged even if the powder is used indiscriminately.

No. 2070

Calcium carbonate	850
Tricalcium phosphate	100

Sodium lauryl sulphate	30
Sodium perborate	20
	1000
Flavouring oils, sweetening material	*q.s.*

No. 2071

Calcium carbonate	780
Kaolin (light)	100
Sodium lauryl sulphate	20
Powdered pumice	100
	1000
Flavouring oils, sweetening material colour	*q.s.*

Powders containing kaolin should be coloured.

A powder containing sodium benzoate is considered to be effective for removing tartar and stains.

No. 2072

Calcium carbonate	500
Kieselguhr	400
Dental soap powder	50
Sodium benzoate	50
	1000
Flavouring oils, sweetening material	*q.s.*

Solid dentifrices provide a convenient and handy form of cleaning for the teeth. Basically they consist of a powder suspended in a base of soap powder, water, and humectant. The soap is first dissolved in a mixture of glycerin and water with the aid of heat. The powder is then incorporated until the whole is a stiff mass. This is dried on trays, and when no more shrinkage occurs, it is cut into blocks and stamped out with a machine. The percentage of flavour is always higher in these products owing to the quantity of soap present, and should not be less than 2 per cent. For large-scale production it is preferable to work very much on the lines of preparing milled toilet soap. The powders are added to dental soap chips, the whole damped and milled. Flavour is added and the whole milled again. The strips are passed through a plodder and stamped out. A typical formula is given:

No. 2073

Dental soap chips	250
Heavy calcium carbonate	700
Glycerol	50
	1000
Flavouring oils, sweetening material, colour	*q.s.*

Liquid dentifrices and mouthwashes

Liquid dentifrices are frequently used on the Continent but have only a small sale in Great Britain. They consist of aqueous-alcoholic solutions of essential oils designed to have a delicate odour and impart a refreshing and pleased flavour during and immediately after use. They may or may not contain either soap or sodium lauryl sulphate as a foaming agent. Antiseptics can also be included in the composition. Formulae are given:

No. 2074

Sodium lauryl sulphate	10
Alcohol	25
Water (softened or distilled)	943
	978
Flavouring:	
Peppermint oil	15·0
Thymol	0·2
Anethol	4·5
Eugenol	1·8
Cinnamon oil	0·5
	1000·0
Colour	*q.s.*

No. 2075

Phenol	10
Boric acid	20
Tincture of myrrh	20
Tincture of quillaia	100
Glycerin	50
Rose	100
Alcohol	687
	987
Flavouring:	
Peppermint oil	5
Anise oil	5
Cinnamon oil	1
Clove oil	2
	1000
Colour	*q.s.*

Mouthwashes can be based solely on an aqueous alcoholic solution of flavouring oils and used for removing bad odours to give a clean refreshing feeling in the mouth. Products containing antiseptic and astringent materials have more therapeutic effects and are used for gingivitis and treatment of sore and tender gums. Such products function by reducing oral bacteria and act as a protein precipitant of mucous membrane.

Mouthwashes are prepared in concentrated form and directions for diluting before use should be given clearly on the label, and packaging designed to distinguish them from preparations intended for internal use. The preparations are in the form of aqueous solutions containing either high or low concentrations of alcohol. If a high proportion of alcohol is used a milky opalescence is formed on dilution with water due to dispersion of oil droplets. It is useful to include glycerin in the formula as a solvent and for its demulcent or conditioning effect on mucous membrane.

Glycerin is also useful as a sweetening and flavouring material, although it is advisable to include a small concentration of saccharin to counteract any characteristic bitterness of the flavouring oils or antiseptic materials. Formulae for mouthwashes published in the pharmaceutical literature are mainly based on active materials which function either as astringents or antiseptics. In many cases no special consideration is given to the flavour appeal of the product. Nevertheless of all the formulae which are available a few have stood the test of time. These are worthy of mention since they indicate some of the properties and types of flavour which the public are accustomed to expect from this type of product. In the following formula the bactericidal properties of phenol are used in the form of sodium phenoxide or phenate. Phenate of potassium is also in common usage as the product known as phenol and alkaline mouthwash.

No. 2076

Phenol	36·5
Solution of sodium hydroxide[1]	95·8
Concentrated orange-flower water	20
Concentrated rose water	10
Glycerin	125
Water (softened or distilled)	712·7
	1000·0
Colouring (solution of bordeaux B)	*q.s.*

[1] Solution containing 3·56 per cent w/v of total alkali calculated as NaOH.

The lotion is diluted with 5 times its volume of warm water before use.

Compound glycerin of thymol contains 0·05 per cent of thymol as the main active ingredient. It is a more powerful antiseptic than phenol and much less toxic. The product also contains certain flavouring oils and includes a small proportion of alcohol as a solvent. On referring to the formula modifications and improvements to the product will be apparent to the cosmetic chemist. The formula is as follows:

No. 2077

Borax	20·0
Sodium bicarbonate	10·0
Sodium benzoate	8·0
Sodium salicylate	5·2
Thymol	0·5
Menthol	0·3
Glycerin	100·0
Alcohol (90% v/v)	25·0
Eucalyptol	1·3
Oil of pumilio pine	0·5

Methyl salicylate	0·3
Water (softened or distilled)	828·9
	1000·0
Colouring (solution of Amaranth)	*q.s.*

Dissolve the salts in 800 parts of the water and add the glycerin; dissolve the menthol, the thymol, the eucalyptol, the methyl salicylate, and the pumilio pine oil in the alcohol, triturate with purified talc or kaolin, add the mixture gradually to the solution of the salts, filter, add the solution of amaranth and sufficient water to produce the required volume. For use the mouthwash is mixed with an equal quantity of warm water.

Antiseptics of potential value as the active ingredient of mouthwashes include chlorhexidine dihydrochloride, parachlorophenol and quaternary ammonium compounds, such as cetyl pyridinium chloride. Some of these compounds are used as lozenges for treatment of bacterial and fungous infections of the mouth and throat.

An aerosol mouthfreshener consists of a solution of flavouring and antiseptic materials dissolved in alcohol and containing a small proportion of water. Suitable flavours can be prepared by referring to the relevant ingredients given in the formulae for compound glycerin of thymol and liquid dentifrices. A formula is as follows:

No. 2078
(Freshener base)

Flavouring oils	10
Alcohol 95%/B.P.	950
Water	40
	1000

Procedure: Dissolve the flavour in the alcohol and add the water. Mix well and filter using a filter aid.

Container charge:

Freshener base	30
Propellent–12/114 (40 : 60)	70
	100

Container: internally lacquered aluminium. Valve: metering type.

CHAPTER FIVE

Depilatories

Depilatories have been used for centuries to remove superfluous hair. Many women today prefer to razor shave but the depilatory has the advantage that it removes the hair at the neck of the hair follicle. Shaving removes hair on a level with the surface of the epidermis and the growth becomes noticeable sooner. Depilation should not be confused with epilation by which is meant permanent removal of hair. This is achieved either by plucking the hair out and removing the root, or using the processes known as electrolysis or diathermy. In this treatment the hair root is destroyed by a high frequency current which is passed through a tiny needle inserted into the hair follicle. This is the most costly way of removing hair, but if the treatment is carried out successfully the hair does not grow again.

Sulphide creams and pastes

The first depilatory used by oriental women was known as rhusma. It consisted of a mixture of orpiment and unslaked lime, and when required for use was made into a paste with water and applied to intimate parts of the body. Orpiment also known as arsenic yellow is a term applied to arsenic trisulphide As_2S_3. Its use has now been abandoned because it is dangerous to use if the skin is not in good condition. The sulphides of calcium, strontium and barium are not so dangerous and have been widely used as depilatories in various forms. Barium sulphide, although very effective, is poisonous, and for this reason, is now seldom used. The strontium salt is considered to be more effective than the calcium salt but it is not as stable. These sulphides are prepared by strongly heating equal parts of the sulphate with powdered charcoal mixed into a paste with linseed oil. The reaction is according to the following equation:

$$XSO_4 + 4C = XS + 4CO$$

After heating, the mixture is allowed to cool and then powdered. Strontium and calcium sulphides are prepared as depilatory creams or pastes and usually marketed in tubes. The vehicle of the paste can be made with a natural colloid such as gum tragacanth, gum arabic, starch, or a synthetic material such as methyl cellulose. Preparation of the pastes is always a difficult procedure because they tend to stiffen a little and even dry out in the tubes. If they are too soft they tend to separate. Sometimes they appear gritty or lumpy in spite of careful and repeated milling. Strontium sulphide gives a whiter product than the calcium salt, which is prone to give greenish products unless whitened by the addition of titanium oxide. Similarly, calcium carbonate, zinc oxide, and osmo-kaolin can be used in the paste to control the discolouration which often develops in sulphide pastes after storage.

The pastes function most efficiently when they are prepared with lime water, and the pH value is adjusted to between 11·0 and 12·0. At this pH value the sulphide odour is less pronounced and the product is less likely to become coloured. Formulae for sulphide pastes are as follows:

No. 2079

Strontium sulphide	300
Titanium dioxide	30
Zinc oxide	70
Calcium carbonate	50
Glycerin	80
Gum tragacanth	50
Lime water	420
	1000

Perfume	0·1–0·2 per cent
Methyl parahydroxybenzoate	0·15 per cent

Procedure: First dissolve the preservative in the glycerin. Mix the gum tragacanth with a part of the glycerin and add sufficient water to prepare a mucilage. Triturate the zinc oxide and titanium oxide with the remainder of the glycerin and add the mucilage. Add the strontium sulphide, calcium carbonate and perfume and the remainder of the lime water. Mix until homogeneous and mill until smooth.

No. 2080

Calcium sulphide	300
Titanium dioxide	50
Zinc oxide	50
Calcium carbonate	100
Methyl cellulose	25
Glycerin	80
Lime water	395
	1000

Perfume	0·1–0·2 per cent
Methyl parahydroxybenzoate	0·15 per cent

Procedure: Dissolve the preservative in the glycerin. Prepare a gel with the methyl cellulose, and triturate the titanium dioxide and zinc oxide with this solution and add to the glycerin. Add the calcium sulphide, calcium carbonate, and perfume and the remaining lime water. Adjust the pH value and mix and mill until smooth.

The pastes are placed on the skin in a thin layer and removed (after about 3 minutes) or as soon as the slightest irritation becomes noticeable. It is then scraped and washed off.

Depilatory wax

Towards the end of the nineteenth century a product was introduced on the continent of Europe which consisted of a mixture of castor and turpentine oils, alcohol and collodion, containing a small percentage of iodine in solution. This was applied daily for four or five days after which time the film was removed leaving the skin free from hair. A modification of this appeared later when the turpentine oil was replaced by Venice turpentine (a mixture of rosin, turpentine, and linseed oils). The application of this mixture differed slightly in that the hairs were pulled out by sharply removing the film. Products designed for epilation or pulling out the hair by the roots in this manner are based on mixtures of waxes and rosin. Formulae are given:

No. 2081

Rosin	700
Beeswax	200
Ozokerite	100
	1000

Perfume 0·2 per cent

Procedure: The ingredients are heated together until a clear solution is obtained. Filter and allow to cool.

A softer product which is easier to apply is made as follows:

No. 2082

Rosin	400
Beeswax	400
Candelilia wax	50
iso-Propyl myristate	150
	1000

Perfume 0·2 per cent

Procedure: Heat to form a clear solution. Filter and allow to clear.

No. 2083

Rosin	520
Beeswax	250
Paraffin wax	180
Petroleum jelly	50
	1000

Perfume 0·2 per cent

Procedure: As before.

To use these depilatory waxes, they are first softened by heating gently and spread evenly over the selected area generally by painting on with a suitable brush. When the mixture has cooled, it is removed quickly and the hair is pulled out at the same time. Epilation in this manner is best carried out by a skilled operator in a beauty salon, although according to some medical authorities it is liable to cause skin trouble especially when used continuously.

Thioglycollates

With most modern products hair is removed by using a controlled chemical reduction method. Several factors contribute to this process. Firstly, it is known that the molecules of dry hair are strongly held in their folded or pleated shape but if the hair becomes wet, molecules of water permeate the hair fibre and cause lateral swelling to take place. The water reduces the strength of the physical and chemical forces that maintain the hair molecules in this normal folded state. This is the reason why wetted hair can be smoothed and more easily shaped into a different position than dry hair. This lateral swelling is also influenced by the pH value and at a pH value of 12·0 dissociation of the keratin takes place. Under these conditions a reducing agent breaks the disulphide linkages and the peptide chains open causing deterioration and solubilization of the hair.

Use of metallic sulphides as the reducing agent has already been described and many other materials have also been suggested. Alkaline salts of thioglycollic acid have now largely replaced the sulphides and in particular calcium thioglycollate has been found satisfactory for us in depilatories. This substance has much less odour than the sulphides and removes hair effectively without causing skin irritation, although this depends upon the balance of thioglycollic acid and the concentration of alkali used. The use of calcium thioglycollate has, however, been the subject of several patents, and the intending manufacturers should ascertain that their product does not infringe any such patents. Although the preparation of a depilatory cream with a thioglycollate gives considerably less trouble than the preparation of sulphide creams, these products do, however, require careful procedure. Several experimental batches need to be made to obtain a product which is both effective and elegant.

A concentration of calcium thioglycollate equivalent to about 3·0 to 4·0 per cent thioglycollic acid should be used and the pH value of the finished product should be adjusted to between 11·5 and 12·5 with calcium or strontium hydroxide. To give the product a cream-like consistency a dispersion of cetyl alcohol can be used, which, together with a suitable wetting agent, provides effective contact of the thioglycollate with the hair shaft. Mucilages of gum tragacanth, gum karaya and dispersions of methyl cellulose of polyvinyl alcohol are also suitable vehicles for the active ingredient. Calcium carbonate is a suitable filler to give body to the preparation and the quantity used can be adjusted to give a product of the desired consistency. Zinc oxide is a useful ingredient to reduce any tendency towards irritation. In depilatories the trihydrate form calcium thioglycollate is used. This is obtainable as a fine white powder. A suitable formula is as follows:

No. 2084

(Cream)

Calcium thioglycollate trihydrate	60
Calcium carbonate	200
Titanium dioxide	20
Cetyl alcohol	50
Sodium lauryl sulphate (powder or needles)	5
Glycerin	50
Water (softened or distilled)	615
	1000

Perfume 0·2 per cent
Calcium hydroxide, sufficient to adjust PH value.

Procedure: Prepare a trituration of the glycerin and titanium dioxide. Melt the cetyl alcohol and add the glycerin mixture blending it with a little water into a paste. Prepare separately a mixture of sodium lauryl sulphate, calcium hydroxide and calcium thioglycollate and add this to a mixture of calcium carbonate in water. Add this mixture to the cetyl alcohol paste with continuous stirring. Adjust the pH value and total weight with calcium hydroxide and sufficient water. Finally mill.

It is considered that any tendency towards irritation is reduced by including an antiseptic, although this does not prevent allergenic reactions. Oxyquinoline sulphate (German Patent 965920. 1958) is useful for this purpose. Addition of 0·25 per cent azulene has also been suggested. Because of the antiphlogistic and protective action of this material it is claimed that higher concentrations of thioglycollates can be used with safety. In such cases the time of contact with the skin is reduced because of the quicker depilatory action. The product should spread easily during the application but must be firm enough to remain localized and in close contact with the hair.

The cream or paste is left on the skin for 10 or 15 minutes according to the strength of the hair growth and then removed by scraping and finally washing. Since moisture is an important factor contributing to the deterioration process from 5·0–10 per cent of glycerin or propylene glycol is generally included in the product to prevent drying out during application. The humectant also helps to prevent the tendency of these preparations to dry out after manufacture and during storage. On the other hand, stability of the thioglycollate is suspect in contact with glycerin or propylene glycol and pastes made with relatively high proportions of glycerin sometimes discolour and develop a sharp acrid odour. A more stable product is based on an emulsion containing a small amount of mineral oil.

No. 2085

Calcium thioglycollate trihydrate	60
Calcium carbonate	175
Mineral oil (cosmetic quality)	50
Cetyl alcohol	50
Sodium lauryl sulphate (powder or needles)	5
Glycerin	35
Water (softened or distilled)	625
	1000

Perfume	0·2 per cent
Calcium hydroxide sufficient to adjust pH value.	

Procedure: Dissolve the sodium lauryl sulphate in about 200 parts of water and mix with the mineral oil/cetyl alcohol mixture to form a cream base. Prepare a mixture of glycerin, calcium thioglycollate and calcium carbonate. Mix to a paste with sufficient water and add to the cream base. Add perfume and adjust the pH value and total weight with water and calcium hydroxide. Finally mill.

Satisfactory depilation occurs using a concentration of 6·0 per cent of thioglycollate but the success of the product is often judged on rapid results which are obtained by using higher concentrations of the order of 8–9.0 per cent thioglycollate. Whatever concentration is used the pH value of the final product must be carefully checked to be in the 11·5 to 12·5 range. A satisfactory product can also be made as follows:

No. 2086

A	Calcium thioglycollate trihydrate	80
	Calcium hydroxide	30
	Sodium sulphite anhydrous	15
	Silicon dioxide[1]	5

B	Urea	60
C	Non-ionic emulsifier[2]	25
	Cetyl alcohol	50
	Water (softened or distilled)	755
		1000

Perfume 0·2 per cent

[1] Aerosil—Degussa.

[2] Ethoxylated fatty alcohol type.

Methyl parahydroxybenzoate 0·15 per cent

Procedure: Mix the ingredients of Part A with a portion of the water. Prepare a solution of urea (Part B) using a suitable quantity of water. Prepare a cream with the remaining water and ingredients of Part C and add the perfume. Mix Parts A, B and C and mill to a smooth cream. The proportion of calcium hydroxide given will normally result in the product having a pH value of between 11·5 and 12·5 but this should be adjusted if necessary.

The following formulae give suitable ingredients for use as the active portion of a depilatory:

No. 2087

Calcium thioglycollate trihydrate	80
Calcium hydroxide	25
Inert filler, Base	895
	1000

No. 2088

Solution of potassium thioglycollate (containing 30 per cent thioglycollic acid)	140
Strontium hydroxide	50
Inert filler, base	810
	1000

All depilatory preparations should be subjected to storage tests and exhaustive user tests should also be conducted to ascertain the 'safe' period for which the product may be allowed to remain on the skin, and also to establish that the product will not give rise to erythema, skin irritation or dermatitis due to incorrect alkali/thioglycollate balance.

Decomposition and discolouration occurs at a high pH value by the action of trace amounts of iron, and as a result of air oxydation. Iron reacts with thioglycollates to form brown and red decomposition products and tin lined tubes must be used for packaging. Since air or oxygen also affects the product it is normally only packaged in tubes where air spaces can be reduced to absolute minimum.

Depilatories are difficult to perfume satisfactorily. Products based on sulphides hydrolyse on storage and in use with liberation of hydrogen sulphide. These require strong floral type perfumes of the rose or violet type excluding ingredients which themselves hydrolyse such as esters, alcohols and aldehydes. Use up to 2·0 per cent of perfume. With improvements in manufacturing methods purer grades of thioglycollic acid and calcium thioglycollate are now available. These have much less odour than those on the market some years ago. Although they are less reactive than the sulphides the materials used must remain stable at a high alkaline pH value. The perfume composition should include materials of strong and powerful odour value such as the ionones, geraniol, citronellol and essential oils such as patchouli, cedarwood, sandalwood and vetivert. Perfume compounds built on diphenyl oxide are particularly effective. Good results are also obtained if a small amount of a raspberry or blackcurrant flavour is included with the perfume, although the amount used need not be discernible in the finished product. Camphor is also useful to cover the basic odour of sulphides and thioglycollates. It can be used at a concentration of from 0·05 to 0·15 per cent and if mixed with an equal proportion of menthol acts as an analgesic and counteracts possible irritation effects of the dipilatory.

An interesting new type of depilatory (British Patents 1030362, 1142090) is based on keratinase. This enzyme has the rare property of digesting keratin to form soluble degration products including peptides and amino acids. It is obtained from cultures of *Streptomyces fradiae* by precipitation with ammonium sulphate, alcohol or acetone. After concentration and purification the activity of the enzyme powder is determined by measuring its capacity to digest wool. One unit of keratinase known as a *k* unit is defined as the amount of enzyme which will digest wool keratin to produce an increase in optical density of 0·04 at 2800 Å, in 3 hours at a pH value of 8·6. The purified enzyme powder prepared on the lines described has an activity of about 200 *k* units per mg, and a depilatory composition should contain from 200,000 to 1 million *k* units per 100 grammes of finished product. The enzyme can be prepared in paste form for direct application or as a dry powder to mix with water when required for use. The depilatory action takes place more slowly than the thioglycollate products but if the purified enzyme is used it is both stable and effective at a pH value of from 7·0 to 8·0. At this range of pH value the product can remain on the skin for long periods (even overnight) without fear of irritating the skin. The enzyme is also odour free so that there is no problem in perfuming the products.

The application of a depilatory preparation in aerosol form depends upon the use of a solution of thioglycollate which remains stable in an aerosol system combined with a foam of suitable density to maintain the thioglycollate in close contact with hair and to remain stable for the period of time required for depilation to take place. Calcium thioglycollate can only be used at concentrations below 7 per cent, since solubility problems arise with higher concentrations and cause valve blocking due to crystallization. This problem is overcome by the use of lithium cations. A suitable foam is obtained using a non-ionic emulsifier of the ethoxylated fatty alcohol type. A formula is given, but manufacturers are advised to check that a proposed formulation for this type of product does not conflict with existing patents (British Patents 1030362, 1142090).

No. 2089

A	Non-ionic emulsifier[1]	140
	Lauric diethanolamine	20
	Lauryl alcohol ethoxylate[2]	20
B	Lithium hydroxide	40
	Water (distilled)	420
C	Thioglycollic acid	60
	Water (distilled)	300
		1000

Perfume	0·5–0·7 per cent
Methyl parahydroxybenzoate	0·15 per cent

[1] Abracol LDS type—Bush Boake Allen.
[2] Empilan KB 3 type—Albright & Wilson Ltd.

Procedure: Heat A, B and C independently to 70°C and add C to A with slow continuous stirring. Continue stirring and add B. Allow to cool and add the perfume at a temperature of 30–35°C. Adjust the pH value to 12, using additional lithium hydroxide.

The usual precautions when handling thioglycollate solutions should be observed, to avoid contamination with iron or dust, otherwise an intense pink colouration will develop.

Container charge:

Cream base	90
Propellent 12/114 (10 : 90)	10
	100

Container: Plastic coated glass on polypropylene.
Valve: Tilt action with no metal parts in contact with the product.

CHAPTER SIX

Hair Preparations

Before referring to hair preparations information of general interest to the cosmetic chemist is discussed. This includes notes on hair growth, a survey of the biochemistry of hair, and other factors which are concerned with preparations for the care and treatment of hair.

Initially it is desirable to define hair as a modified epithelial structure formed as a result of the keratinization of germinative cells. A hair follicle is formed as a result of a down-growth from the epidermis into the dermis or subcutaneous tissue. At the depth of the down-growth a cluster of cells known as the germinal matrix forms over a papilla of connective tissue which acts as a source of tissue fluids to the area. The part of the epidermis which has grown downwards becomes canalized between the germinal matrix and the surface and forms the external root sheath of the hair follicle. Proliferation of the cells of the germinal matrix forces the cells up the external root sheath and as they become further away from their source of nourishment, transition into keratin takes place. Growth of hairs is therefore the result of continuous proliferation of epidermal cells of the matrix. In addition, these cells form a sheath of tissue known as the internal root sheath, which extends partly up the follicle, and separates the hair from the external root sheath. During the development of the follicle some of the cells forming the external root sheath grow out into the dermis and form sebaceous glands, with their ducts opening into the follicle. The hair follicles are usually in a slanting position

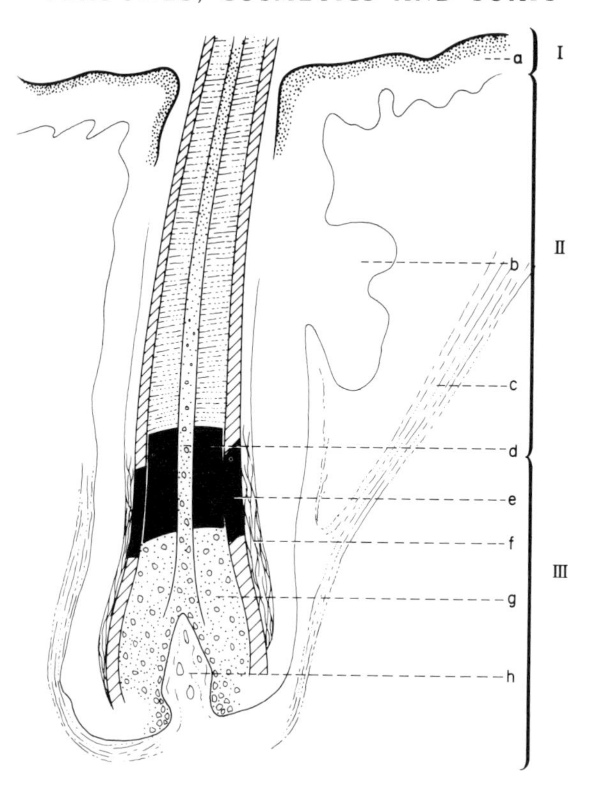

Fig. 6.1 Diagram to illustrate structure of hair follicle in relation to Epidermis I, Corium II, and Connective Tissue III showing epidermis (a), sebaceous gland (b), arrector pili (c), medulla (d), internal root sheath (e), external root sheath (f), hair bulb (g), and papilla (h).

and the sebaceous glands are generally on the obtuse angle of the follicles. The arrector pili is a small bundle of muscle fibres extending from the connective tissue sheath of the hair follicle on the obtuse side, upwards towards the dermis at a short distance from the opening of the hair follicle. Contraction of these muscles give the characteristic 'goose pimple' appearance on the skin and at the same time promotes the flow of sebum into the hair follicle and onto the skin.

A single hair consists of the central medulla of soft keratin which is surrounded by the cortex of hard keratin. The outer surface consists of very thin flat overlapping scales which interlock to form the cuticle. The medulla which always contains air spaces is present in what is known as coarse hair. There is no medulla present however in fair or blond types of hair. The colour of hair depends upon the amount of pigment present in the cortex. This pigment melanin is responsible for the colour of brown and black hair and other hair colours are also probably due to varying proportions of melanin type pigments. The melanin is formed in and by the epithelial cells of the matrix and the pigment is carried along by these cells as they are pushed up the follicle to become keratinized. If there is only a small amount of pigment present the hair appears white even though it may contain a dark medulla. Thus hair becomes white when melanin is no longer formed due to a change in the metabolism. Grey hair, therefore, is a mixture of white hairs and hairs containing varying amounts of pigment.

The process of keratinization of the epithelial cells which forms the hard keratin of the cortex and cuticle is similar to that which forms finger nails and the horny layers of epidermal tissue. Animal horn, feathers, and wool are similarly formed. The keratins consist of a large group of compounds which are obtained from keratinous structures, and in relation to hair they are often loosely referred to as 'hair keratin'. Hair, however, contains at least the four keratins of the inner and outer cuticle, the cortex and the medulla. These vary slightly in composition and, for this reason, they are separately involved in the chemical reactions and physical behaviour of hair. The keratins belong to the group of fibrous proteins which includes fibrin, collagen, elastin and silk protein and are composed of polypeptide chains. The building of a peptide chain is illustrated by reference to a basic amino acid structure thus:

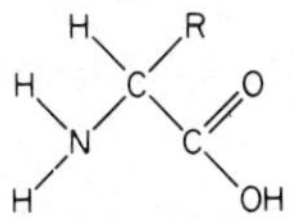

Such a structure is joined to another amino acid by elimination of a molecule of water to give a *di*-peptide with the formation of a peptide linkage.

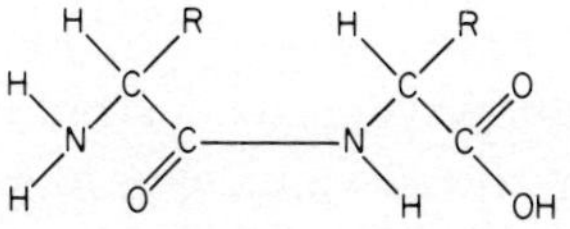

In similar fashion, a number of amino acids join together by peptide linkages to give a *poly*-peptide chain.

The polypeptide chains may take up different configurations depending upon the type of hydrogen bond which is formed

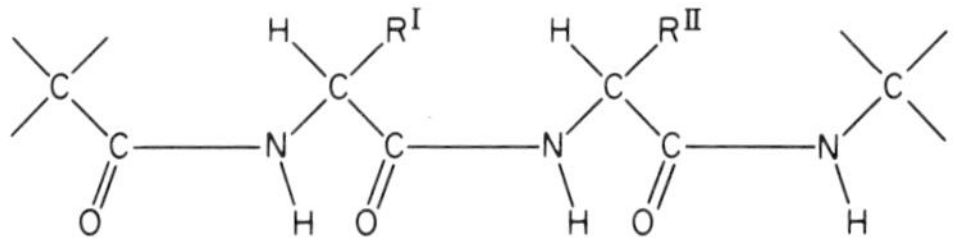

between the –CO and –NH groups. Hair keratins occur naturally in the alpha form, in which much of the hydrogen bonding takes place between the –CO and –NH groups of the same chain, as shown in the following diagram:

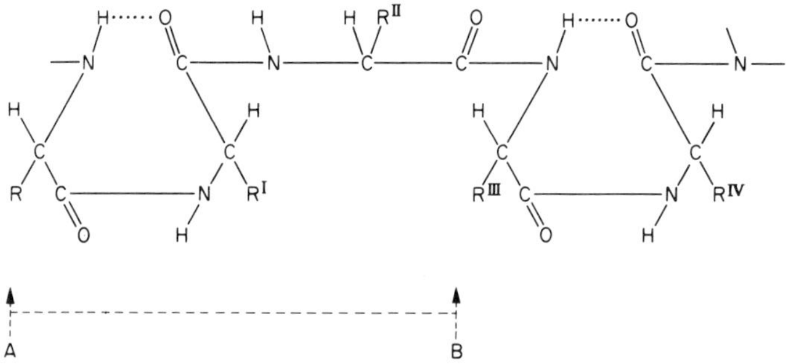

This diagram illustrates the chain-bundle formation of alpha-keratin in the unstretched form described by Astbury.

If the hydrogen bonds are broken, the structure A to B can be stretched to give the beta form of keratin which is of similar configuration to that of silk fibrogen. The stretched form can be illustrated as follows:

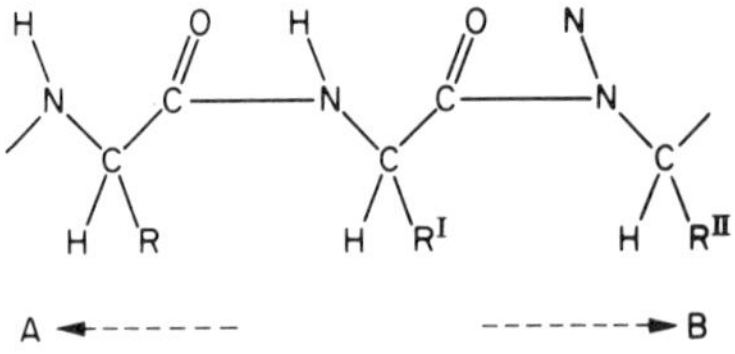

This formation will, however, give a strained structure on account of the resistance due to cross linkages, in particular the disulphide linkages of the cystine residues present, and although hydrogen bonding has been broken between groups of the same polypeptide chain, resistance will also occur due to the hydrogen bonding between neighbouring molecules of parallel chains. This stretched form, therefore, will contract to the stable alpha formation when the tension caused by stretching is released. The hydrogen bonding of the polypeptide structure A to B, with an adjacent chain, is shown below:

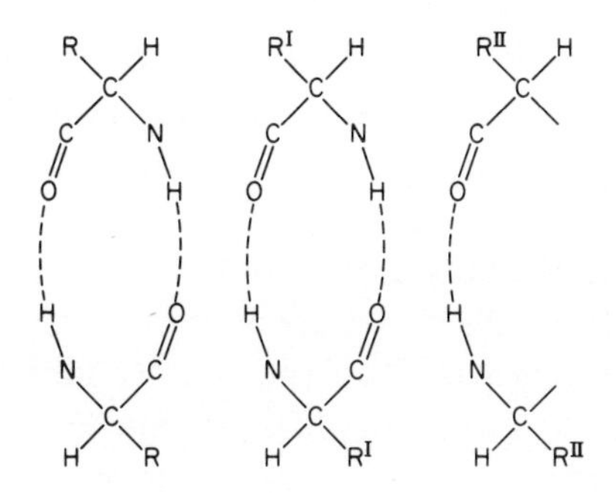

Permanent hair waving

There is evidence to suggest that a form of permanent hair waving was first practised by the Egyptians by tightly winding a mesh of hair on a cane stick, covering it with liquid mud and allowing it to bake dry in the sun. Wigmakers of old discovered that when hair was wound on a curling rod and immersed in boiling water, or steam for some hours, a permanent curl was obtained. They found that the time of immersion could be reduced considerably if a mild alkali such as borax was added to the water. The commercial possibilities of this process were first realized by Nessler in the early part of the twentieth century who claimed that hair can be stretched by 30 per cent when dry, 60 per cent when wet and up to 100 per cent when steamed, without rupture. In actual practice the safety margin is usually about two-thirds of these figures. Since stretched moist hair contracts to its original length when it dries, to wave hair by this method it must be dried before contraction can take place. A waving process based on this principle consists of winding the hair onto wooden rods, and covering the mesh with a paste of borax spread on muslin. The treated mesh is then protected with a tube of brown paper and the whole cylinder enclosed by a heating iron. This is left on the head until all the moisture has been driven off and the brown paper is scorched.

Opinions vary as to the most efficient alkali to use for this process and also as to its strength, and hairdressing specialists have their own fancies according to their experience. Two types of alkalis are employed; the non-volatile group containing borax, sodium and potassium carbonates and sodium bicarbonate, and the volatile group consisting of liquid ammonia and its salts. The main objections to the latter is their odour and incompatibility with dyed hair. Such solutions usually contain about 30 per cent of liquid ammonia, 0·880 strength, together with approximately 5 per cent of borax. So-called wave 'oils' are usually an almost saturated solution of potassium carbonate.

During the 1914–18 war Eugene Suter developed a process for permanent waving based on the Nessler ideas. The hair is first saturated with a solution of an alkali, wound onto curlers and then wrapped in sachets. In this arrangement the heating element is close to the root of the hair and steam is generated inside the sachet. During this period several similar processes were developed, and in 1923 Sartory obtained patents covering the use of exothermic materials as a means of heating the hair. These methods are long and often painful but are, nevertheless the basis of the falling heat system still in use. A machine is used for the process. It consists of several metal bars mounted on a chandelier type stand, clamps are attached to the bars which are electrically heated, the temperature being controlled by a thermostat. In the waving process the hair is first wetted with the wave solution and then wound onto curlers. The heated clamps are taken from the bars, placed over the curlers, and left for approximately ten minutes. The operating temperature is governed by the strength of the lotion being used and the tightness of curl required. The lotions are prepared with soluble sulphites or sulphides together with alkalis. Solutions of ammonium thioglycollate are also used with or without soluble sulphite. The success of the process depends entirely on controlling the amount of heat and the strength of solution. As a general rule a stronger solution of reducing substance requires less heat than a weak solution of reducing agent. Over heating in either case results in damage to the hair.

When the hair is wetted and wound on a curler, the strain produced causes the hair to form in the stretched and unstable beta-form. If the hair is left in this position and during the time the hair dries, it seems likely that some new linkages will be formed in the polypeptide structure which will be in line with the curled shape of the hair. When the curling pins are removed the hair will hold temporarily in its new stretched form. These curls are unstable and soon lose their new shape particularly if the hair is allowed to become damp.

Cold waving preparations

Cold waving preparations produce new permanent cross linkages in hair whilst it is held in the altered shape on the curlers, generally without applying heat. It can be said that the present methods of cold waving are based on work carried out on hair and wool by Goddard and Michaelis, 1934 (*J. Biol. Chem.* **106**, 605), who showed that alkaline solutions of thioglycollic acid reduce the di-sulphide linkages in the keratin of wool. From analysis of the amino residue of hair keratin it can be shown that hair contains about 17 per cent of cystine. The cystine molecule is made up of two amino acid groups linked together by two atoms of sulphur

$$\underset{|}{C}H_2{-}S{-}S{-}H_2\underset{|}{C}$$

These form cross linkages in the polypeptide structure as shown

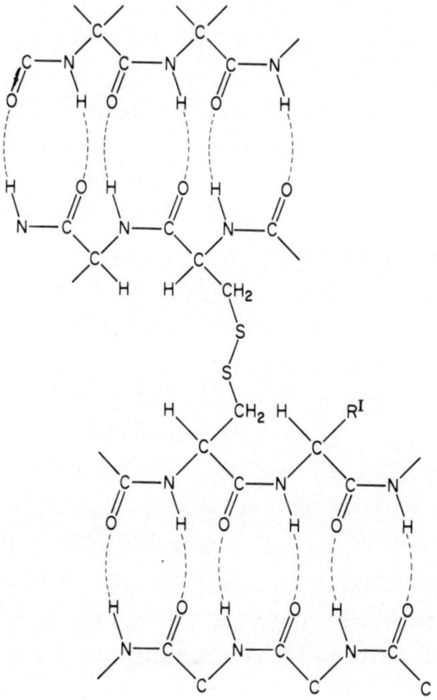

The content of cystine sulphur can readily be estimated and it is probably for this reason that the chemistry of waving has concentrated on the reduction processes which cause the breakage of the —S—S cross linkages of cystine. The general shape, elasticity and stability of the hair depends to a large extent on the cystine cross linkages since complete reduction of these structures results in breakdown of the fibre. The waving solution must break the disulphide linkages sufficiently to change their shape without destroying the fibres. An active reducing agent splits the cystine molecule into two molecules of cysteine as illustrated:

$$\begin{array}{ccc} & \text{cystine} & \\ CH_2S & \text{———} & S\cdot CH_2 \\ | & & | \\ CH\cdot NH_2 & & CH\cdot NH_2 \\ | & & | \\ COOH & & COOH \\ & \downarrow & \\ CHSH & & CHSH \\ | & & | \\ CH\cdot NH_2 & & CH\cdot NH_2 \\ | & & | \\ COOH & & COOH \\ \text{cysteine} & & \text{cysteine} \end{array}$$

and the second half of the process consists of reconstituting the cystine while the hair is still on the curlers by means of a suitable oxidizing agent.

This explanation is now considered to be a simplified version of the mechanisms which take place in the cold waving process. Reference has already been made to the hydrogen bonding which occurs in the polypeptide structure (see Fig. 6.5) and it would seem that these bonds must be involved in the waving process. This is confirmed by observing the effects of concentrated solutions of urea on hair. This substance, which is known to break hydrogen bonds, gives a tighter curl to hair when it is used together with a reducing agent. It is also possible to curl hair by using a solution of a hydrogen bond breaking substance only. Speakman's work (Speakman 1936, *J. Soc. Dyers & Colourists* **52**, 335) on hair and wool shows that in addition to the changes involved in the breakdown of disulphide linkages the salt linkages across the peptide chains are also affected in the waving process.

Cold wave lotions for permanent waving are mainly based on thioglycollic acid, generally as ammonium thioglycollate. Monoethanolamine and isopropanolamine are also used as alkalis for neutralization of the acid. Lotions prepared by ammonia neutralization are considered to give a uniform curl and not have any adverse effect either on the appearance or control of the hair after processing. On the other hand, ammonia is constantly being lost by evaporation during the process and this is considered by some workers to contribute to a non-uniform result. Monoethanolamine thioglycollate gives a curl which is more difficult to control in the final set. Both monoethanolamine and isopropanolamine however, do not increase the odour level of the thioglycollate to the same extent as ammonia neutralized lotions. Sodium and potassium hydroxides give thioglycollates which are slower acting than the amine or ammonia neutralized preparations, and make the hair very soft to handle. For this reason they are only used occasionally for neutralizing and then only as a small proportion of the total alkali. It is claimed that a more uniform and natural curl can be obtained with monoethanolamine thioglycollate if a monoethanolamine salt such as the hydrochloride, sulphate or nitrate is included in the formulation together with a small proportion of a polyhydric alcohol (U.S. Patent 3039934). Thioglycollates are prepared by straightforward neutralization of thioglycollic acid with the selected alkali and final adjustment of the required pH value.

$$HS\,.\,CH_2\,.\,COOH + NH_4OH \rightleftharpoons HS\,.\,CH_2\,.\,COONH_4 + H_2O$$

There are however, several precautions to be observed. Heat is evolved during the mixing process and both materials should be diluted before mixing, using iron free water for all the dilutions. The strength of concentrated solutions of ammonia should always be checked before dilution so that allowance can be made for any loss of strength which may have occurred during storage. It is also advisable to add ammonia solution below the surface of the acid to reduce losses of ammonia gas to the atmosphere.

All containers and mixing vessels should be made from, or lined with, glass or polythene, to avoid contact with iron and it is a good plan to provide loose covers for all vessels. Even contamination by particles of rust will cause the thioglycollate solution to become pink.

Lotions are prepared with varying concentrations of thioglycollic acid depending upon whether the final product is to be sold as a home-perm kit, or for professional use in the hairdressing salon. A general guide to the concentrations for particular requirements is as follows:

(1) Lotions based on from 5·0 to 6·0 per cent thioglycollic acid are suitable for easy-to-wave hair or for bleached and tinted hair. It is advisable to recommend a test application on a small mesh of hair in the latter case.
(2) Lotions based on 6·0 to 7·0 per cent thioglycollic acid are used for so-called normal hair. Concentrations of acid up to 8·0 or 8·5 per cent are used for hair which is generally described as difficult or hard-to-wave.
(3) For professional use lotions are prepared containing up to a maximum of 10 per cent thioglycollic acid. At concentration above this figure there is a risk of causing irritation of the scalp and permanent damage to the hair.

Except in certain special types of formulations the effective performance of the thioglycollate at any of the concentrations mentioned depends upon the pH value. For optimum waving effect this should be in the pH range of 9·5 to 9·7. If the pH value should fall to within a range of from 9·0 to 9·2 owing to loss of free ammonia during storage, the lotion will not give a satisfactory result even if the required concentration of thioglycollate is present. At pH values above 9·7 permanent damage of hair can occur by careless use of the lotion and there is also a possibility of depilation taking place. Control of pH value therefore, governs the effective performance of a waving lotion but does not necessarily control the degree of tightness of the curl. Soft or casual wave effects can be obtained with a lotion having a pH value below 9·5 but the wave is less 'permanent' than that obtained at the correct range of pH value. It is the method of winding the hair over the curlers and the number and size of the curlers used for the process which controls the appearance of the finished style. It is therefore most important to determine the pH value accurately by using a pH meter. The quantity of 'free' ammonia present should also be estimated as a further check of the pH measurement. This is made by titration with deci-normal sulphuric acid, using methyl red as the indicator. From this determination a content of 1·40 grammes of free ammonia (NH_3) per 100 ml of solution corresponds to a pH value of 9·5, and 2·20 grammes of free ammonia per 100 ml of solution is equivalent to a pH value of 9·7. Concentrations of free ammonia between these two limits are satisfactory.

To prepare a cold wave lotion therefore, the basic strength of thioglycollic acid must first be decided according to the type of hair or type of market for which the product is intended. The diluted acid is neutralized with the selected alkali by mixing in the manner described. Whichever alkali is used for neutralization the final adjustment of pH value should be made with ammonia solution.

Special additives

There are several special additives which can be included in the formulation. Some of these are intended to maintain the stability and performance of the lotion in use whereas others are added to improve the appearance of the finished product. From 1·0 to 1·5 per cent of a surface-active agent should be included as a wetting agent. For this purpose sodium lauryl sulphate or any other material normally used for shampoos can be used. Non-ionic surface-active agents are often preferred because these are more likely to be compatible with any other additives present. It is advisable to make practical use tests with all formulations to check the effect of the surface-active material. If too much is included the hair becomes slippery and difficult to control during the winding process.

Polyvinyl pyrrolidone has also been suggested as a useful additive. The amount used should be from one to three per cent of the total volume of solution. It is considered that a proportion of the polymer penetrates the hair shaft during the process and gives the wave extra strength.

Clouding agents are often used to give an opaque or milky appearance to the finished product. This effect conforms with a popular notion implying that an opaque product has a gentle action on the hair. Clear lotions are also sold and manufacturers have suggested that this type of lotion also has special properties. Several materials are available for use as clouding agents, but before deciding which is the most suitable for a particular formulation, tests should be made for compatibility with other ingredients of the lotion and also to check that they do not settle on the bottom of the container during storage. Satisfactory clouding or opacifying effects can be obtained with a mixture of an ethylene oxide type of surface-active agent and cetyl alcohol. From 0·5 to 2·5 per cent of such a mixture is used according to the degree of opacity required. A similar proportion of a water-soluble lanolin is also effective. Several speciality products are also available for use as clouding agents and details for using these can be obtained from the manufacturers.

A quaternary ammonium compound is sometimes included in the product as a conditioner provided this is compatible with the other ingredients of the composition. Suitable materials are:

(*a*) Cetyl trimethyl ammonium bromide.
(*b*) Stearyl dimethyl benzyl ammonium chloride.
(*c*) Dodecyl trimethyl ammonium bromide.

About 1·0 per cent gives wetting properties to the lotion and because these materials are substantive to hair give some softening effect. There is also less risk of the hair becoming frizzy and unmanageable on account of careless technique or over processing. The choice of suitable dyestuffs is limited, and those offered by manufacturers should be carefully checked to ensure that they are stable during storage and not affected by the reducing action of thioglycollates.

Perfumes are also affected and must be tested for stability. The quality and purity of thioglycollic acid and thioglycollates manufactured during the last few years has improved considerably. As a result, these materials are now available with much less odour and are consequently not so difficult to perfume. Perfumes stable to thioglycollates can be obtained from several perfumery houses. Those of a strong floral character such as sweet-pea, violet, or wallflower are the most effective, and should be used at a fairly high concentration (up to 1 per cent) to provide effective cover. Perfumes based on flavouring materials such as raspberry are also used by some manufacturers of professional lotions in an attempt to mask the prevailing odour of thioglycollates in the hairdressing salon.

Practical application

Practical application of cold wave lotions generally follow a standard method of procedure but the results obtained can be varied according to the number and size of the curlers used. The hair is first shampooed, rinsed, and towel-dried. It is then divided into sections and wound onto curlers starting with the section at the nape of the neck. The lotion is applied to a mesh of hair with cotton wool or a sponge, combed through the length of the mesh and wound round a curler of the selected size. The hair should not be wound too tightly or fastened too near the scalp so as to cause tension at the hair root. Tight winding also prevents penetration of waving lotion and results in failure of the process. Winding is continued at the back and sides, and finally the top front, large size curlers generally being used for this section. The wound hair is then redamped with lotion taking care not to allow an excessive amount to come into contact with the scalp.

About 10 to 30 minutes is allowed to complete the process but a guide to the length of processing time required can be obtained by examination of a test curl. A curler in the middle back section is carefully half unwound and the hair pushed towards the head. If a good wave movement is shown this indicates that the front section only requires about 5 to 10 minutes total processing. A poor waving movement of the test curl shows that a longer processing time is required and in this case a second test curl should be examined after a further 5 or 10 minutes processing time. When the hair has obtained the required amount of curl the lotion is thoroughly rinsed off with lukewarm water and the curlers blotted dry with a towel. The neutralizing lotion is then applied.

Lotions used by the professional hairdresser which are based on concentrations of up to 10 per cent of thioglycollic acid are prepared by neutralizing the acid either with monoethanolamine or isopropanolamine and the pH adjustment made with ammonia solution. With these stronger solutions a different technique can be used for the method of application. After the hair has been shampooed, rinsed, and towel dried the wave lotion is applied with a sponge or cotton wool and distributed evenly throughout the hair by combing. The hair is divided into sections and wound on to the curlers as already described, each mesh of hair being re-wetted with lotion before winding. Hot water is then sprayed over each curler whilst the head is held over the wash basin, the water being as hot as possible without causing discomfort. Spraying is continued until all the lotion is completely removed from the hair. The curlers are dried with a towel and a neutralizing solution applied in the usual manner. The procedures of applying thioglycollate solutions and the removal of the lotion by rinsing complete the first half of the waving process. At this stage the hair is in a softened and swollen condition, and in its new shape.

The second half of the process consists of neutralizing the reducing agent to reconstitute or stabilize the hair keratins in their new shape. This is also referred to as the hardening process. Neutralization therefore can be carried out by a procedure which involves oxidation. If the hair is unwound before oxidation has taken place, it will not take up the new shape. The hair will, in fact be straight, and this principle forms the basis of preparations used for straightening curly hair.

Neutralizers consist of mild chemical oxidizing agents such as sodium perborate, sodium or potassium bromate or sodium percarbonate. When supplied as part of a home-perm outfit they are packed sealed in separate sachets and the powder is dissolved in lukewarm water immediately before use. This is poured over the curlers for several minutes in order to saturate the hair thoroughly. After application it is advisable to leave the neutralizing lotion on the hair for ten or fifteen minutes. This allows time for oxidation to take place and the success of the whole process depends upon complete and effective oxidation. The hair is then rinsed with lukewarm water, and the curlers removed. Finally the hair is given a further rinsing with lukewarm water.

Neutralizers for sale in powder form are simply made by preparing sealed envelope-type packages each containing about 5.0 grammes of the oxidizing chemical. Potassium bromate is the material often used since it has a mild action and does not have any tendency to cause bleaching of the hair. Mild bleaching can occur if any alkali is present when the oxidizing agent is applied. To counteract alkalinity of the waving lotion, a buffer salt with an acid reaction in solution can be mixed with the oxidizing agent. Sodium dihydrogen Phosphate ($NaH_2PO_4 . H_2O$) is used as a buffering agent in the following formula:

No. 2090

Potassium bromate	500
Sodium dihydrogen phosphate	500
	1000

10 grammes of the mixture is used for each package of neutralizer.

These two chemicals react together if moisture is present and they also tend to cake. To prevent this they should be dried separately before mixing until the moisture content is about 0·01 per cent. After mixing pack in sealed moisture-proof containers. Neutralizers are also prepared as creams or lotions which do not require a preliminary mixing procedure. They are supplied as part of a home-perm kit and are also used by professional hairdressers. A typical formula for a lotion or thin pourable cream is prepared as follows:

No. 2091

A	Cetyl alcohol	100	
	Water (softened or distilled)	400	
		500	500
B	Sodium bromate	100	
	Water (softened or distilled)	400	
		500	500
			1000

Perfume	0·3–0·5 per cent
Methyl parahydroxybenzoate	0·15 per cent

Procedure: Warm A until homogenous and B in a separate container until the bromate has dissolved. Add B to A with stirring. Add any perfume and continue stirring until the mix is cold.

A neutralizer of a firm cream consistency with an attractive pearly sheen suitable for home or professional use can be prepared to the following formula:

No. 2092

S.D.M.B.A.C. (Stearyl dimethyl benzyl ammonium chloride)	100
Cetyl alcohol	20
Non-ionic emulsifier[1]	50
Sodium bromate	100
Water (softened or distilled)	780
	1000

Perfume	0·3–0·5 per cent
Methyl parahydroxybenzoate	0·15 per cent

[1] Emulsene 1220 type—Bush Boake Allen

Procedure: Dissolve the S.D.M.B.A.C. and sodium bromate in water by warming gently to 70°C with constant stirring. (Stirring is essential to prevent the bromate settling.) In a separate vessel melt together the cetyl alcohol and emulsifier at a temperature of 70°C and add the bromate solution with continuous gentle stirring. Mix and allow to cool and add the perfume.

Hydrogen peroxide solution is an effective neutralizer and a dilute solution is often used by hairdressers as a final rinse after the application of a more conventional preparation.

Foam neutralizers can be prepared by mixing hydrogen peroxide solution with an emulsion or detergent base immediately before they are required for use. An emulsion base is prepared from a mixture of waxes such as cetyl alcohol or stearic acid together with a suitable surface-active material. Use of a surface-active agent is indicated as follows:

No. 2093

1.	Sodium lauryl ether sulphate	50
	Sodium chloride	50
	Water (softened or distilled)	400
		500
2.	Solution of hydrogen peroxide (acid stabilizer)	20 volume

Procedure: Dissolve the sodium chloride in part of the water. Mix the sodium lauryl ether sulphate with the remaining water and add the sodium chloride solution.

When required for use one part of the 'foam neutralizer' is added to an equal amount of hydrogen peroxide solution, and mixed vigorously until a light foam-like preparation is obtained. Foam neutralizers are effective because they contain a high proportion of surface-active material which helps to penetrate the hair wound on the curlers. They are also easy to control during use. Foam type neutralizers are also prepared by mixing a solution of sodium bromate (10 per cent) with a suitable liquid or paste type shampoo base.

No. 2094

Shampoo base[1]	700
Sodium bromate	100
Water (softened or distilled)	200
	1000

Perfume, preservative *q.s.*

[1] Empicol CL Type–Albright & Wilson Ltd., Marchon Division.

Procedure: Dissolve the sodium bromate in the water without the aid of heat and mix with the shampoo base.

A comparatively new idea in the technique and application of cold waving preparations is to combine the waving process with a hair colourant. This is carried out by incorporating an acid or basic dyestuff in the neutralizer. Any of the formulae for liquid or cream type neutralizers can be used although formula No. 2104 which contains a quaternary ammonium compound is particularly recommended.

Cold wave lotions are also sold without a neutralizer in which case the directions for use are modified. After the thioglycollate lotion has been removed by rinsing, the curlers are left on the hair for four to six hours or overnight until it is dry. Neutralization in this case is by slow and gradual air oxidization. More permanent results are generally obtained when a chemical oxidation process is used.

Hair straighteners are also related to cold waving preparations and are a thioglycollate treatment 'in reverse'. They are used to straighten kinky or negroid type hair but in this case the reduction process breaks down the –S–S linkages of the curly hair and permits them to reform in the uncurled shape. The curlers are removed after thioglycollate treatment when the hair is in a softened condition before application of a neutralizer. After rinsing, a firm cream similar to that given under Formula No. 2104 is applied since this helps to set the hair firmly in its new shape in addition to its neutralizing action.

Shampoos

Shampoos continue to have a tremendous sale and the development of synthetic detergents most probably accounts for the growth of this market. The modern shampoo is designed to provide sufficient cleansing power together with adequate foam, to remove soil from the hair and scalp without seriously reducing natural oiliness–so that the hair is left with a natural gloss and in a soft and manageable condition.

The function of a shampoo has been variously defined by

several workers. The author considers that an acceptable shampoo should have the following characteristics:

1. It should remove the build-up on the hair and scalp of surface soil, excessive sebum, and residues of setting lotions and dressings.
2. The degree of foam should be sufficient to satisfy the psychological requirements of the user.
3. The shampoo should be removed easily by rinsing, and leave the hair in soft and lustrous condition, with good manageability and a minimum of fly-away.
4. The shampoo should impart a pleasant fragrance to the hair during use, in order to mask the odour of wet hair. It is, however, not essential for the perfume to be substantive.

It should be noted that the cleansing properties defined above give an indication of the required degree of detergency, since over degreasing leaves the hair dry, frizzy and unmanageable. New synthetic detergents are constantly being offered as suitable bases for shampoos but it is probably true to say that the majority of shampoos on the market are based on sulphated fatty alcohols —known commercially as lauryl or alkyl sulphates. These may include monohydric alcohols of chain length C_{10} to C_{18}. It is generally understood that a high degree of sulphation is necessary to obtain good detergency, and this is often used as the criterion when good solubility and low cloud point are required. On the other hand, a certain amount of foam is desirable and foam stability is achieved to some extent by the presence of some unsulphated alcohol. Development of foam, however, does not always indicate good detergency.

The properties of these detergents depends upon the chain lengths of the fatty alcohol used. Thus an alkyl sulphate based on C_{12} (lauric) and C_{14} (myristic) alcohols gives a wet and billowy type of lather with good cleansing properties at lower temperatures than those based on C_{16} (palmitic) and C_{18} (stearic) alcohols. Since the solubility of the sulphated alcohol decreases as the chain length of the fatty alcohol increases, shampoos based on C_{16} and C_{18} alcohols, have poor cold storage properties. Because of this they are less soluble in water and they give less foam at normal shampoo temperatures. The hardness of water has also more effect on alcohol sulphates of higher molecular weight.

Another factor to consider is that C_8 (caprylic) and C_{10} (capric) alcohol sulphates tend to be irritant, and it is usual for fractions to be carefully cut to eliminate all the C_8 alcohols and nearly all the C_{10} alcohols. Alkyl sulphate based on C_{12} anc C_{14} alcohols are non-irritant, have good storage properties, and give the most acceptable type of lather at normal shampoo temperatures.

The properties of a sulphated fatty alcohol are also influenced by the cation present. Thus, sodium alkyl sulphate has more degreasing action than the ammonium or triethanolamine compound, and because of its harsh cleansing action sodium alkyl sulphate is now generally only used as the base of compositions for the treatment of greasy hair. This treatment can nevertheless be criticized, since complete degreasing can promote an increase in the flow of sebum. In addition after a degreasing treatment the hair is dry, frizzy and unmanageable and has a dull appearance. The modern conception of a successful shampoo accentuates the manageability of the hair immediately after shampooing and in this context the most popular detergents are triethanolamine or monoethanolamine compounds, which have a mild cleansing effect. Triethanolamine alkyl sulphate which is probably the most widely accepted cleansing agent used in shampoos has the added advantage of good storage stability. Ammonium alkyl sulphate which also has a mild cleansing effect, is seldom used alone because of its high solidifying point. It is usually used in blended compounds such as ammonium-monoethanolamine, and ammonium-triethanolamine alkyl sulphates. Shampoos formulated with these bases have good foaming and cleansing properties but still leave the hair manageable.

In addition to the sulphated fatty alcohols, shampoos with particular characteristics can be prepared from sulphated fatty ethers. The sodium salts of sulphated lauryl alcohol ether are available as colourless liquids of different viscosities—and give profuse and stable foam. The viscosity of the diluted product can also be easily adjusted and controlled by the addition of a predetermined amount of sodium chloride or any other suitable electrolyte.

Shampoo additives

The search for the perfect shampoo has also led to investigation of the effect of additive materials to the basic detergent. Fatty acid alkylolamides are used to increase stability of the lather and control the viscosity. These are condensation products of fatty acids with either mono- or diethanolamine or the corresponding isopropanolamine. These materials also disperse insoluble lime and magnesium soaps produced during washing and prevent their deposition on hair and scalp.

The correct choice of detergent base plus additive, which governs the performance characteristics of a shampoo is closely correlated to the manageability and final appearance of the hair. To this end many additives claimed to have conditioning properties have been suggested. These include lanolin or one of the lanolin derivatives, cholesterol, oleyl alcohol and acetoglycerides. Additives such as lanolin or egg powder are frequently used and special claims are made that these materials will give the hair lustre and manageability. Although it is questionable whether such materials can have any influence on the final condition of the hair, they may assist by reducing the degree of degreasing, and in a two-application shampoo some deposition on the hair fibres may increase gloss or lustre. Materials of this type, however, invariably reduce the amount and stability of the foam if they are added to standard compositions.

An interesting observation based on chemical investigation claims that if amino acids are added to a shampoo, some deposition remains on the hair and scalp after shampooing. It is considered that there is a hygroscopic substance present in hair which is required to maintain the hair in good condition. This hygroscopic material mainly consists of amino acids. Thus, if amino acids do remain on the hair and scalp after shampooing it is possible that the hair will benefit by including a suitable hydrolysate containing all the essential amino acids. Formulations with this type of additive form the basis of protein shampoos.

Other additives examined for their effect on foam formation and stabilization include the glycols and polyvinyl pyrrolidone. Carboxymethyl-cellulose has been used with detergents as a soil suspending agent; and it has also been found that liquid silicones, particularly at a concentration of about 4 per cent, increase foam stability, improve setting properties, and give gloss to the hair.

Evaluation of shampoos

In the formulation of shampoos, the physical properties must be related to the performance characteristics, and several laboratory techniques have been developed for this purpose. Some success in the measurement of foams has been achieved by using simple stirring devices in graduated temperature controlled vessels and the effects of adding selected soil such as lanolin, mineral oil or vegetable oil can be determined. Ross and Miles (Miles, G. D., Ross, J. and Shedlovsky, L. *J. Am. Oil Chemists Soc.*, **27**, 26-273) developed a satisfactory method for evaluation of foam, and although this method does not give results which can be correlated to practical application, is useful for screening purposes. Myddleton designed an apparatus for the measurement of foam quality which could be related to the results of actual shampoo tests made with the same source of detergent (The Evaluation of a Shampoo. Myddleton, W. W., *J. Soc. Cos. Chem.* **Vol. IV**, 150-156).

The final assessment of a shampoo, or an individual detergent, however, is probably best determined by practical demonstration. For this purpose, the half-head technique—practised according to a standardized procedure—is often employed. This can be used to provide instructions which will be issued with the finished product. The data obtained should include the amount used, details of dilution, the number of applications, information on temperature, time of lather formation, volume and texture of lather, and ease of rinsing. The following is the normal procedure:

(*a*) During the application of a test shampoo or detergent it is advisable to note the texture and volume of the lather, together with its degree of stability and odour.

(*b*) The condition of the wet hair should be examined after rinsing for information on feel and effect of combing. Information regarding the odour value of a perfume can also be obtained at this stage.

(*c*) A similar examination should be made after the hair has been towel-dried. Needless to say, for the purpose of these tests, setting lotions should not be used.

After some experience of the half-head technique, minor variations in the performance of a shampoo or detergent can be identified and small differences in the final condition of the hair noted. The important data finally obtained includes: the degree of gloss or shine, development of highlights; ease of combing; assessment of fly-away; and degree of manageability. A similar standardized technique also can be used for evaluating perfumes in standard shampoo compositions.

Shampoo formulations

Shampoo formulations can be classified into the following main groups:

Powders.
Liquids.
Liquid creams.
Lotions.
Creams.
Jellies.

Shampoo powders

The earliest types of shampoo powders were prepared by the simple admixture of diluent or extender with soap powder. Sodium carbonate, bicarbonate, sodium sesquicarbonate, disodium phosphate or borax are suitable diluents.

No. 2095

Sodium bicarbonate	500
Disodium phosphate	200
Soap powder	300
	1000

Perfume 0·5 per cent

Shampoos of this type continue to be of some interest. They can be used with henna or chamomile to give products which are supposed to give a slight colouring effect to the hair. A henna shampoo can be prepared as follows:

No. 2096

Henna powder	50
Borax	150
Sodium carbonate	250
Potassium carbonate	50
Soap powder	500
	1000

Perfume 0·5 per cent

The following formula can be used to prepare a camomile shampoo:

No. 2097

Powdered camomile flowers	50
Borax	250
Sodium carbonate	200
Soap powder	500
	1000
Perfume	0·5 per cent

Partly because of their poor performance in hard water, powder shampoos based on soap have now been superseded by synthetic detergents. Sodium lauryl sulphate is generally used as the base for these products. This material can be obtained from the manufacturer as a white, free flowing spray dried powder in various concentrations. About 5 grammes of the pure compound is required to give a satisfactory shampoo and to increase the bulk it is mixed with a suitable diluent. Manufacturers also supply the alkyl sulphate in diluted form (perfumed if required) ready for packaging. These are prepared by spray drying the alkyl sulphate generally with sodium sulphate as the diluent.

Dry shampoos

Dry shampoos are sold in powder form although these products have a limited sale. They are useful for use between wet shampooings or when it is inconvenient to use water and are particularly effective when the hair is greasy. Provided all the powder is removed by brushing they give a quick method of improving the appearance of lank or drooping hair. They are made from powders with absorbent properties such as talc, kieselguhr or starch.

No. 2098

Talc	400
Starch	150
Kieselguhr	450
	1000
Perfume	0·5 per cent

No. 2099

Talc	250
Borax	50
Starch	700
	1000
Perfume	0·5 per cent

Liquid dry shampoos

Liquid dry shampoos are also used as a method of cleaning the hair and scalp when time is not available for the usual lather type shampoo. They are mainly designed to remove grease by use of suitable solvents such as alcohol, and when applied with gentle massage of the scalp they have a pleasant refreshing effect. They can also be used as a means of leaving the hair pleasantly perfumed. This type of product is also sold as a lacquer removing shampoo but since there is no actual washing or rinsing procedure in the treatment any lacquer on the hair must be dissolved rather than removed. On the other hand if hair and scalp is massaged with a lotion which contains a high proportion of alcohol and a suitable plasticizer, some of the lacquer on the hair will be softened or dissolved and removed during towel drying. The effects are likely to be most noticeable when a rigid film forming type of lacquer has been used on the hair. A suitable product can be made using the following:

No. 2100

Glycerin	20
Diethyl phthalate	50
Alcohol	650
Water	280
	1000
Perfume	0·5 per cent

Clear liquid shampoos

Clear liquid shampoos have already been referred to in the general discussion dealing with behaviour characteristics of detergents. The detergent and additive used in formulating are selected to give the required performance. The viscosity of the product and correct choice of auxiliary additives such as perfume and colour are also important factors to consider. The basic detergents are available from the manufacturers as solutions which differ either in physical characteristics including viscosity or in content of active material. So called 'built' detergents are available which are preformulated and contain an alkylolamide.

There is a general method of manufacture which can be followed for all clear liquid shampoos. If an alkylolamide is being used this should first be dissolved by heating gently with about half the total amount of detergent until solution is effected. The remaining quantity of detergent is then added gradually. The perfume is dissolved in the cold concentrated detergent. Any colour is dissolved in an adequate amount of water and added to the mix. The remaining water is finally added with gentle stirring to avoid incorporating air bubbles and the formation of an excessive amount of froth.

A preservative should be included in all the formula which are given and 0·2 per cent of solution of formaldehyde (40 per cent) is recommended. Phenyl mercuric salts are also effective, at a concentration of 0·001–0·002 per cent but as these are poisonous their use should comply with Poisons regulations. Phenyl mercuric borate is preferred because it is more soluble in water.

Perfumes are used in shampoos at concentrations of from 0·3 to 1·0 per cent—0·5 per cent being a generally accepted level. It is also usual to add a suitable proportion of a water soluble dyestuff.

The formulae which follow have been selected to give shampoos with the characteristics and acceptable performance already described:

No. 2101

Triethanolamine lauryl sulphate[1] (40% active matter)	600
Lauric isopropanolamide[2]	20
Water	380
	1000

[1] Empicol TL 40 type—Albright & Wilson Ltd.
[2] Empilan LIS type—Albright & Wilson Ltd.

No. 2102

Preformulated triethanolamine alkyl sulphate[1]	500
Water (softened or distilled)	500
	1000

[1] Empicol TLP type—Albright & Wilson Ltd., Marchon Division.

No. 2103

Monoethanolamine lauryl sulphate[1]	550
Lauric isopropanolamide[2]	10
Water (softened or distilled)	440
	1000

[1] Empicol LQ 27 type—Albright & Wilson Ltd., Marchon Division.
Cycloral SA type—Cyclo Chemicals Ltd.
[2] Empilan LIS type—Albright & Wilson Ltd., Marchon Division.

Several variations of this formula can be made by increasing the amount of alkylolamide and decreasing the detergent without affecting the viscosity of the product to any marked degree.

No. 2104

Preformulated monoethanolamine lauryl sulphate[1]	500
Water (softened or distilled)	500
	1000

1 Cycloryl SA/N type—Cyclo Chemicals Ltd.

Optical brighteners (optical bleaches) are useful additives particularly for use with clear liquid shampoos although they can also be used in creams and lotions. These materials increase visible light reflection and are considered to give the hair more lustre. Fluorescent materials are also used for the same purpose. A Swiss* patent describes a similar effect by using a colourless hair-substantive organic coumarin derivative.

Concentrated clear liquid shampoos are prepared for sale either to manufacturers who simply dilute the concentrate with water before packing in bottles for retail distribution or to hairdressers who dilute the product as required for use in the salon. They are particularly useful for hairdressing premises since they help to solve the problem of storage of large amounts of liquid. Concentrated clear liquid shampoos can be made as follows:

No. 2105

Monoethanolamine lauryl sulphate[1] (27% active)	750
Coconut diethanolamide[2]	20
Water (softened or distilled)	230
	1000

* Swiss pat. 326772.

[1] Empicol LQ 27 type–Albright & Wilson Ltd., Marchon Division.
Cycloral SA type–Cyclo Chemicals Ltd.
[2] Empilan CDE–Albright & Wilson Ltd., Marchon Division.

The sulphated fatty alcohol ethers are clear colourless almost water-white liquids and have outstanding foaming properties. Chemically they are obtained by the reaction of one molecule of a fatty alcohol with one or more molecules of ethylene oxide, thus introducing hydrophilic groups into the compound. They are stable over a wide range of pH values, and are compatible with most of the builders or special additives which have been mentioned. They are more soluble and less affected by hard water than the fatty alcohol sulphates. Several of these materials are available commercially varying either in viscosity or content of active matter. A typical formula for a clear liquid shampoo is as follows:

No. 2106

Sulphated lauryl alcohol ether[1] (sodium salt)	400
Water (softened or distilled)	600
	1000
Perfume, colour	*q.s.*

[1] Empicol E.S.B. 30 type–Albright & Wilson Ltd., Marchon Division.

This formula gives a shampoo of a thin consistency but an interesting property of this group of raw materials is the manner in which the viscosity of the final product can be altered.

Viscosity, together with perfume and colour are important characteristics which often contribute towards a successful product. The viscosity of solutions of the lauryl alcohol ether sulphates can be controlled by adding electrolytes and a convenient way of doing this is to add a solution of sodium chloride. Prepare a solution containing about 1 part of sodium chloride to 3 parts of water, and add small portions of this solution into the shampoo stirring well after each addition until thc mix is uniform. As more salt solution is added the shampoo becomes more viscous until a maximum viscosity is obtained. After this stage has been reached, further additions of salt solutions lower the viscosity.

Many perfumes also affect the viscosity of shampoos based on these materials and the perfume should be mixed with a small proportion of detergent and added to the bulk before the main dilution with water is made. The final viscosity of the product can then be carried out by adding salt solution. The quantity of salt solution used should be carefully noted so that the process can be repeated successfully when subsequent batches are prepared. Usually it will be found that between 1·0 and 2·0 per cent of sodium chloride is required based on the total weight of the shampoo. This amount does not greatly affect the detergent concentration of the final product.

Among recent raw materials developed for use in shampoos ditertiary amines have been mentioned (Brit. Pat. 867874). In addition to good foaming and cleansing properties it is claimed that the materials impart gloss and also have anti-static properties which promotes ease of combing.

Liquid cream or lotion shampoos

Liquid cream or lotion shampoos are pourable preparations prepared either from fatty alcohol sulphate pastes, or from clear liquid detergents by mixing with a stabilizer and suitable opacifying agent. Concentrated preformulated materials can be obtained from detergent manufacturers which only require dilution and addition of colouring and perfume. Liquid creams are often used as the basis of a large variety of shampoos, sold by many names, such as protein, lanolin, egg, brandy, milk, lemon cream and even strawberry, depending upon the additive used. Their attractive pearly appearance and texture suggests they have a mild cleansing action, and since the additives used generally reduce the cleansing properties of the detergent there may be some truth in this suggestion. A thick but mobile pearly cream can be prepared as follows:

No. 2107

Fatty alcohol sulphate paste[1]	500
Sodium chloride	3
Water	497
	1000

[1] Empicol C. L. Paste—Albright & Wilson Ltd., Marchon Division. Cycloryl M.1—Cyclo Chemicals Ltd.

Procedure: Dissolve the sodium chloride in the water and add slowly to the paste with continuous stirring.

In common with many other detergents of this type the viscosity of the final product is controlled by the concentration of sodium chloride in the final dilution. As previously described electrolytes should always be added in solution and preferably to the diluted detergent since this helps the mix to be of uniform viscosity. The viscosity of liquid cream shampoos is also affected by the perfume so these should always be tested by preparing a laboratory scale batch before finally adjusting the viscosity.

Liquid cream shampoos with an attractive pearly sheen can be prepared as follows:

No. 2108

Monoethanolamine lauryl sulphate[1] (27% active)	400
Ethylene glycol mono stearate	50
Water (softened or distilled)	550
	1000

[1] Empicol L.Q. 27–Albright & Wilson Ltd., Marchon Division.

Procedure: Heat and mix the ethylene glycol monostearate with a small quantity of the detergent to form a homogeneous mixture. Gradually add further quantities of detergent and water mixing well after each addition.

An alkylolamide can be added as a foam booster:

No. 2109

Monoethanolamine lauryl sulphate[1] (27% active)	400
Lauric isopropanolamide[2]	15
Ethylene glycol monostearate	50
Water (softened or distilled)	535
	1000

[1] Empicol L.Q. 27–Albright & Wilson Ltd., Marchon Division.
[2] Empilan L.I.S.–Albright & Wilson Ltd., Marchon Division.

The following formula is for an egg shampoo

No. 2110

Fatty alcohol sulphate paste[1]	275
Lauric isopropanolamide[2]	10
Ethylene glycol monostearate	30
Egg powder	2·5
Water (softened or distilled)	682·5
	1000

[1] Empicol L.M. 45–Albright & Wilson Ltd., Marchon Division.
[2] Empilan L.I.S.–Albright & Wilson Ltd., Marchon Division.

Either dried whole egg powder or dried egg yolk can be added to this shampoo. It should be mixed into a paste with a small quantity of water or glycerin and then diluted with some of the detergent before adding to the mix. The amount of egg powder used is not sufficient to colour the shampoo and a suitable quantity of a water soluble egg colour is necessary.

Shampoo pastes or creams

In addition to liquid cream or lotion shampoos thick viscous products are also in demand. These are now mainly used by the professional hairdresser. In the trade they are often called cream shampoos but a more accurate description is the alternative name of shampoo pastes. They are based on sodium alkyl sulphates made from a medium cut alcohol which gives products of a firm consistency with more tendency to pearl.

To prepare the pastes use a wax such as cetyl alcohol as the builder and sodium alkyl sulphate in needle or paste form.

No. 2111

Ingredient	Amount
Sodium lauryl sulphate needles[1] (84% active matter)	400
Cetyl alcohol	75
Water (softened or distilled)	525
	1000

[1] Empicol LZV type—Albright & Wilson Ltd., Marchon Division.

No. 2112

Ingredient	Amount
Sodium lauryl sulphate paste[1] (34% active matter)	800
Cetyl alcohol	50
Water (softened or distilled)	150
	1000

[1] Empicol LZ 34 type—Albright & Wilson Ltd., Marchon Division.

No. 2113

Ingredient	Amount
Sodium lauryl sulphate paste[1] (45% active matter)	750
Cetyl alcohol	75
Water (softened or distilled)	275
	1000

[1] Empicol LM45 type—Albright & Wilson Ltd., Marchon Division.

No. 2114

Sodium lauryl sulphate needles[1] (82% active matter)	450
Cetyl alcohol	50
Water (softened or distilled)	500
	1000

[1] Empicol LMV—Albright & Wilson Ltd., Marchon Division.

An alkylolamide such as coconut diethanolamide or lauric isopropanolamide can be included in any of the above formulae as a foam stabilizer. About 1·0 to 2·0 per cent only should be included, otherwise the consistency of the final paste may be affected. A similar procedure is followed for the manufacture of these products whether they are made either from the needle or paste form of the alkyl sulphate.

The concentrated detergent is heated together with the water to a temperature of about 80°C, in a steam jacketed pan fitted with a stirrer. Direct heat should not be used. The wax is added and stirring continued for about 15 minutes. The mix is allowed to cool to about 40-45°C. Any perfume or colour is then added and stirring continued until the mix is uniform. The mixes should always be stirred gently and prolonged stirring avoided as aeration makes the finished product spongy and less firm. It is also advisable to fill into containers whilst still warm since this helps to give a firm product with an attractive pearly appearance.

The necessity for heating sodium alkyl sulphate in paste or needle form and the incorporation of waxes or other modifiers can be a laborious and tedious process. For this reason it is sometimes preferable to obtain one of the several proprietary products which are available from the manufacturers of detergent products. These are preformulated or built products and can be purchased containing about 35 per cent active material suitable for filling into jars, tubes or sachets.

Concentrated bases can also be purchased which can be coloured and perfumed to prepare an original and distinctive product.

No. 2115

Sodium lauryl sulphate paste (50–60% active matter)	450
Water	550
	1000

Although some processing is required in this case a formula made from concentrated paste is easier to handle than one prepared from the basic detergent. All the ingredients, with the exception of the perfume are heated together in a steam jacketed vessel and processing continued in a similar manner to that already given. Although these pastes are finished products ready for use as shampoos they can be sold as raw materials, provided they do not contain perfume and the word 'shampoo' does not appear on the label.

Aerosol shampoos

Shampoos have to be specially formulated for aerosol packing. They should emerge from the aerosol foam head as a soft manageable foam, but have enough staying power to shampoo the hair thoroughly and efficiently. In the following formula coconut diethanolamide is used as a foam builder and stabilizer:

No. 2116
(Basic shampoo)

Shampoo base:

Ammonium/Triethanolamine lauryl sulphate	600
Coconut diethanolamide	15
Water	385
	1000

Perfume	0·3–0·7 per cent
Phenyl mercuric borate	0·01 per cent

Procedure: Gently warm the diethanolamide with part of the lauryl sulphate. Add the remainder of the lauryl sulphate and perfume. Stir gently. Finally add the water and preservative, and mix.

Container charge:

Shampoo base	92
Propellent—12/114 (50 : 50)	8
	100

Container: internally lacquered aluminium or tin plate. Valve: standard with lacquered cup fitted with a foam button.

The shampoo foam is sprayed on to the palm of the hand and thoroughly rubbed into the pre-wetted hair and the hair shampooed in the usual way.

The following formula, containing a lanolin derivative, gives very good results:

No. 2117

(Lanolin shampoo)

Shampoo base:

Ammonium/Triethanolamide lauryl sulphate	600
Coconut diethanolamide	10
Ethoxylated lanolin alcohols[1]	25
Water	365
	1000

Perfume	0·3 to 0·7 per cent
Phenyl mercuric borate	0·01 per cent

[1] Solulan 16 and Solulan 25 types—American Cholesterol Products Inc.

Procedure: Gently warm the coconut diethanolamide and lanolin derivative with part of the lauryl sulphate. Add the remainder of the lauryl sulphate and perfume. Stir gently. Finally add the water, preservative, and mix.

Container charge:

Shampoo base	92
Propellent—12/114 (50 : 50)	8
	100

Container: internally lacquered aluminium or tin plate. Valve: standard with lacquered cup fitted with a foam button.

Dandruff shampoos

Dandruff or scurf is the dry form of seborrhoea capitis known as seborrhoea sicca in which the epidermis is shed in dry flaky scales. It is generally considered that this is due to a functional disorder which causes some alteration of the skin keratinization process. Cell division in the deeper layer of the epidermis results in the cells being pushed towards the surface, those farthest from the dermis being transformed into keratin. This results in the formation of an invisible film of dead cells which is continually being cast off at the surface of the skin. If for any reason the normal functions do not synchronize properly, and keratinization and casting off of cells increases to an abnormal rate, the cells become visible and give the characteristic dry flaky scales of dandruff.

Another common form of seborrhoea capitis is due to an

abnormal secretion of sebum. This condition is known as seborrhoea oleosa. For a sebaceous gland to secrete sebum the processes of the epidermis and desquamation from the surface must also be in progress. The sebaceous glands develop from the hair follicles and normal secretion of sebum is sufficient to oil the hair and lubricate the surface of the skin. An abnormal secretion of sebum therefore causes the hair to become excessively greasy. In a few sites in the body, for example in the eyelids and corners of the eyes, sebaceous glands develop without hairs; in others, particularly in the skin around the nose, the sebaceous glands which develop from a hair follicle become much more prominent than the hair. Any disturbance of the normal epidermal function can therefore cause either a greasy or dry condition of hair, scalp and skin. In some cases excessive casting off of dead epidermal cells can occur together with abnormal secretion of sebum and when this takes place the dead cells become entangled and matted by the sebum and remain on the scalp. The degree of itching in these cases is often quite severe and if the condition is not treated inflammation and seborrhoeic dermatitis can occur.

The abnormal conditions of the scalp which have been mentioned are accompanied by an increase in growth of bacteria and fungi, the predominating species being staphycoccus aureus and pityrosporum ovale. There is also evidence to suggest that a greater variety of yeasts is present than in normal conditions, but it is not known whether these are in any way responsible for the altered condition. Although the etiology of the various forms of the disease have not been clearly defined several suggestions have been made. Some of these are:

1. Disturbance of hormonal balance.
2. Biochemical changes in the epidermis or dermis of the scalp.
3. Excessive use of strong alcoholic lotions or strong alkalis.
4. Excessive or abusive use of other irritating hair preparations, such as cold-wave preparations and hair dyes.
5. Vitamin and mineral deficiencies.

Whatever the cause of the disturbance may be, it is clear that if it is due to any physiological condition, this cannot be removed simply by use of a preparation which is applied externally. Nevertheless, it should be remembered that thorough and regular cleansing is essential to maintain the skin in a healthy condition and it is also essential to maintain the scalp in good condition in order to have a healthy skin.

Treatment of seborrhoea capitis is important because although it can be considered an entity in its own right, it is frequently associated with a predisposition to some other skin disease. Dandruff is associated with loss of hair, excluding male and female pattern baldness for which there is as yet no specific treatment; with acne, which is a general term denoting inflammation of the sebaceous glands; and psoriasis which is a chronic scaly skin disease. Some of the conditions are often attributed to an hereditary disturbance in the metabolism of fat, even though they are frequently confined to the scalp. Infantile eczema is generally considered to be of an allergic origin and yet this condition is also associated with the scalp condition known as 'cradle cap'. Cosmetic treatment preparations are those containing 'active' ingredients for the treatment of impaired skin conditions. In the field of hair preparations, shampoos, and hair dressings in particular are prepared containing active materials for the treatment of seborrhoea. It seems highly probable that these will continue to develop as research reveals the causes and gives the key to the prevention and treatment of such problems as greying hair and baldness.

Medicated shampoos are thus prepared as a sensible approach towards control of the various forms of dandruff. The function of a medicated shampoo can be defined as follows:

1. It must clean the hair and scalp without leaving the hair greasy or dry and unmanageable.
2. It should not irritate the sebaceous glands which would only serve to increase their activity.
3. It should contain an effective germicide, fungicide or antiseptic material to counteract increased bacterial growth and prevent infection for a period after the shampoo.
4. The concentration of the active material used, should not increase sensitivity of the scalp.
5. By virtue of the above qualities it should reduce the degree of itching, scaling and inflammation associated with the disease.

Several types of active materials are used as indicated in the formulae which follow. The recommendations previously given concerning use of preservatives and colouring should be followed in all cases.

No. 2118

Thymol	2
Triethanolamine lauryl sulphate	500
Water (softened or distilled)	498
	1000

Thymol is an effective bactericide but its use is limited on account of low solubility in water. It can however, be dissolved in a suitable proportion of alcohol or alternatively dissolved in the perfume and dispersed in the concentrated detergent before dilution with water.

Suitable perfumes for medicated shampoos are often based on dipentene, oils of pine and wintergreen, or suitable coal tar fractions. A mixture of thymol and camphor is a useful active material, the latter being useful as a counterirritant.

No. 2119

Thymol	1·5
Camphor	1·5
Ammonium/triethanolamine lauryl sulphate[1]	600·0
Water (softened or distilled)	397·0
	1000·0

[1] Empicol TCR type—Albright & Wilson Ltd., Marchon Division.

Menthol can be used in either of the above formula for its cooling effect and consequent tendency to reduce irritation of the scalp.

No. 2120

Thymol	0·5
Menthol	1·0
Camphor	1·0
Triethanolamine lauryl sulphate	500·0
Water (softened or distilled)	497·5
	1000·0

Procedure: Mix together the thymol, menthol and camphor. Add the perfume oils and mix with a small quantity of the detergent. Continue to add the detergent with continuous gentle stirring. Finally add water to volume.

Bithional (2,2′ thiobis (4,6-dichlorophenol)) is a white or greyish white crystalline powder with bactericidal and fungicidal properties, and is suitable for use as the active ingredient in medicated shampoos. It is insoluble in water but solution can be effected by dissolving the compound in a small quantity of alcohol or alkali before adding to the shampoo base. The compound which is substantive to skin and hair is effective at a concentration of 1·0 to 1·5 per cent. Small amounts retained on the scalp after shampooing reduce resident bacteria by 97·5 per cent. The compound is also used as the active material of antiseptic or deodorant soaps.

2,4,4′-trichloro-2′-hydroxydiphenyl ether (Irgasan DP 300—Geigy) is a white powder with a faintly aromatic odour, and is an effective bacteriostat active against bacteria and certain fungi. It is insoluble in water but solutions can be prepared by dissolving the compound in a small quantity of alcohol, or the surface-active base, before mixing with the remaining ingredients of a shampoo.

Hexachlorophane ((Di-3:5:6-trichloro-2-hydroxyphenyl) methane) is an effective bactericide which retains its activity in the presence of soap and detergents, and daily use of soap containing hexachlorophane reduces the bacterial flora of the skin. It has been used successfully in shampoos at a concentration of 0·5 to 2·0 per cent. Recent experimental studies in the United States have indicated that in certain circumstances hexachlorophane can induce brain damage in animals. It is suggested that a potential health hazard may exist, particularly to infants, when total body bathing or widespread application of a product containing hexachlorophane is carried out without subsequent rinsing.

The Committee on Safety of Medicines have recommended that products containing hexachlorophane for medicinal use on infants, such as talcs, powders, creams, lotions, and emulsions, should only be recommended on medical advice. There is, however, no evidence to date that preparations containing hexachlorophane are harmful to human adults, and attendant risks would appear to be minimal in a shampoo, when adequate rinsing forms part of the normal shampooing procedure.

Prepared coal tar, B.P. (Pix Carbonis Praeparata), is a well-known antiseptic used for the treatment of pruritus, psoriasis, and eczema, and is used to alleviate itching. The most convenient way of using coal tar is in the form of coal tar solution (liquor picis carbonis B.P.)—a solution prepared by macerating 20 per cent of prepared coal tar and 10 per cent quillaia in alcohol. The quillaia forms a milky dispersion of coal tar when it is added to water, but is readily soluble in surface-active materials. Clear alcoholic solutions can be prepared by macerating prepared coal tar in alcohol only and quillaia, filtering out insolubles after maceration.

Medicated shampoos are prepared containing either one or more of the antiseptic materials given and several heterogeneous compositions have appeared on the market.

Typical mixtures of active ingredients for use with the shampoo bases already mentioned are as follows:

No. 2121

	(per cent)
Bithional	1·5
Thymol	0·05
Menthol	0·10
Camphor	0·10
	1·75

No. 2122

	(per cent)
Bithional	1·0
Coal tar solution (B.P.)	2·5
	3·5

No. 2123

	(per cent)
Bithional	0·5
Coal tar solution (B.P.)	2·0
Thymol	0·1
Camphor	0·1
Oil of rosemary	0·1
	2·8

No. 2124

	(per cent)
Hexachlorophane	1·0
Methyl salicylate	0·25
	1·25

No. 2125

	(per cent)
Hexachlorophane	0·5
Prepared coal tar B.P.	2·0
Thymol	0·1
Camphor	0·1
Oil of rosemary	0·1
	2·8

No. 2126

	(per cent)
Hexachlorophane	1·00
Methyl salicylate	0·25
	1·25

In the above formula, oil of rosemary and methyl salicylate are included for their odour value only. In similar fashion suitable proportions of dipentine or oil of pine can be included.

The following are also used as active ingredients of medicated shampoos:

Resorcinol (*m*-dihydroxybenzene) occurs as a colourless powder and is a useful material to relieve itching. In shampoos it is usually used at a concentration of from 0·05 to 0·10 per cent. When the powder is exposed to light and air or used as an ingredient of alcoholic solutions it becomes pinkish in colour. An alcoholic solution of resorcinol also has a tendency to discolour fair hair slightly. It is not likely to give any trouble at the concentrations normally used in shampoos.

Oil of eucalyptus is used in medicated shampoos, generally at a concentration of from 1·0 to 2·5 per cent. It has a characteristic aromatic camphoraceous odour and when applied externally has a cooling effect and counter-irritant properties. Eucalyptol or Cineole is one of the main constituents of eucalyptus oil and has similar properties to oil of eucalyptus. Use from 0·25 to 0·5 per cent in medicated shampoos.

Cade oil also known as juniper tar oil or juniper tar is a dark reddish brown oily liquid with a tarry odour. It is sometimes used (at about 0·25 to 1·0 per cent) to stimulate circulation.

Ichthammol or ammonium ichthosulphonate obtained by distillation from certain bituminous shales occurs as a black viscous liquid with a strong characteristic odour. It is slightly bacteriostatic and reduces inflammation. Use from 0·5 to 2·0 per cent. Although both cade oil and ichthammol effect the colour of a product they are nevertheless materials which are useful to prepare a speciality treatment preparation.

Precipitated sulphur is used externally as a mild antiseptic and parasiticide. It is usually applied in lotion or ointment form in order to maintain intimate contact with the skin. Salicylic acid is a bacteriostatic and fungicide also used in lotions and ointments for treatment of the scalp. Up to 2·0 per cent of these two materials can be used in medicated shampoos nreferably in a cream or paste type base. A micropulverized sulphur is recommended.

Cationic surfactants are so termed because the surface active portion of the molecule is contained in the positively charged cation. On this account they are substantive to hair and skin and have antistatic properties. Unlike the anionic surfactants many cationic compounds have achieved commercial importance in applications where surface activity as such is a minor consideration. Thus where the cation is present as a quaternary ammonium group attached to a long chain alkyl group, the compounds can be used as detergents or emulsifying agents and also for their bactericidal properties. As bactericides their properties vary according to the chain length of the alkyl group. Compounds of high molecular weight have lower bactericidal properties. Care must be exercised in selecting a particular quaternary ammonium compound when this is to be used for regular application to the hair and scalp since a material with high bacteriostatic properties may cause defatting of the skin and give rise to irritation. Several cationic surfactants are used to prepare medicated shampoos, as a specific treatment for dandruff and associated irritation of the scalp. The following are particularly recommended:

Cetrimide B.P. (CTAB) consists of a mixture of dodecyltetradecyl–and hexadecyl-trimethylammonium bromides. It is non-irritant and non-toxic in the dilutions generally used.

Benzalkonium chloride is a mixture of alkyl dimethyl benzyl ammonium chlorides, and has similar bactericidal and bacteriostatic properties to cetrimide. At high concentrations of the order of 12 per cent it is slightly less toxic than cetrimide.

Stearyl dimethyl benzyl ammonium chloride is another typical quaternary ammonium salt and although this material has less bactericidal action than either cetrimide or benzalkonium chloride its use also improves the appearance of the hair.

Medicated shampoos based on quaternary ammonium compounds are prepared as follows:

No. 2127

Cetrimide B.P.	175
Alcohol	100
Diethyl phthalate	20
Water (softened or distilled)	705
	1000

Procedure: Dissolve the cetrimide in water using gentle heat. Cool to 35°C and add the mixture of diethyl phthalate and alcohol. Add perfume and stir gently until homogeneous.

Cetrimide B.P. is soluble in two parts of water but to maintain clarity of solution during storage a small proportion of alcohol is included in the above formula. Diethyl phthalate increases the viscosity of the shampoo. Other quaternary ammonium compounds such as hexadecyl trimethyl ammonium bromide can be used in the above formula if required in place of cetrimide. For this type of product use a concentration of 15 to 20 per cent active material. Any perfumes used in these mixtures should be carefully checked for stability.

A cream shampoo for treatment and control of seborrhoea capitis and seborrhoeic dermatitis can be prepared as follows:

No. 2128

Alkyl trimethyl ammonium bromide[1]	150
Cetyl alcohol	150
Water (softened or distilled)	700
	1000

[1] Cetrimide B.P. or hexadecyl trimethyl ammonium bromide.

Procedure: Warm gently the cetyl alcohol and alkyl trimethyl ammonium bromide with 200 parts of water. Stir gently until homogeneous and add the remaining water at about 40°C. Add perfume and colour and mix.

For use, the hair is first thoroughly wetted with warm water and about a teaspoonful of cream or lotion is massaged well into the scalp with the fingers. An extensive lather is not obtained. The hair is then rinsed thoroughly and the application repeated and on this occasion the shampoo can be worked up to produce a lather. Finally the hair is thoroughly rinsed with warm water. Care should be taken to keep the preparation away from the eyes. It is claimed that weekly treatment controls seborrhoeic conditions and maintains the scalp and hair in a healthy condition.

A liquid cream shampoo with an attractive pearly sheen can be prepared to the following formula:

No. 2129

Monoethanolamine lauryl sulphate[1]	500
Stearyl dimethyl benzyl ammonium chloride	25
Water (softened or distilled)	475
	1000
Perfume, colour	*q.s.*

[1] Empicol LQ 27 type–Albright & Wilson Ltd., Marchon Division.

Selenium sulphide SeS_2 is a bright orange powder which was developed by Norlander (U.S. Patent Nos. 1711742, 1860336). It is used in the form of an emulsion or with a surface-active agent in persistent cases of seborrhoea capitis. Sometimes during the treatment, the hair becomes oily and occasionally hair loss occurs. Hair loss of this nature very often takes place after shampooing a seborrhoeic condition and is generally only of a temporary nature. It has been reported however, that use of selenium sulphide appears to bring about changes in the hair roots which results in weakening of the hair. When persistent hair loss occurs, treatment with selenium should be stopped.

Selenium sulphide preparations are generally described as creams or suspensions and are considered primarily as a medicated treatment rather than a shampoo. Formulation with selenium is difficult on account of its insolubility and it is more easily handled if mixed with a suitable inert diluent before adding to a surface-active agent. Norlander found that by heating equal parts of selium disulphide and bentonite to a temperature where fusion of the selenium occurred, the resultant mixture could be easily dispersed in water. This mixture was shown to consist of selenium monosulphide SeS. A typical selenium preparation for treatment and control of seborrhoea capitis can be prepared as follows:

No. 2130

Selenium disulphide	25
Bentonite	50
Sodium lauryl sulphate (paste form)	400
Water (softened or distilled)	525
	1000

Procedure: Mix the selenium disulphide and bentonite until the selenium is evenly dispersed. Heat the sodium lauryl sulphate and water in a steam jacketed vessel fitted with a stirrer. Stir gently to a temperature of 90°C. In a separate vessel add a suitable proportion of the detergent mix to the selenium dispersion and stir until homogeneous. Add to the main mix and continue stirring. Cool to 40 to 45°C and fill.

A firmer cream can be prepared by adding 0·5 to 1·5 per cent of cetyl alcohol to the formula. This helps to disperse the selenium if added during the preparation of the concentrated selenium detergent mix.

Selenium preparations are generally applied twice a week for the first two weeks of the treatment and once a week for a further two weeks. After this they are used occasionally between the use of ordinary shampoos to maintain control of the seborrhoric condition. During application continuous massage of the scalp is desirable and after treatment thorough rinsing should be carried out to remove all traces of the preparation.

Undecanoic acid or undecylenic acid is a fungicide used for the treatment of athlete's foot and other superficial fungal infections of a dermatological nature. It is usually used in conjunction with zinc undecanoate either as ointments, emulsions or dusting powders, the total concentration of active fungicide being about 10 to 15 per cent. Undecanoic alkanolamides derived from undecanoic acid and similar aliphatic acids of medium chain length are fungicidal and bactericidal compounds suitable for treatment of seborrhoea capitis. These materials which are non-allergenic and non-sensitizing are substantive to keratin and on this account are useful materials for use in shampoos, since some retention of antiseptic properties occurs on the hair and scalp after shampooing. They have a characteristic musk-like odour and although this is less pronounced than the odour of undecanoic acid this factor must be considered when selecting a perfume.

Floral perfumes of a rose or violet type generally blend satisfactorily.

Several compounds are available commercially. Those derived from mono-alkanolamides of fatty acids of medium and short chain lengths are waxy solids with varying melting points (Loramine U.185, 1PU.185). They are almost insoluble in water, but soluble in alcohol, aqueous-alcoholic mixtures, organic solvents, and oils and fats. Shampoos can be formulated by first dissolving the compounds in alcohol or the perfume oil and mixing with a suitable detergent base such as triethanolamine lauryl sulphate. The surface-active material of the shampoo base increases the activity promoting intimate contact of the compounds with fungi and bacteria. The compound derived from undecanoic acid and diethanolamine is available commercially in the form of a loose complex containing a percentage of amine soap (Loramine DU.185). On this account it has a greater solubility in water than the waxy materials. Aqueous solutions containing 2 per cent of the compound have sufficient wetting and foaming properties to be used as detergents on their own but a more satisfactory product is obtained by using a base of triethanolamine lauryl sulphate. Undecanoic mono-alkalolamide has also been prepared as a sodium sulpho succinate derivative (Loramine SB. U185). This compound is available as a cream coloured paste with a weak but characteristic odour containing 50 per cent active matter. It has been shown to be an efficient, non-toxic fungicide and bacteriostatic and is substantive to keratin.

In water the compound gives a slightly opalescent solution but it is readily soluble in luke-warm water. Clear solutions can be prepared by mixing with surface-active agents or by dissolving in a mixture of alcohol and water using 30 to 60 per cent of alcohol. Optimum solubility can be obtained by using aqueous *iso*-Propyl alcohol. The compound can thus be used for the treatment of seborrhoea capitis as a medicated shampoo or a scalp lotion. The usual dosage is from 0·5 to 2·0 per cent. (Undecanoic alkanolamides (alkylolamides) are available from (1) Rewo G. m.b. H. Germany, (2) Dragoco, Germany, (3) Dutton and Reinisch, England).

Hair conditioners

Hair conditioners are considered to improve the appearance of the hair by giving body and lustre, and to achieve this effect two types of hair conditioning creams are used. The most popular are those used as rinses and are applied to the hair after shampooing and rinsing, whilst the hair is still wet. The second category includes hair dressings or creams, and brilliantines. These are applied to dry hair either immediately after setting and drying, or at any time when extra gloss or light fixing properties are required.

Preparations for wet hair

Hair conditioning preparations for use on wet hair after a shampoo are prepared either as creams or lotions. Probably the most successful type are those which contain a quaternary ammonium compound as the active ingredient. Following a shampoo or permanent wave the cream is applied with the hands over towel-dried hair and distributed evenly by combing. If the product is in lotion form this can either be poured over the hair followed by combing, or diluted with water and poured repeatedly over the hair whilst the head is held over a basin. The cream or lotion is left on the hair for a few minutes and then rinsed off with warm water.

Reference has already been made to the bactericidal properties of the quaternary ammonium compounds, but their conditioning properties are caused by the relatively large complex cation which is substantive to protein. The quaternary ammonium compound is absorbed onto the hair fibre from a cream or lotion and is retained on rinsing. This removes the static charge and reduces friction between the individual hair strands. As a result the hair is more manageable and easier to handle during setting. Cationic softening agents are used in the textile industry to improve handling. In a similar manner a thin layer of quaternary ammonium compound acts as a lubricant, gives a feeling of body and improves the handling of hair. These products are used to give lustre to dull looking hair, to overcome frizziness and unmanageability caused by faulty or careless permanent waving, use of harsh shampoos, or excessive bleaching.

Satisfactory preparations can be prepared using stearyl dimethyl benzyl ammonium chloride as the active material. Although compounds containing a stearyl group are less bacteriostatic than for example, lauryl compounds, they have better substantive properties, probably on account of their lower solubility. Dosage per application should be carefully controlled and directions for use should also be given clearly. These are best determined by salon trials which are necessary because a high concentration softens the hair too much so that it becomes limp and does not retain a set. Most products are made as emulsions or creams with small proportions of fatty materials and an emulsifier. The instructions for use should emphasize the importance of thorough rinsing after the application, or otherwise deposition remaining on the scalp has a similar effect to overdosing.

No. 2131

Stearyl dimethyl benzyl ammonium chloride	250
Ethylene glycol monostearate	50
Lactic acid	1
Water (softened or distilled)	699
	1000

Perfume	0·3–0·5 per cent
Methyl parahydroxybenzoate	0·15 per cent

Procedure: Add the water at a temperature of 75°C gradually to the quaternary compound and monostearate previously heated to the same temperature. Stir slowly and continuously until all the water has been added. Allow to cool and add the lactic acid and perfume when the temperature is about 35°C. The cream can either be applied direct to wet hair or mixed with water before use. In either case the dosage should be calculated to give from 0·5 to 0·75 grammes of active material per application. The above formula gives a thin, pourable cream. The consistency can be modified by varying the proportion of ethylene glycol monostearate.

A firmer product can be made to the following formula. Although cetyl alcohol gives body to the cream, the solids content should be kept to a minimum in order to assist ease of removal of the cream during the rinsing process.

No. 2132

Stearyl dimethyl benzyl ammonium chloride	50
Ethylene glycol monostearate	20
Cetyl alcohol	30
Water (softened or distilled)	900
	1000

Perfume	0·3–0·5 per cent
Methyl parahydroxybenzoate	0·15 per cent

No. 2133

Stearyl dimethyl benzyl ammonium chloride	80
Ethylene glycol monostearate	10
Water (softened or distilled)	910
	1000
Perfume	0·3–0·5 per cent
Methyl parahydroxybenzoate	0·15 per cent

10 grammes of the above cream is sufficient for one application.

A further typical formulation is given below:

No. 2134

Stearyl dimethyl benzyl ammonium chloride	35
Cetyl alcohol	25
Mineral oil (cosmetic quality)	5
Water (softened or distilled)	935
	1000
Perfume	0·3–0·5 per cent
Methyl parahydroxybenzoate	0·15 per cent

Procedure: Melt together the fatty materials and heat to 70-80°C. Dissolve the quaternary compound in water in a separate container and heat to 70-80°C. Add the aqueous solution to the fatty materials whilst slowly stirring to avoid incorporation of air. Continue stirring until cold adding perfume during cooling at about 35°C.

Directions for use can be given so that one part cream added to 5 parts of warm water before use gives the desired quantity of 0·7 grammes of active material per application.

Several quaternary ammonium compounds for use as hair conditioning agents are available commercially. Triton X-400 is based on stearyl dimethyl benzyl ammonium chloride together with an alcohol. A formula using this material is given:

No. 2135

Triton X-400	75
Cetyl alcohol	3
Potassium chloride	8
Water (softened or distilled)	914
	1000
Perfume	0·3–0·5 per cent
Methyl parahydroxybenzoate	0·15 per cent

Procedure: Heat part of the water with the cetyl alcohol. Heat the Triton X-400 and add to the mixture with stirring. Stirring is continued for 30 minutes to stabilize the emulsion. Finally dissolve the potassium chloride in the remainder of the water and add to the batch.

Potassium chloride acts as a thickener of Triton X-400. A concentrated form of this compound is also available known as Triton X-400 concentrate (Rohm & Haas Co., Philadelphia 5, Pa.). It contains a minimum of 67 per cent of stearyl dimethyl benzyl ammonium chloride and other cationic materials to give a total of 82 per cent active material.

Emcol 61 (Emulsol Corp., Chicago 3, Ill.) is a condensation product of a fatty acid and amide. This compound is available as an amber coloured paste containing about 10 per cent active material. It is compatible with sodium alkyl sulphate and can be used in shampoos at a concentration of about 5 per cent.

Arquads (Armour Hess Chemicals Ltd.) is the trade mark for a series of quaternary ammonium salts containing alkyl groups with varying chain lengths. Several compounds in this series are water soluble whilst others are either oil-soluble or water-dispersable.

Acid conditioning creams have largely been replaced by those based on quaternary ammonium compounds. Although the former are still popular with certain sections of the public, this type of cream is probably mainly used in the professional salon, particularly after permanent waving. The cream is also applied with the hands or a comb after washing and rinsing, to towel dried hair. It is left on the hair for a few minutes and finally rinsed off with warm water. The cream is considered to give body to the hair, this effect being largely obtained by the thin residual film of fatty material which remains on the hair. A formula is as follows:

No. 2136

Cetyl alcohol	150
Sodium lauryl sulphate	5
Citric acid	20
Water (softened or distilled)	825
	1000

Perfume	0·3–0·5 per cent
Methyl parahydroxybenzoate	0·15 per cent

Procedure: Melt together the cetyl alcohol with a small proportion of the water. Prepare a solution of sodium lauryl sulphate in water. Warm and add with gentle stirring to the mix. Continue stirring to form a stable cream and finally add a solution of the citric acid in the remaining water.

Dressing-type conditioners

Dressing-type conditioners are creams applied to dry hair before setting. Used on dull or brittle hair their purpose is to apply a thin film of oil in order to give a natural looking shine or gloss without any appearance of oiliness. A similar effect is obtained with an aerosol spray of a light oily material such as iso-propyl myristate.

Thus, although often referred to as hair conditioners they are more correctly described as dressings. Conditioning properties are claimed because when used in the correct manner the creams give gloss and shine to dull or brittle hair. Only a small quantity of the cream is applied to the hair from the hands. Alternatively a small amount of cream is liquefied by rubbing on the palm of the hand, and the hairbrush is dipped lightly into the liquefied cream and brushed over the hair. Either method is designed to avoid using an excess of cream which gives the hair an oily or greasy appearance. The amount of cream required to give the best results varies. Fine hair does not need as much cream as coarse hair and it is advisable to draw attention to this point. It follows that water-in-oil emulsions are most suitable for application in this manner. Creams suitable for tube packaging can be made as follows:

No. 2137

A	Wool alcohols	50
	Petroleum jelly	450
	iso-Propyl myristate	200
	Mineral oil	50
B	Water (softened or distilled)	250
		1000

Perfume	0·5 per cent
Methyl parahydroxybenzoate	0·15 per cent
Propyl parahydroxybenzoate	0·02 per cent

Procedure: Melt together the wool alcohols and petroleum jelly. Add the *iso*-Propyl myristate and mineral oil and heat to 70°C. Add the preservatives. Heat the water to 70–75°C and add to the oil mixture with stirring. Homogenize and allow to cool.

No. 2138

A	Ethoxylated cetyl oleyl alcohol[1]	80
	Polyethylene glycol 400 monostearate	80
	Mineral oil	200
	Silicone fluid[2]	10
	Paraffin wax	40
	Petroleum jelly	80
B	Water (softened or distilled)	510
		1000
	Sorbic acid	0·1 per cent
	Perfume	0·5 per cent

[1] Empilan KL 6 type—Albright & Wilson Ltd., Marchon Div.
[2] MS 200 type—Midland Silicones Ltd.

Procedure: Melt together the ingredients of part A to a temperature of 75°C and add part B to part A slowly with continuous stirring. Stir until cool adding the perfume when the temperature has fallen to about 35°C.

Aerosol conditioners

An aerosol hair conditioner for use as a rinse is prepared as follows:

No. 2139

Hair conditioner rinse base:

A	Non-ionic emulsifier[1]	20
	Alkyl myristate[2]	5
	Stearic acid	4
B	Water	969
	Cetyl trimethyl ammonium bromide (or Cetrimide B.P.)	2
		1000
	Perfume	0·2–0·3 per cent
	Methyl parahydroxybenzoate	0·15 per cent
	Colour (water-soluble)	*q.s.*

[1] Emulsene 1220 type—Bush Boake Allen
[2] Bush Boake Allen

Procedure: Heat part A and part B independently to 75°C. Add B to A slowly with continuous stirring. Cool with stirring, adding the perfume at 30°C.

Container charge:

Hair conditioner rinse base	87
Propellent—12/114 (10 : 90)	13
	100

Container: plastic coated glass. This type of product is known to attack tin plate, therefore rigorous shelf testing is essential before marketing in a metal container. Valve: tilt action fitted with a foam head.

Setting lotions

Setting lotions are used both at home and at the hairdressing salon to style the hair. After permanent waving use of a setting lotion enables the hair to hold its shape for several days, and between visits to a hairdressing salon they are often used to pin the hair at night to help brushing out and styling the following morning.

Lotions based on gums and mucilages continue to be popular. The most favoured gum is tragacanth, followed by the cheaper substitute—karaya gum also known as Indian tragacanth. Acacia or gum arabic is less popular, although it has the advantage of easy solubility. Mucilages prepared by decoction of quince seeds, psyllium seeds, and Irish moss can also be used. The mucilages prepared from gums are generally prepared as weakly alcoholic solutions, and the decoctions are used as aqueous solutions. Both types of lotion are coloured and perfumed, and must contain an adequate amount of preservative. It is not desirable to lay down any hard and fast rules as to the strengths of the various setting lotions because a good deal depends upon the type of hair and individual preference. In particular, mucilages prepared with gum tragacanth vary considerably depending on the quality of the gum used. The percentage margins are, however, approximately as follows:

	(per cent)
Alcohol	1–10
Gum tragacanth in powder or flakes	0·1–1·5
Gum karaya	0·8–3
Gum arabic	0·2–0·5
With borax or sodium carbonate	0·5–2
Quince seeds	0·5–1·5
Psyllium seeds	0·5–1·5
Irish moss	1–2
With borax	2–5
Perfume	0·1–0·5

Procedure for preparing mucilages: in the case of tragacanth and karaya gums the alcohol, preservative and perfume are mixed with the powdered gum and the water added in a continuous stream to facilitate even distribution and swelling. In the case of quince and psyllium seeds hot decoction for ten minutes is sufficient followed by straining and cooling. Irish moss is infused cold for a few hours after being freed from extraneous matter by washing. All the other constituents are merely dissolved or mixed as the case may be.

The alcohol used for wetting purposes to prepare the mucilaginous products also reduces drying time of the lotion in use. From 2 to 5 per cent of glycerin can also be included in the formula and other suitable additives include 0·1 per cent of cholesterol or 1 to 2·5 per cent of a water-soluble lanolin or lanolin derivative. A formula for a setting lotion of this type is as follows:

No. 2140

Tragacanth	15
Glycerin	50
Water-soluble lanolin[1]	20
Cholesterol	1
Water	914
	1000

Methyl parahydroxybenzoate	0·2 per cent
Water-soluble perfume	0·3 per cent

[1] Solulan 98 type—American Cholesterol Products Inc.

Procedure: Dissolve the preservative in the glycerin using gentle heat. Add the tragacanth and cholesterol; mix to a smooth paste. Mix the water, water-soluble lanolin, and perfume separately and add to the tragacanth mixture using vigorous stirring to commence. Allow to stand before filling out. Alcohol (about 5·0 per cent) can be included to disperse the tragacanth before adding the glycerin. Add a suitable water-soluble dyestuff.

A carboxy vinyl polymer is used to prepare an attractive transparent gel used for setting the hair. The product is easy to handle and more attractive than a mucilage. Polyvinyl pyrrolidone or a co-polymer is included as a film former. A suitable product is made to the following formula:

No. 2141

Carboxy vinyl polymer[1]	6
Triethanolamine	7·5
Alcohol	100
Water (softened or distilled)	866·5
Polyvinyl pyrrolidone	20
	1000·0

Methyl parahydroxybenzoate	0·2 per cent
Water-soluble perfume	0·3–0·5 per cent

1 Carbopol 940 type—B.F. Goodrich Chemical Co.

Procedure: Dissolve the polyvinyl pyrrolidone and preservative in the alcohol. Disperse the polymer in the triethanolamine and gradually add the water with stirring. Add the alcohol solution and perfume.

Polyvinyl pyrrolidone co-polymers and other resinous materials already mentioned have, to a large extent, replaced setting lotions based on gums and colloids. Aqueous lotions of polyvinyl pyrrolidone are quite pliable after application so that the hair can easily be set before drying. On drying a transparent film is formed which gives body and firmness depending upon the concentration used in the lotion.

In order to prevent the dried PVP film from flaking a plasticizer is used such as a water soluble lanolin or a glycol. Diethylene glycol is particularly suitable since it helps to form a dry film. The concentration of plasticizer should be 10 per cent of the PVP content of the lotion. Suitable formulae are as follows:

No. 2142

Polyvinyl pyrrolidone	20
Diethylene glycol	2
Alcohol	220
Water (softened or distilled)	758
	1000
Methyl parahydroxybenzoate	0·2 per cent

Use a water-soluble perfume and add a suitable water-soluble dyestuff.

This lotion has a slightly acidic pH value and addition of an alkali to a pH value of about 9·0 causes softening and slight swelling of the hair fibre and increases penetration of the PVP. An alkaline lotion is thus considered to give a more firm and lasting set. With this type of product, on account of the residual PVP film, if the hair is wetted slightly with water on subsequent days after use it can be re-styled and set without a further application of the lotion.

No. 2143

Polyvinyl pyrrolidone	25
Diethylene glycol	2·5
Triethanolamine	1·0

No. 2143 (*continued*)

Alcohol	200
Water (softened or distilled)	771·5
	1000·0

Methyl parahydroxybenzoate 0·2 per cent
Water-soluble perfume and pigment

When using aqueous alcoholic solutions of PVP the dried PVP film has a tendency to flake when the hair is subsequently combed, and although the degree of flaking is reduced when the correct glycol is used as the plasticizer, even less flaking occurs when a small proportion of an ethylene oxide condensate is included in the formula.

No. 2144

Polyvinyl pyrrolidone	30
Diethylene glycol	3
Ethylene oxide condensate of cetyl/oleyl alcohol[1]	1
Alcohol	185
Water (softened or distilled)	781
	1000

Methyl parahydroxybenzoate 0·15 per cent

[1] Empilan KL 6 or KL 10 type—Albright & Wilson Ltd.

Use a water-soluble perfume and include a suitable water-soluble dyestuff.

Polyvinyl pyrrolidone co-polymers are used in similar fashion. Their use is sometimes preferred because the residual film remains more stable in conditions of high humidity. A good product is prepared as follows:

No. 2145

Polyvinyl pyrrolidone:	
Vinyl acetate 60 : 40 co-polymer	10
iso-Propyl myristate	1·5
Alcohol	700
Water (softened or distilled)	283·5
	1000·0

Perfume 0·3–0·5 per cent

[1] Badische Anilin & Soda Fabrik Ag.

Add an alcohol and water-soluble perfume.

Procedure: Dissolve the PVP:VA in the alcohol with stirring. Add the perfume and iso-propyl myristate. Slowly add the water, stirring constantly.

A pearly setting lotion based on stearic acid is made as follows:

No. 2146
(Base)

Stearyl dimethyl benzyl ammonium chloride	160
Stearic acid	80
Water (softened or distilled)	760
	1000

Procedure: Gently heat all the ingredients together and stir until a smooth gel is formed. To prepare the setting lotion dilute 5 parts of the base with 95 parts of water. Include methyl parahydroxybenzoate 0·15 per cent. Use a water-soluble perfume and a suitable water-soluble dyestuff.

Good results are also obtained with dimethyl hydantoin formaldehyde resin, as indicated in the following formula:

No. 2147

A	Dimethyl hydantoin formaldehyde resin	20
	Alcohol	700
B	*iso*-Propyl alcohol	50
	Diethyl phthalate	10
C	Glycerin	5
	Water (softened or distilled)	215
		1000
	Perfume	0·3–0·5 per cent

Procedure: Dissolve the resin in the alcohol with stirring. Add the *iso*-Propyl alcohol and diethyl phthalate. Add the perfume, followed by the glycerine and water, and mix. A water-soluble dyestuff can be added.

Coloured setting lotions used as temporary colourants are made by adding certified soluble dyestuffs to a suitable setting lotion base. The dyestuffs used are mainly alcohol-soluble types although certain water-soluble dyestuffs and others which are both oil and alcohol soluble are also suitable. The solvent of the setting lotion base must therefore be adapted to suit the solubility characteristics of a particular dyestuff used to achieve the desired colour effect. Coloured setting lotions are used by the professional hairdresser and also sold in one application packs for the home user. A note regarding their preparation is also given in the next section dealing with aerosol temporary colours.

Aerosol setting lotions

Aerosol setting lotions are convenient to handle as they prevent dripping and spilling of the liquid in use. The dried film on the hair must not be too brittle otherwise when the set is combed out the film breaks, causing dust particles to cling to the comb and make the hair look dull and lifeless. The following formula gives very good results:

No. 2148

	Setting lotion base:	
A	Dimethyl hydantoin formaldehyde resin[1]	20
	Alcohol 96% v/v	700
B	*iso*-Propyl myristate	5
	Diethyl phthalate	5
	iso-Propyl alcohol (IPS/C grade)[2]	50
C	Glycerin	5
	Water (softened or distilled)	215
D	Pluronic L. 103[3]	50
		1000

Perfume 0·2–0·3 per cent

[1] Glyco Products–U.K. Agent: Rex Campbell & Co. Ltd.
[2] Shell Chemical Co.
[3] Jacobsen Van Den Berg & Co. (U.K.) Ltd.

Procedure: Dissolve A with continuous stirring, and add B. Mix together C and add to the bulk. Mix with stirring, and finally mix the perfume with D and mix.

Container charge:	
Setting lotion base	50
Propellent–12/114 (50 : 50)	50
	100

Container: internally lacquered aluminium or tin plate. Valve: standard fitted with a button similar to that used for hair sprays.

If a conditioning agent is required in the above formula then the addition of 0·5 per cent cetyl trimethyl ammonium bromide is recommended. In this case, however, it is not advisable to pack in a metal container without shelf testing, as cetyl trimethyl ammonium bromide can attack tin plate in certain conditions.

Container: Polypropylene or plastic coated glass

Container charge:

Base	45
Propellent–12/114 (10 : 90)	55
	100

Valve: tilt action with a micro mist actuator.

Hair lacquers or sprays

Hair lacquers or hair sprays are used on the hair after setting and are intended to hold the style firmly in shape without detracting in any way from the artistic finish. Apart from a satisfactory formulation the appearance of the finished pack, the use characteristics and choice of perfume often contribute to the success of a product. They consist of a film-forming material in a suitable solvent together with a plasticizer or modifier to promote flexibility of the resin. The basic composition is given as follows:

1. Film-former
2. Plasticizer (modifier)
3. Solvent
4. Perfume.

Shellac was invariably used as the film-former in early formulations. This resin is secreted from a scale insect and deposited on the twigs of various trees in India, Siam, and Indo-China. The resin encrusted twigs which are collected from the trees are known as 'stick lac'. When this has been ground and washed it is known as 'seed lac' and the shellac of commerce is prepared from this material by melting and straining or by solvent extraction. It contains about 95 per cent of resinous substances including alevritic, shellolic, kerrolic, and butolic acids. For use in hair lacquers, the refined quality known as palest dewaxed flake shellac should be used. The flakes are prepared by pouring the melted resin over plates or cylinders and as the resin hardens it scales off in thin flakes. Shellac is insoluble in water, but dissolves readily in warm alcohol. *iso*-Propyl alcohol is the best solvent and can be used to prepare solutions containing up to 25 per cent of the resin. A solution of this strength is a convenient method of storing the resin in the form of a concentrate available for dilution as and when required. Although shellac is considered by many to be inferior to other film formers which are now available, it is still used for fantasy hair styling particularly by the professional hairdressers, when a high degree of fixation is required. It gives a rigid film which tends to break when handled even though plasticizers or modifiers are used. The resin is also difficult to remove from the hair and often causes flaking when the hair is subsequently brushed or combed. In addition, when a shellac film is used on the hair it cannot be wetted with water and re-styled. On the other hand a shellac film is not affected by humidity and holds the hair in position for long periods.

The film obtained with a shellac-type hair lacquer can be controlled to a large extent by the amount and type of modifier or plasticizer used.

A typical formula is as follows:

No. 2149

Shellac	40
Castor oil	2
Diethyl phthalate	2
Alcohol	956
	1000
Perfume	0·5 per cent

A shellac film which is more water-dispersible can be obtained by the addition of an alkali illustrated by the following formula.

No. 2150

Shellac	50
Alcohol	300
Triethanolamine	10
Water	640
	1000
Water-soluble perfume	1·0 per cent

Procedure: Dissolve the perfume in the alcohol and dissolve the shellac in the mixture. Mix the triethanolamine with the water and add the alcoholic solution slowly with constant stirring.

It is important to consider the time of drying when formulating a hair spray, and this will naturally take longer when the product includes water.

Selection of a suitable perfume plays an important part in the success of all hair sprays. It must remain stable in note and intensity of odour. Unless the hair spray is one of a range of perfumed products containing a specific perfume it should be of a light and fugitive character so that it does not clash, but preferably blends with the users normal perfume.

With formulations which require a water-soluble perfume, the properties of the spray should be checked both with and without the perfume to determine the effect of the solubilizer which has been used. Some perfume solubilizers act as plasticizers and affect the properties of the film on the hair. Mention has already been made of the effect of both amount and type of modifier or plasticizer on the properties of a resinous film on the hair. This applies particularly with shellac based sprays and the table which follows gives an indication of some of the different effects which are obtained.

In assessing the properties of the films on the hair with the formulations given in Table 6.1 the extent of gloss is given as determined by visual examination. A standard procedure of combing the dried film is used to give a measure of the degree of flaking. Maximum flaking is indicated by the figure '100' and occurs when no plasticizer is used.

The results of the practical tests illustrated in Table 6.1 indicate the value of carrying out tests with various concentrations of shellac and modifier before deciding on a particular formulation.

Polyvinyl pyrrolidone or PVP is a popular material for the formulation of hair sprays. It is a synthetic polymer first used during World War II as a blood plasma extender, as was initially prepared by the action of acetylene on formaldehyde at high pressure. It is non-toxic and neither a primary skin irritant nor a sensitizer.

The polymer is used to prepare aqueous solutions of different viscosities, the grades available being denoted by a *K* value. This is a function used to define the molecular weight of the product as related to the viscosity of a 1·0 per cent aqueous solution. Three viscosity grades are available—*K*30, *K*60, and *K*90, but the material normally used for the preparation of hair sprays is the low viscosity grade with a *K* value of 30 (30–35). In solid form the

TABLE 6.1

Concentration of shellac	*Modifier used*	*Content of modifier*	*Type of film*	*Degree of gloss*	*Degree of flaking on combing**
6·25%	Propylene glycol monolaurate	0·625%	Soft	Poor	5
	Diethyl phthalate		Soft	Poor	25
	Silicone MS 555[1]		Rigid	Poor	50
	Glycerol		Firm	Good	25
	None		Firm	Good	100
6·25%	Alcohol-soluble lanolin deriv.	2%	Soft (sticky)	Medium	25
	Propylene glycol monolaurate		Soft	Good	5
	Diethyl phthalate		Firm	Medium	50
	iso-Propyl myristate		Soft (pliable)	Poor	25
4·0%	Propylene glycol monolaurate	2%	Firm	Good	25
	iso-Propyl myristate		Firm	Good	50
8·0%	Propylene glycol monolaurate	2%	Rigid	Good	50
	iso-Propyl myristate		Rigid	Good	none

* 100 represents maximum degree of flaking
[1] Midland Silicones Ltd.

material is a white or off-white hygroscopic powder and as generally supplied contains about 5 per cent water. If the material is required for products which must not contain appreciable amounts of water this should be specified to the suppliers. In this case PVP is packed hot in heat sealed polythene containers immediately it is recovered from the drying plant. This form normally contains from 1·0 to 1·5 per cent of water. Although packaged in moisture-proof containers it readily picks up moisture once the container is opened, and can become sticky and difficult to handle. For this reason it is often convenient to obtain supplies in solution form. These solutions keep satisfactorily, provided they contain a suitable preservative. Aqueous solutions which are quite stable can be obtained containing about 30 per cent of solid material. This solution has a specific gravity of 1·07, is pale yellow and slightly viscous in appearance and has a slight characteristic odour. In addition to being miscible with water, polyvinyl pyrrolidone is soluble in many organic solvents including some esters and fatty acids and a wide range of alcohols.

The solubility properties make PVP a versatile raw material and its use is referred to in the formulation of shampoos, creams, after-shave lotions, and shaving cream, in addition to its use in hair setting lotions and hair sprays.

Hair lacquers based on PVP can therefore be prepared either with all alcohol or water or aqueous/alcoholic solutions as the solvent, and they can easily be removed by shampooing. Lacquers prepared without water enhance the appearance and give a so-called conditioning effect which is more apparent than those containing water. This is probably due to a more even distribution and greater penetration of PVP into the hair shaft during the particular condition of usage. They give a soft and pliable set which is firm but not stiff or rigid like the shellac film and after use the hair can be reset by wetting with a small amount of water using a brush or comb. The PVP film is, however, somewhat brittle when used on its own and use of a suitable modification or plasticizer is recommended. The fact that the polymer is hygroscopic is a disadvantage in conditions of high humidity. These cause the set to fall and the hair to become quite sticky. Better results are obtained by including in the formula a small proportion of a material to act as a water repellent. Several materials have been recommended for use as plasticizers with PVP including dimethyl, diethyl and dibutyl phthalates, *iso*-Propyl myristate and certain lanolin derivatives.

When formulating a PVP lacquer, it is important to examine the effect of various concentrations of the plasticizer not only in relation to the pliability of the lacquer film, but also to the effect obtained when the hair is subsequently brushed or combed. Used on its own the PVP film causes flaking, so the best plasticizer is one which can be used at a concentration which does not affect the particular rigidity required and at the same time will give the least amount of flaking. In this case the plasticizer is not used primarily to soften the film but rather to improve its adherence to the hair and prevent flaking. If too much plasticizer is used the PVP film tends to become sticky.

Table 6.2 indicates the effect of several plasticizers when they are used with PVP at various concentrations and the results are shown in a similar manner to those given by shellac films.

TABLE 6.2

Concentration of polyvinyl pyrrolidone	*Modifier used*	*Content of modifier*	*Type of film*	*Degree of gloss*	*Degree of flaking on combing**
5%	Alcohol-soluble lanolin deriv.	0·5%	Firm	–	100
	Propylene glycol monolaurate		Firm	Good	50
	Diethyl phthalate		Firm	Good	50
	iso-Propyl myristate		Firm	–	100
	Silicone MS. 555[1]		Rigid	Fair	50
	None		Rigid	–	100
5%	Alcohol soluble lanolin deriv.	2·0%	Rigid	–	100
	Propylene glycol monolaurate		Soft	Fair	25
	Diethyl phthalate		Soft	Fair	25
	iso-Propyl myristate		Soft	–	50
2·5%	Propylene glycol monolaurate	2·0%	Firm	Good	None
	Diethyl phthalate		Firm	Medium	None
7·5%	Propylene glycol monolaurate	2·0%	Firm	Good	50
	Diethyl phthalate	2·0%	Firm	Medium	None

* 100 represents maximum degree of flaking
[1] Midland Silicones Ltd.

It will be seen that 2·5 per cent PVP and 2·0 per cent propylene glycol monolaurate gives a firm film with a better gloss than that obtained using diethyl phthalate. No flaking occurred on combing in either case. If the concentration of PVP is increased to 7·5 per cent, the same proportion (2·0 per cent) of propylene glycol monolaurate still gives a firm film with a good gloss but causes flaking on combing, whereas no flaking occurs in this case with diethyl phthalate.

The results do indicate that effective lacquer systems can be prepared with a range of concentrations of PVP provided the effect of the plasticizer is checked carefully. To increase the water repellency of PVP lacquers small proportions of shellac, silicone oil, or certain modified resins can be included. As shown in the table it will be seen that by using 10 per cent of silicone oil based on the PVP content a rigid film is obtained although this flakes when combed. The degree of flaking can be reduced if a phthalate is added as the plasticizer to the system.

It does not follow that water repellency is always required. Where climatic conditions are dry a small percentage of a humectant is used to prevent any drying effect on the hair caused by using PVP (Brit. Pat. 747806). Examples of formulations for lacquers based on PVP are given:

No. 2151

Polyvinyl pyrrolidone	25·0
Dimethyl phthalate	20·0
Silicone	0·5
Alcohol	954·5
	1000·0
Perfume	0·3–0·5 per cent

Preparation of the lacquers is a straightforward procedure. The PVP is dissolved in the alcohol and any additives added and mixed until solution is effected. The perfume is then added and the lacquer is allowed to stand and finally filtered before filling.

The above formula gives a firm but flexible film which does not flake. It is particularly effective on fine hair.

No. 2152

Polyvinyl pyrrolidone	75
Propylene glycol monolaurate[1]	20
Silicone oil	1
Alcohol	904
	1000
Perfume	0·3–0·5 per cent

[1] Or propylene glycol monomyristate.

No. 2153

Polyvinyl pyrrolidone	75
Diethyl phthalate[2]	20
Water	100
Alcohol	805
	1000
Perfume	0·3–0·5 per cent

[2] Or dimethyl phthalate

This lacquer is suitable for use on coarse or 'wiry' hair which are more difficult to control, or on fine hair if a good fixative is required.

Reference has already been made to the effect of a solubilizer when a water-soluble perfume is used in a lacquer system. It is equally important to check the properties of a lacquer formulation with and without perfume when a perfume concentrate is being used in an alcohol based lacquer. Some perfumes contain phthalates and will affect the properties of the film. In some circumstances a phthalate is also required to be used as a denaturant for the alcohol and adjustment of the formula must be made to allow for this.

Precipitation of PVP can also occur due to reaction with certain raw materials used in compounding perfumes, and it is advisable to check that this does not occur with any other cosmetic products containing PVP.

No. 2154

Polyvinyl pyrrolidone	40
Alcohol soluble lanolin derivative	1
Dimethyl phthalate	4
Alcohol	955
	1000

Perfume	0·3–0·5 per cent

Although it is useful to include lanolin or a lanolin derivative in a lacquer as a selling point, tests show that when these materials are used as the sole plasticizer they tend to improve the flexibility of a PVP film and reduce the hygroscopicity but do not prevent flaking. For this reason they are best used in small amounts only together with a phthalate.

Suitable alcohol soluble derivatives include Acetulan,[1] Amerchol L-101,[1] and Lanethyl.[2] It should be noted that the total amount of plasticizer used is usually from 10 to 12 per cent of the resin content.

Lacquers are also presented to give special effects with soft flexible films for casual wave effects. One way of obtaining this effect is to use a higher concentration of plasticizer. The following formula gives a soft film with a good gloss.

[1] American Cholesterol Products Inc.
[2] Croda Ltd.

No. 2155

Polyvinyl pyrrolidone	50
Propylene glycol monolaurate	5
Diethyl phthalate	2
Alcohol	943
	1000
Perfume	*q.s.*

When an aqueous/alcoholic solution is used as the vehicle this naturally extends the drying time of the lacquer. This type of product is less expensive to produce and is useful for the professional hairdresser when the hair is dried by heat after setting. It is also sold as a lacquer and setting lotion combined and the properties and uses are described according to the proportions of water and alcohol used in the formula. In particular the lotion is suitable to apply to damp hair after shampooing. In this case the hair is set with rollers or by pin-curlers and after setting is allowed to dry and then combed out. The spray can then be used again to keep the hair in place. A suitable product can be prepared as follows:

No. 2156

Polyvinyl pyrrolidone	35
Lanolin derivative	1
Diethyl phthalate	2
Water	100
Alcohol	862
	1000
Perfume	0·3–0·5 per cent

Co-polymers of polyvinyl pyrrolidone and polyvinyl acetate have different properties to PVP partly due to the plasticizing effect of the acetate. These materials, generally referred to as PVP/VA co-polymers, give films suitable for use as hair sprays, particularly because they do not pick up moisture as readily as PVP. The resins are prepared with different proportions of acetate as shown:

PVP/VA	70 : 30	co-polymer
PVP/VA	60 : 40	co-polymer
PVP/VA	50 : 50	co-polymer
PVP/VA	30 : 70	co-polymer

Since the acetate exerts a plasticizing effect the proportion of acetate present in the co-polymer affects the type of film obtained on the hair. If the proportion of acetate is too high the lacquer film does not bind onto the hair shaft. The PVP/VA 60:40 co-polymer gives a satisfactory firm film which is made more flexible by using a suitable plasticizer. Although the film on the hair is not affected by humidity to the same extent as a PVP film the lacquer will become soft and sticky if the proportion of modifier used in the formula is too high.

Diethyl and dimethyl phthalates, and polyethylene glycol 400 are suitable plasticizers for the PVP/VA film. These materials are used at a concentration of 10 per cent of the amount of co-polymer.

No. 2157
(Soft film)

Polyvinyl pyrrolidone/vinyl acetate 60 : 40 co-polymer	25·0
Dimethyl phthalate	2·5
Alcohol	500·0
Water	472·5
	1000·0
Perfume	0·3–0·5 per cent

No. 2158
(Hard film)

Polyvinyl pyrrolidone/vinyl acetate 60 : 40 co-polymer	50
Dimethyl phthalate	4
Silicone fluid[1]	1
Alcohol	945
	1000
Perfume	0·3–0·5 per cent

[1] Silicone fluid MS. 555 type—Midland Silicones Ltd.

A soft flexible film can be obtained by including *iso*-Propyl myristate, in this instance to give softness and added gloss or lustre to the hair.

No. 2159
(Soft film)

Polyvinyl pyrrolidone/vinyl acetate 60 : 40 co-polymer	25
Dimethyl phthalate	2

iso-Propyl myristate	1
Alcohol	972
	1000
Perfume	0·3–0·5 per cent

Solutions of the resin or the completed spray solution should be filtered before packaging, preferably after standing overnight in suitable closed containers.

Dimethyl hydantoin formaldehyde resin (DMHF) (Rex Campbell & Co. Ltd.) is a light coloured brittle material soluble in both alcohol and water. Used as the film-former of hair sprays the compound gives soft, firm or rigid films, depending upon the concentration used. It is less hygroscopic than either PVP or PVP/VA co-polymers and when used with the correct amount and type of plasticizer gives a non-tacky film and a good gloss to hair. Solutions in alcohol are more easily prepared with the aid of gentle heat.

Suitable plasticizers for use with DMHF resin are as follows:

1. Alcohol-soluble lanolin derivatives: (or water-soluble depending upon the solvent system used)
2. Propylene glycol monomyristate:
3. Diethyl or dimethyl phthalate:
4. *iso*-Propyl myristate.

Using 10 per cent of any of these plasticizers based on the content of resin, firm lacquer films are formed which show good gloss and do not flake when the hair is combed. If the amount of plasticizer is increased to 20 per cent the films obtained with soluble lanolin derivatives and diethyl phthalate become sticky, and the film obtained with propylene glycol monomyristate has less gloss. The types of lacquers obtained using various concentrations of resin with different plasticizers are shown as follows:

No. 2160
(Hard film)

DMHF resin	50
Soluble lanolin	5
Alcohol	945
	1000

No. 2161
(Hard film)

DMHF resin	50
Dimethyl phthalate	5
Alcohol	945
	1000

No. 2162
(Hard film)

DMHF resin	50
iso-Propyl myristate	10
Alcohol	940
	1000

No. 2163
(Soft film)

DMHF resin	50
iso-Propyl myristate	20
Alcohol	930
	1000

Lacquers based on the proportions of resin and plasticizer given above all give a good gloss to the hair and do not cause flaking. The lacquers are prepared by first dissolving the resin in the solvent. The remaining ingredients are then mixed with the solution which is allowed to stand overnight in suitable closed containers. The solution is filtered before filling out.

DMHF resin is used with a small proportion of shellac to prepare a spray suitable for use on difficult or 'hard-to-hold' hair. The resin is also useful to include in sprays based on PVP or PVP/VA to reduce hygroscopicity.

No. 2164
(Hard film)

DMHF resin	50
Shellac (25 per cent solution in *iso*-Propyl alcohol)	5
Silicone fluid[1]	1
iso-Propyl myristate	4
Alcohol	940
	1000

Perfume 0·3–0·5 per cent

[1] Silicone fluid MS. 555 type—Midland Silicones Ltd.

Dantoin 693 (Rex Campbell & Co. Ltd.) is a modified dimethyl hydantoin terpolymer resin free from formaldehyde and is used in a similar way to dimethyl hydantoin formaldehyde resin. It is non-hygroscopic and forms a good flexible film which is completely removed from the hair by shampooing.

Resin 28-1310 (National Adhesives and Resins Ltd.) is a vinyl acetate co-polymer and Resin 28-3307 is a carboxylated vinyl acetate co-polymer including a co-polymerized ultra-violet light absorber. These materials are available as fine transparent beads which are soluble in alcohol and chlorinated solvents, the latter being used as a co-solvent for concentrated solutions. Partial or complete alkali neutralization of alcoholic solutions with amino hydroxycompounds determines the final water solubility, complete water solubility being obtained with 100 per cent neutralization. Thus hard spray films with poor water solubility are obtained by 10 to 70 per cent neutralization and softer more water-soluble films by 80 to 90 per cent neutralization of the resin. Consequently the properties of the resin film can be varied to suit conditions or provide specific properties by controlling the degree of neutralization. The most suitable neutralizers are AMPD, 2-amino-2 methyl-1,3 propanediol and AMP, 2-amino-2 methyl-1 propanol.

Further modifications of the film are obtained by using plasticizing additives such as lanolin, dimethyl phthalate and iso-propyl myristate, as already indicated in the previous text.

It seems likely that new resins will become available to provide specific properties to hair sprays. Amongst the several available at the present time polyvinyl imidazole (PVI) (U.S. Patent 2953498) is a resin of similar chemical structure to PVP. This material is soluble in both alcohol and water, and gives films which do not become tacky at high relative humidities. The films are flexible and can be used without special additives and modifiers.

Polymers prepared from N-vinyl oxazolidones (Devlex—Dow Chemical Co.) are available either as water or alcohol soluble materials suitable for formulation of hair sprays and setting lotions. Gantrezhan (General Aniline & Film Corp.) is a co-polymer of vinyl methyl ether and maleic anhydride, suitable for application in water based hair sprays and setting lotions.

Consumer research inevitably controls the type of product required by the public and this applies particularly to the properties of hair sprays. Products should be carefully formulated and evaluated to provide the particular properties required. As a general guide to performance the hair-spray should provide sufficient adhesion and control and be easy to remove from the hair; it should not become sticky or greasy in conditions of high humidity and should not flake, dry, or dull the hair. The spray can be prepared for its hair controlling properties only, as a dual purpose setting and fixing lotion or for conditioning effect to give the hair lustre and gloss and maintain a soft, natural looking set.

Complete formulations for aerosol hair sprays are now given based on the materials discussed in the previous text. The following formulation has been found to give good results and washes out completely by shampooing:

No. 2165
(PVP type)

Hair spray base:	
Polyvinyl pyrrolidone	30
Propylene glycol monomyristate[1]	3
Alcohol 99·5% v/v	967
	1000
Perfume	0·2–0·5 per cent

Procedure: Dissolve the polyvinyl pyrrolidone in the alcohol. Stir well and add the propylene glycol monomyristate and perfume. Mix well.

Container charge:	
Hair spray base	40
Propellent–11/12 (50 : 50)	60
	100

Container: internally lacquered tin plate or aluminium. Valve: standard with lacquered cup.

The following formula gives good holding properties to the hair with a good gloss and is easily washed out by shampooing:

No. 2166
(PVP : VA type)

Hair spray base:	
Polyvinyl pyrrolidone: Vinyl acetate 60 : 40 co-polymer	50
Diethyl phthalate	2·5

iso-Propyl myristate	2·5
Alcohol 99·5% v/v	945
	1000·0

Perfume 0·2–0·5 per cent

Procedure: Dissolve the PVP:VA in the alcohol with stirring. Add the rest of the raw materials and the perfume. Mix well.

Container charge:	
Hair spray base	40
Propellent–11/12 (50 : 50)	60
	100

Container: internally lacquered tin plate or aluminium. Valve: standard with lacquered cup.

Dimethyl hydantoin formaldehyde resin can be used in a range of formulae from 'soft hold' sprays to 'firm hold' sprays, and is readily removed by shampooing. A 'firm hold' spray is obtained with the following formula:

No. 2167
(DMHF type)

Hair spray base:	
Dimethyl hydantoin formaldehyde resin	50
iso-Propyl myristate	15
Alcohol 99·5% v/v	935
	1000

Perfume 0·2–0·5 per cent

Procedure: Dissolve the DMHF resin in the alcohol. Add the perfume and *iso*-Propyl myristate and mix well.

Container charge:	
Hair spray base	40
Propellent–11/12 (50 : 50)	60
	100

To increase the hardness of the film on the hair 0·5 per cent of palest dewaxed shellac can be added to the above formula. The film will still shampoo out of the hair satisfactorily. To give a softer film on the hair reduce the percentage of DMHF to about 3·0 per cent.

Container: internally lacquered tin plate or aluminium. Valve: standard with lacquered cup.

No. 2168

(Dantoin resin 693 type)

Hair spray base:	
Dantoin resin 693[1]	30
iso-Propyl myristate	3
iso-Propyl alcohol	967
	1000

Perfume 0·5–0·75 per cent

[1] Rex Campbell & Co. Ltd.

Procedure: Dissolve the Dantoin resin 693 in half the amount of *iso*-Propyl alcohol with gentle heat, stirring occasionally. Cool, add the remainder of the *iso*-Propyl alcohol, the *iso*-Propyl myristate and the perfume. Mix well.

Container charge:	
Hair spray base	40
Propellent—11/12 (50 : 50)	60
	100

Container: internally lacquered tin plate or aluminium. Valve: standard with lacquered cup.

This formula can also be made using ethyl alcohol instead of *iso*-Propyl alcohol, and the perfume content can then be reduced to between 0·2 and 0·5 per cent.

No. 2169

(Resin 28–1310 type)
(Soft to medium hold)

Hair spray base:	
Resin 28–1310[1]	25
2-amino-2-methyl-1: 3-propanediol (AMPD)[2]	0·1
Alcohol 99·5% v/v	972·9
	1000·0

Perfume 0·2–0·5 per cent

[1] National Adhesives & Resins Ltd.
[2] Honeywill & Stein Ltd.

Procedure: Dissolve the AMPD and Resin 28-1310 in the alcohol with stirring. Add the *iso*-Propyl myristate and perfume. Mix well.

Container charge:	
Hair spray base	40
Propellent–11/12 (50 : 50)	60
	100

Container: internally lacquered aluminium or tin plate. Valve: standard with lacquered cup.

The above formula is also suitable for use as a man's hair fixative.

The most rigid film former used in hair spray formulations is, of course, shellac. It has the biggest disadvantage, that of being insoluble in water. A hard and rigid hair spray film based on shellac is made as follows:

No. 2170

Hair spray base:	
Shellac–palest dewaxed[1]	40
Diethyl phthalate	4
Alcohol 99·5% v/v	956
	1000

Perfume 0·2–0·5 per cent

[1] A. F. Suter Ltd.

Procedure: Dissolve the shellac in the alcohol with continuous stirring, care being taken not to stop stirring until all the shellac has been dissolved, otherwise the shellac will sink to the bottom of the reactor and stick as a resinous mass. Finally, when dissolved add the diethyl phthalate and perfume.

Container charge:	
Hair spray base	40
Propellent–11/12 (50 : 50)	60
	100

Container: internally lacquered aluminium or tin plate. Valve: standard with lacquered cup.

Use of Resin 28-3307 and the effect of neutralization with 2-amino-2 methyl-1,3 propanediol (AMPD) is shown in the following three formulations. Hair sprays of 'soft', 'regular', and 'hard-to-hold' residual films are obtained by varying the proportion of resin neutralizing agent and plasticizer or modifier. The sprays give good gloss to the hair, do not show any dust after combing, and are easily removed by shampooing.

No. 2171
(Soft hold)

Resin 28–3307[1]	36
2-amino-2-methyl-1:3-propanediol (AMPD)[2]	4
iso-Propyl palmitate	3
Alcohol 99·5% v/v	957
	1000

Perfume 0·2–0·5 per cent

[1] National Adhesives & Resins Ltd.
[2] Honeywill & Stein Ltd.

No. 2172
(Medium hold)

Resin 28–3307[1]	27
2-amino-2-methyl-1:3-propanediol (AMPD)[2]	2·5
iso-Propyl palmitate	2
Alcohol 99·5% v/v	968·5
	1000·0

Perfume 0·2–0·5 per cent

[1] National Adhesives & Resins Ltd.
[2] Honeywill & Stein Ltd.

No. 2173
(Hard-to-hold)

Resin 28–3307[1]	49
2-amino-2-methyl-1:3-propanediol (AMPD)[2]	4
iso-Propyl palmitate	2
Alcohol 99·5% v/v	945
	1000

Perfume 0·2–0·5 per cent

[1] National Adhesives & Resins Ltd.
[2] Honeywill & Stein Ltd.

Procedure: Weigh out the alcohol into a mixing vessel fitted with a variable speed stirrer. Add the perfume, AMPD, and *iso*-Propyl palmitate. Commence stirring fairly rapidly and slowly add the Resin 28-3307 beads. Continue stirring until the beads have completely dissolved.

Container charge:	
Hair spray base	40
Propellent–11/12 (50 : 50)	60
	100

Container: internally lacquered aluminium or tin plate. Valve: standard.

CHAPTER SEVEN

Hair Colourants

Hair colourants today form an important group of products and are prepared for use at home or at the hairdressing salon. Colouring or tinting of hair is, however, by no means a modern cosmetic refinement since it was practised in a crude form from earliest antiquity. It is, for instance, well-known that Egyptian women used kohl, probably a naturally occurring lead sulphide (galena), to give a black colour to the hair, eyebrows and eyelashes. Henna was also used to obtain auburn tints. During the Roman era, several other useful plants had been discovered which were used for obtaining a wider range of hair tints. For instance, Lysimachia was used for imparting a blond tint, by far the most esteemed colour amongst the Romans. This plant was discovered by King Lysimachus of Thrace, a contemporary of Alexander the Great. Botanists have since identified this plant as the purple willow-herb, *Lythrum salicaria*. L. Hypericon, known also as Corisson, was used to dye the hair black. This plant is believed to be the perforated St. John's Wort, *Hypericon perforatum L.* Other plants yielding a black dye were Ophrys, the eyebrow plant, now believed to be *Ophrys ovata* or *bifolia, L.*, and *Polemonium caeruleum, L.* This latter plant was boiled in oil and the extract used to impart blackness to the hair.

A modern conception of hair colourants includes products of the following types.

(*a*) Lighteners.
(*b*) Temporary colourants.

(*c*) Semi-permanent colourants.
(*d*) Permanent colourants.

Hair lighteners or bleaches

Hair lighteners or bleaches are included as hair colourants since they alter the natural colour. More correctly they may be described as decolouring preparations since they lighten or remove colour.

The melanin type pigments which are responsible for the colour of hair are formed in the living cells of the matrix of the follicle and as the cells grow the pigment becomes distributed in the dead keratinized cortex and cuticle. Bleaching of the pigment is carried out by using a peroxide, generally hydrogen peroxide at an alkaline pH value. Three to 6 per cent solutions of hydrogen peroxide (10 volumes or 20 volumes) are mainly used for home bleaching and the solution is made alkaline immediately before use by adding a few drops of strong ammonia solution. The bleaching effect is obtained by the oxidation effect of the oxygen as it is released by the action of the alkali. During oxidation some damage occurs to the sulphur linkages of the hair, depending upon the length of time the hair is exposed to the effect of alkali.

Hair is soluble to some extent in alkalis and if over-bleached it has a soft gelatinous feel when wet and is often difficult to comb. After drying it feels coarse and brittle. Because of the damage which affects the hair structure in these conditions, special care is required if thioglycollate solutions are subsequently used for permanent waving. Hair which has been previously softened by alkali is more readily affected by thioglycollates and special weak solutions are used to prevent further damage taking place. It has been shown that bleached hair is also affected by ordinary shampooing—when it appears that some colloidal material is removed from the hair causing turbidity in water, whereas unbleached hair treated under the same conditions does not show similar turbidity (Flesch *Proc. Toilet Goods Assoc., Sci.* Sec. No. 32, 1, 1959).

With increasing popularity of producing temporary colours the hair is often lightened or bleached before a colourant is applied. Dark shades of hair cannot be made lighter by treatment with most of the dyestuffs used as colourants so that pre-bleaching becomes part of a colouring process when pastel shades, light auburn or golden effects are required. When an alkaline bleaching is to be preferred avoid damage to the hair shaft by prolonged contact with the alkali. At the same time when a strong solution of ammonia is used as the activator rapid and excessive evolution of oxygen takes place and the comparatively large bubbles of oxygen give most rapid bleaching on the outside of the hair shaft, before penetration has occurred. If this type of bleaching solution is re-applied to treat any regrown hair, the rapid evolution of oxygen tends to migrate to that portion of the hair which has already been treated and the effects of over bleaching soon become noticeable. For this reason strong ammonia solutions are now seldom used as activators.

Bleaching pastes

Bleaching pastes or creams are made to provide a more progressive form of bleaching. These are made by mixing hydrogen peroxide solution with an absorbent powder immediately before use. The paste or cream so formed does not run, is easier to apply, and because it acts only where applied is suitable for treating the roots of previously bleached hair. A suitable powder which forms a paste when mixed with water or hydrogen peroxide was at one time sold as 'white henna'. This term is, however, a misnomer for light magnesium carbonate and although the paste obtained with this material is easier to control than hydrogen peroxide solution on its own, it still requires the addition of ammonia as an activator.

Bleach powders

Bleach powders for converting into paste form with either water or hydrogen peroxide are derived from the original white henna. They usually contain solid oxidizing agents and alkalis in an inert filler which can consist wholly or partly of light magnesium carbonate. A formula for a bleach powder to be mixed with hydrogen peroxide before use, is given below, using ammonium bicarbonate as the alkali:

No. 2174

Ammonium bicarbonate	200
Ammonium bisulphate	100

Light magnesium carbonate	500
Light calcium carbonate	200
	1000

Ammonium bicarbonate or ammonium hydrogen carbonate NH_4HCO_3, dissociates slowly during storage with the formation of ammonia, but it has reasonable shelf life properties in the type of inert base indicated. Ammonium bisulphate NH_4HSO_4 is included in the formula to act as stabilizer for the bicarbonate. Other materials can be incorporated in this basic composition to increase the bleaching effect or the rate at which bleaching takes place.

Sodium perborate or sodium borate perhydrate $NaBO_2 . H_2O_2$ is a white free flowing powder which dissolves in water to give an alkaline solution of peroxide.

Ammonium persulphate $(NH_4)_2S_2O_8$ occurs as white crystals or powder. It is a strong oxidizing agent and decomposes gradually during storage but remains sufficiently stable to retain activity if protected from moisture. Up to 20 per cent of this material is used in bleach powders particularly those used by professional hairdressers, to give rapid effects or platinum shades.

Sodium peroxide Na_2O_2 is another strong oxidizing agent used in bleach powders to give rapid effects and very light shades. In water the compound develops heat and decomposes to form sodium hydroxide and hydrogen peroxide. It is claimed that melamine perhydrate is a stable oxidizing agent suitable for use as a bleach powder. This compound is considered to be more effective than sodium perborate (German Pat. 1141749). Alternative formulae are as follows:

No. 2175

Ammonium bicarbonate	100
Ammonium bisulphate	100
Sodium peroxide	100
Sodium sulphate (anhydrous)	100
Calcium carbonate	200
Light magnesium carbonate	400
	1000

Sodium peroxide should be handled with care. It can cause ignition with volatile solvents particularly in the presence of water.

No. 2176

Ammonium persulphate	200
Ammonium bicarbonate	30
Ammonium bisulphate	30
Sodium perborate monohydrate	25
Light calcium carbonate	200
Magnesium silicate	515
	1000

All bleach powders are prepared by grinding (if necessary) and mixing all the ingredients together, with the exception of any oxidizing agent. Oxidizing agents are first triturated with a small proportion of the base before mixing with the bulk. The powder is finally sieved before packaging.

Sodium sulphate anhydrous (Na_2SO_4), light magnesium oxide (MgO), and magnesium silicate ($3MgSiO_3 . xH_2O$), are all suitable inert powders for use as a vehicle for the bleach powder or alkali. These materials do however, vary in density or specific volume and because of this it is necessary to experiment with the selection and proportion of fillers to give a product of the correct specific volume. This depends upon the directions which should be issued in detail with the finished product. The powder is generally measured by tablespoonful measure or by using a plastic scoop supplied with the package. For use the user requires to know the amount of powder and the quantity of hydrogen peroxide solution which is to be mixed with the powder to produce a paste of the correct consistency. The quantities prescribed should be detailed to give a thick cream-like mass and sufficient mix should be obtained with one mixing either for application to the complete head of hair or for bleaching regrowth at the roots, as the case may be.

As with many preparations for use on the hair, trials should always be carried out with the final product. In this case these are required to determine the length of time the bleach can safely be left on the hair and the bleaching effects which are obtained on different colours and types of hair by the treatment. One of the less attractive results of bleaching hair is the bright yellow or brassy effects obtained with the original ammonia-peroxide solutions and by over bleaching. In some cases reddish tints are also obtained. These colours can be toned down by treating the hair with a rinse containing a water soluble blue dyestuff, which makes the hair slightly blue in colour and whiter in appearance. In a similar way, yellow and red tints can be eliminated or reduced by adding a blue dyestuff to a bleach powder. For this purpose either methylene blue or phthalocyanine blue can be used.

Methylene blue is soluble in water and a sufficient quantity should be added to the powder bleach so that it forms a blue tinted cream when mixed with peroxide. Phthalocyanine blue is a copper complex which is considered to have a catalytic action on the bleaching mixture in addition to preventing development of yellow and reddish tints. The pigment is insoluble in water and should therefore be used as a finely ground powder so that it is easily dispersed in the paste. A sulphonated water soluble form of the dyestuff can also be obtained. About 0·5 to 1·0 per cent of either form of phthalocyanine included in the powder mix is sufficient to give an attractive blue colour to the finished paste. Powders of this type are used to obtain highlights by mixing with water only, and the paste is applied to towel-dried hair after shampooing. A mild lightening effect is also obtained by adding one or two ounces of hydrogen peroxide and mixing to a paste with a liquid shampoo. The mix is used as a shampoo and the lather allowed to remain on the hair for a few minutes before rinsing.

The powders are however, most frequently used with solution of hydrogen peroxide to obtain a stronger effect. Either 3 or 6 per cent solutions (10–20 volume) hydrogen peroxide should be recommended when the product is intended for home use. The powders are mixed with thirty or even 40 volume hydrogen peroxide by professional hairdressers when very light shades are required. Considerable skill is required when handling these mixtures in order to apply the paste rapidly and evenly without over bleaching or causing damage which will result in breakage of the hair. A powder suitable for professional use is as follows:

No. 2177

Ammonium persulphate	200
Sodium percarbonate	200
Ammonium bicarbonate	30
Sodium perborate monohydrate	25
Light magnesium carbonate	545
	1000
Blue colour	*q.s.*

Liquids and cream bleaches

Hair lighteners are also prepared as liquids and creams. In both cases the products provide a convenient way of applying hydrogen peroxide.

Liquid bleaches are generally sold as a two solution pack for mixing immediately before use. One bottle contains either 20 or 25 volume solution of stabilized hydrogen peroxide and the second bottle contains the activator. When the solutions are mixed a thin cream is formed of a consistency which can be controlled during application, and does not run down the hair. The activator generally contains a concentrated soap solution and forms a cream or lotion when it is mixed with the peroxide. An activator lotion based on oleate soaps is often referred to as an oil bleach by the professional hairdresser. A similar product is also sold for home-use and is claimed to have 'conditioning' properties. A formula for this type of preparation is as follows:

No. 2178

Oleic acid	454
Alcohol	183
Ammonium hydroxide (0·880)	227
Triethanolamine	136
	1000
Perfume	0·5 per cent

Procedure: Dissolve the perfume in the alcohol and add the oleic acid. Mix thoroughly. Slowly add the solution of ammonium hydroxide with constant stirring. Finally add the triethanolamine.

When required for use one part of the oil bleach or activator lotion is mixed with two parts of hydrogen peroxide solution (20 volume). The two solutions are supplied in these proportions if the product is for home use. The product is used either as a brightener or to give very light effects according to the length of time the mix is allowed to remain on the hair. Tests on hair switches should always be carried out before marketing this type of product. From these tests it is possible to include instructions so that the user can obtain a particular effect.

A cream bleach is prepared by incorporating a surface-active material with an oil bleach as indicated in the following formula:

No. 2179

Oleic acid	409
Alcohol	165
Ammonium hydroxide (0·880)	205
Triethanolamine	126
Sodium lauryl ether sulphate[1]	95
	1000

[1] Empicol E.S.B. 30 type—Albright & Wilson Ltd.

The bleach is prepared by mixing one part of the lotion with four parts of hydrogen peroxide, and the mix gives a cream which can be easily controlled as it is applied to the hair.

A liquid preparation with an acid pH value is used to give controlled or gradual lightening. The following formula has a pH value of 5·0–5·5:

No. 2180

Hydrogen peroxide solution[1]	800
Stearyl dimethyl benzyl ammonium chloride	10
Water (softened or distilled)	190
	1000

[1] Acid stabilized (30 volume).

Procedure: Dissolve the S.D.M.B.A.C. in the water and add the hydrogen peroxide solution.

When this type of lotion is used, bleaching is continued until the hair becomes dry. The lotion can be allowed to dry on the hair in this manner as there is no risk of alkaline damage. The degree of lightening is controlled by the amount used and the number of applications. The hair is not damaged to the same extent when peroxide bleaching is carried out at an acid pH value.

Bleach creams are convenient since the preparation is controllable on the hair and the degree of bleaching can be regulated. They are prepared as emulsions with glyceryl monostearate or cetyl alcohol, usually with the addition of a non-ionic surface-active agent. A suitable lanolin derivative can also be included ((*a*) Solan, Polychol 10—Croda Ltd. (*b*) Solulans, Amerchols—American Cholesterol Products Inc.). Preparations of various types can be obtained depending upon the emulsifying agent or thickener which is used. An interesting product can be made using Carbopol 934 as the thickening agent.

No. 2181

A	Carbopol 934	20
	Hydrogen peroxide solution[1]	880
B	Triethanolamine (10 per cent solution in distilled water)	100
		1000

1 Acid stabilized (30 volume).

Procedure: Disperse the Carbopol 934 in the peroxide by stirring with blades having good 'shear' properties. Allow to stand for 15 minutes then add solution of triethanolamine to a final pH of about 5·5.

Bleach preparations are best applied either with a toothbrush or a piece of cotton wool round an orange stick.

If the product consists of two solutions which require mixing, this is carried out in a suitable glass or porcelain container, avoiding use of metal containers or spoons. In most cases, bleaching preparations are affected by the temperature conditions. Heat accelerates the process and for the best results it is advisable to wrap the head in a hot towel or warm the treated hair with a hair dryer. This also reduces time of contact of the hair with the bleaching agent and hair damage is less likely to occur.

Solutions of hydrogen peroxide are prepared from commercial concentrates containing 35 and 50 per cent H_2O_2. Only deionized or distilled water is suitable to prepare the diluted solutions since impurities such as heavy metal ions present in natural or 'softened' waters cause catalytic decomposition. Mixing is carried out in stainless steel or glass lined containers. Cleanliness is essential since contamination affects stability and consequently shelf life. To prevent decomposition, stabilizers are added to peroxide concentrates and it is advisable to add an additional stabilizer to diluted solutions. About 450 parts per million or phenacetin or acetanilide is suitable for this purpose. Stability is also improved by adjusting the solution to a pH value of 4·0 with phosphoric acid.

Temporary colourants

Before the introduction of shampoos based on synthetic detergents it was common practice after shampooing with soap, to rinse the hair with either vinegar or lemon juice. Soap which remains on the hair has a dulling effect and the rinse was considered to give highlights as a result of neutralizing the alkali. A similar effect is obtained by rinsing with a solution of citric or tartaric acid and this principle is still used to prepare temporary colourants sometimes referred to as water-rinses. These consist of a mixture of a suitable dyestuff with an acid, and are prepared as either powders or liquids. Powder products are made by mixing the dyestuff with citric acid or tartaric acid in a sachet or capsule. To prepare the rinse the contents of the package is dissolved in about half a pint of warm water and the solution poured repeatedly over the wet head of hair immediately after shampooing. Some substantivity and absorption of dyestuff on to the hair cuticle takes place from the acid solution. The colourant does not penetrate to the cortex or medulla and the dye can easily be removed with a shampoo. The most successful water-rinses are those formulated to give blue, pink or light golden or auburn shades. These give good effects on bleached hair or very light shades of blonde or brown hair. Strong colour effects cannot be obtained with water rinses on dark hair, neither are they effective for covering grey hair. A black water rinse however, gives a satisfactory result on a dark-haired person.

No. 2182

Certified colour	50
Tartaric acid	950
	1000

Several manufacturers supply suitable acid dyestuffs. It is interesting to note that an acid rinse as used for giving pastel shades to bleached hair assists in preventing the alkali degradation of peroxide bleaching. Such a treatment can therefore be considered to have a conditioning effect. The powder form of acid rinse is generally used as a home application and each package contains sufficient dye to prepare a single rinse.

Liquid acid rinses consist of dye solutions in an aqueous-alcoholic vehicle, using a basic solvent of the following type:

No. 2183

Citric acid	300
Alcohol	200
Water (softened or distilled)	500
	1000

Deionized or distilled water should always be used for all types of hair colourants and it is also advisable to add a sequestering agent to prevent colour changes which may arise from accidental contamination with metallic ions.

Liquid rinses are most frequently used by professional hairdressers as concentrated solutions. A dropper is usually provided to dispense the concentrate, which is diluted with a suitable quantity of warm water when required for use. Experiments with acid dyes show that different dyes give varying colour yields on hair according to the type of acid used as the vehicle. It is not unusual to find that a particular dye gives a satisfactory colour yield with a citric acid base but is ineffective when used with acetic acid. This is because a sorption of a specific dye on hair varies according to the affinity of the dye and acid at the pH value given by a particular acid.

To prepare a range of rinses, individual dyestuffs must be tested in solutions of different acids to determine which particular vehicle gives the optimum colour yield. Basic formulae for the tests are given:

No. 2184

Acid dyestuff	60
Alcohol	100
Acetic acid (30 per cent)	100
Water (softened or distilled)	740
	1000

No. 2185

Acid dyestuff	60
Alcohol	100
Citric acid	400
Water (softened or distilled)	440
	1000

Tests can also be carried out using sulphuric and phosphoric acids in solutions adjusted to a pH value between 4·0 and 4·5.

Certain basic dyestuffs are occasionally used as rinses. These include methylene blue, rhodamine, and methyl violet (gentian violet). Solutions of these materials give pastel shades, and white or platinum blonde effects on bleached hair. A few drops of solutions of these dyestuffs are often mixed with a bleaching mixture to reduce the harsh red and yellow tones which are often obtained as a result of strong alkaline-peroxide treatment.

Colour shampoos

Colour shampoos can be regarded as falling into two main groups depending upon the extent of absorption or substantivity of the colour. Both types are considered as temporary colourants as distinct from the oxidation complexing involved in a permanent colouration process. One type combined in a shampoo base acts in a similar way to a water-rinse giving a slight colouring effect which is removed after one or two shampoos. The second type is more substantive and gives a higher degree of colouring. These are also described as semi-permanent colourants and are not generally used for their function as a shampoo but are applied after shampooing when the hair is in a degreased condition so that maximum absorption of colour takes place. Temporary and semi-permanent colourants can be presented to give various degrees and intensity of colour effects, but the dyestuff is essentially water soluble and is eventually removed by shampooing. Temporary colour shampoos are prepared with certified water-soluble dyes. Some non-certified dyes are also used and suitable selections can be obtained from dyestuffs manufacturers. An indication of the performance of a dyestuff in a detergent base can be obtained from their behaviour as acid-rinses.

Usually from 0·5 to 2·0 per cent of the dyestuff is used depending upon the particular effect required. This is dissolved in water and mixed with a shampoo detergent base such as triethanolamine or monoethanolamine lauryl sulphate. To assist absorption the pH value of the shampoo is adjusted to about 5·0. In a similar way to the behaviour of dyes in an acid-rinse, the behaviour of a dyestuff in a detergent base depends to some extent on the affinity of a dye for a particular acid. The most suitable acid to use for adjusting the pH value should therefore be determined by experiment for each dyestuff. It is not wise to adjust the pH value lower than 5·0 because the dyestuff could damage the skin or eyes by a vigorous shampoo application. The results of tests with individual dyestuffs is used as a basis for blending a range of shades.

Semi-permanent colourants

Semi-permanent colourants are distinguished from the colour shampoos described since they are designed to give a stronger and more permanent colouration to the hair. The colour effect persists for six to eight subsequent shampoos although some of the colour is removed during each shampoo. This type of colourant is easy to apply, and compared with permanent dyeing processes there is less risk of damage to hair and skin. This is sufficient reason to account for the popularity with the professional hairdresser and the home user.

Before attempting to prepare a range of colour shades it is also essential to examine the effect of each dyestuff using hair switches under specific conditions. They are mainly based on basic dyestuffs or nitro-amino dyes. These materials give highly coloured solutions which have some cation-active properties and are consequently absorbed to some extent by the hair fibre. To determine the best results of a particular dyestuff the following factors must be considered.

1. Solubility in water.
2. The composition of the vehicle or base.
3. The effect of pH value on the base.
4. The effect of solvents added to the basic composition.

The water solubility of a dyestuff is a factor which influences selection of the vehicle. Thus, a nitro-amino type dyestuff with low water solubility does not give satisfactory results when applied in an ionic surface-active agent. In such cases the effect of the dyestuff is increased by adding 2 to 5 per cent of an alkylolamide. It does not follow, however, that the same mixture will be the best vehicle for a basic dyestuff. The basic dyes being cation-active agents or in mixtures of these two materials.

Sorption of quaternary ammonium compounds has already been mentioned with reference to their so called conditioning effects. It follows that basic dyestuffs will also colour the hair by a sorption effect when they are applied in a cation-effective vehicle. In many instances, however, the dyestuff becomes too finely dispersed in the solution with the result that most of the colour rinses off the hair immediately. This phenomenon leads to the application of this type of semi-permanent colourant in bases prepared by mixing anionic and/or non-ionic surface-active agents with a cation active material. The mixtures form anion-cation colour complexes, and these are prepared in various ways according to the behaviour of a particular dyestuff. Correct formulation gives optimum conditions for transferring the colour onto the hair.

Colour complexes can be prepared using the following procedures:

1. Addition of a cationic surface-active material to a solution of anionic dyestuff.
2. Addition of an anionic surface-active material to an aqueous solution of a basic dyestuff.
3. Addition of an anionic surface-active material to a solution of a dyestuff in a cationic surface-active agent.

Experiments on these lines indicate the best complexing mixture to use for a particular dyestuff. The objective of the formulation experiments is to obtain the colour complex in the form of a fine precipitate. In some cases, the complexes are most effective if they are in solution form, but as already mentioned, if they are completely dissolved in the vehicle they are not necessarily as effective in transferring colour to hair.

The colourants are applied to towel-dried hair after shampooing and allowed to remain on the hair for a period of from 20 to 30 minutes to allow maximum absorbtion of colour. The hair is finally rinsed well with warm water to remove the excess of colour. The following basic formula illustrates a method of preparing an anion-cation colour complex:

No. 2186

Quaternary ammonium compound	100
Anionic surface-active material	80
Acid	40
Alkylolamide	100
Dyestuff	10
Water (softened or distilled)	500
	830

To prepare the complex first dissolve the dyestuff in a mixture of the alkylolamide and anionic surface-active material. Next dissolve the acid and quaternary ammonium compound in the water. Add this solution gradually to the colour solution whilst stirring.

It will be seen that a number of materials can be used as constituents of the basic formula in order to determine the most suitable and effective vehicle for use with a particular dyestuff. An extension of the basic formula is given:

No. 2187

Quaternary ammonium compound	100
Nonyl phenol ethylene oxide condensate	80
Lactic acid	40
Coconut diethanolamine	100
Dyestuff	10
Water (softened or distilled)	500
	830

In this formula a nonyl phenol ethylene oxide condensate is used as the basic solvent of the dyestuff. Coconut diethanolamide is used as the foaming agent to control distribution of the colourant in use.

Addition of an acidified solution of a quaternary ammonium compound forms a colour complex as it reacts to decrease the solubility of the dyestuff. As a general guide the colour complex which is formed shows maximum sorption properties before floculation or precipitation occurs. A preliminary test indicates the effect of the colour complex on hair and modifications are then made to the basic solution to determine whether the colour absorption can be improved. Modifications include:

1. Effect of different quaternary ammonium compounds.
2. Variation of the pH value of the system.
3. Effect of addition of solvents such as aldehydes, amyl, butyl and benzyl alcohols (French Patent 1138955; U.S. Patent 2940902).

Results of tests made with individual colours are used as a basis for blending to prepare a range of shades. Finally, if necessary, the viscosity of the product is adjusted, since a thin water solution is difficult to control in use. Hydrophilic colloids such as methyl cellulose, carbopol, or one of the natural gums can be used provided these are compatible with the particular colour complex.

Permanent colourants

Permanent colourants are preparations which are generally accompanied by a chemical process. They are used nowadays either to intensify or completely change the natural hair colour, although with few exceptions, they do not make the hair lighter than its natural colour. They are most successful when used to give a shade near to the natural colour of the hair. Recently, permanent colourants have become popular to give dramatic colour effects. Light brown or dark ash blonde hair for example can be successfully dyed to red, auburn or any dark colour including jet black. One of the most popular applications is for concealing grey hair and a shade nearest to the natural hair colour should be used.

Permanent colourants can be prepared from materials of vegetable origin, or salts of heavy metals, but those of most commercial importance are the oxidation dyes based on synthetic organic chemicals.

Vegetable dyes

Probably the most important of the vegetable hair dyes is Henna which when used on its own gives a brownish-chestnut shade. Henna powder is prepared by grinding the dried leaves and stems of the shrub *Lawsonia alba* (=*inermis*) known also as Egyptian privet. For use as a colourant the powder is made into a thin smooth paste with boiling water (henna pack) and brushed into the hair while hot with a small stiff brush and combed to ensure even distribution. Hot towels are wrapped round the head and replaced as they cool for half an hour. When the vegetable mass is removed by shampooing the hair has a characteristic bright chestnut colour. A small amount of solution of ammonia is sometimes added to the paste to enhance the effect. As with other dyes of vegetable origin henna is non-toxic, does not cause irritation or sensitization, one of the main disadvantages of other types of permanent colourants. On the other hand, the preparation and application of a henna pack is a laborious and messy process and causes staining of the hands and finger nails. Repeated applications of henna also give an unnatural reddish-auburn colour. This effect can be modified to some extent by mixing the henna with other materials. Synthetic indigo or powdered natural indigo leaves obtained from various species of *Indigofera* is sometimes used for this purpose. Mixtures of henna and indigo also give brown, auburn and black colour effects depending on the proportion of indigo. Pyrogallic acid and metallic salts are other additives used with henna powder. Typical formulae for auburn shades are as follows:

Light

No. 2188

Powdered henna	900
Pyrogallic acid	50
Copper sulphate	50
	1000

Medium

No. 2189

Powdered henna	880
Pyrogallic acid	60
Copper sulphate	60
	1000

Dark

No. 2190

Powdered henna	860
Pyrogallic acid	80
Copper sulphate	60
	1000

If an aqueous extract or infusion of henna is prepared the dyeing properties are available in a more convenient form. A liquid henna dye can be made as follows:

No. 2191

Powdered henna leaves	250
Powdered camomile flowers	80
Pyrogallic acid	2
Citric acid	1
Alcohol	30
Glycerin	10
Distilled water	1000
	1373

Procedure: Dissolve the citric acid in the water. Bring to the boil and infuse the camomile flowers in this solution. Then add all the other ingredients except the alcohol and maintain the source of heat gently for an hour. When cold add the alcohol and filter. Apply by brushing into the hair every hour until the desired tint is obtained.

An infusion of henna powder extracts the active principle, Lawsone (2-hydroxy-1, 4-naphthaquinone). This orange and red pigment is substantive to keratin and has a colouring effect 100 to 150 times greater than that of henna powder. It is available commercially in a pure and stable form. Use a concentration of 0·5 per cent in a suitable detergent base to prepare a henna shampoo, and a concentration of from 0·5 to 0·10 per cent for a powder or liquid acid rinse.

Camomile is another vegetable dye of interest. The active principle is present in the double or semi-double flowering heads obtained from *Anthemis nobilis*, also known as 'Roman' camomile or Manzanilla romana. The variety known as German or Hungarian camomile is derived from the species *Matricaria chamomilla.* Camomile powder is used as a paste and applied in a similar manner to a henna pack, particularly for giving a lighter or brighter effect to fair hair. An infusion prepared from the flowers is used as a rinse and can also be added to a suitable base to prepare a camomile shampoo. The colouring effect is due to a viscous deep blue volatile oil of which about 0·4 per cent is present in the flowers. The chief constituents of the oil are esters of butyric, angelic and tiglic acids and the colouring material apigenin also known as azulene. It has now been shown that there are several colouring materials of a similar chemical structure present in many other essential oils. They have an intense blue, violet, reddish-violet or bluish-violet colour according to the position of the methyl and iso-propyl groups in the basic ring structure. These compounds are very expensive when isolated from camomile oil but synthetic azulene is now commercially available. Although interest in the synthesis of azulene is mainly due to investigation of pharmacological properties, this work has shown that all azulenes having an intense blue colour are more effective than those with a red or violet colour. Azulene can be used in cosmetic products instead of camomile flowers or powder, and it is possible to obtain more positive effects because a standardized product is being used. To prepare shampoos or colour rinses use from 0·01 to 0·025 per cent in the basic composition.

Metallic hair colourants

Metallic hair colourants are based on various compounds of lead, silver, iron, cobalt, cadmium and copper. They are frequently referred to as 'colour restorers' and as such are claimed to restore the natural colour of grey hair. The compounds used do, however, act as dyes by depositing a coloured metallic salt on the hair shaft.

Lead dyes are generally based on solutions of lead acetate and it is considered that the insoluble lead salts are formed as a result of the reducing action of keratin. The reaction takes place slowly, the rate of colouring being influenced by the quantity of lead present in the solution and also by the effect of air and light. This results in a gradual colouration of grey hair and gives an illusion of restoring the natural colour. It is for this reason that these compositions are known as 'progressive' hair dyes. Successful results from lead colour restorers depends upon fresh preparation of the lead salt. This can be achieved by using a mixture of lead acetate with sodium thiosulphate or precipitated sulphur. Solutions prepared with sodium thiosulphate are probably the more elegant because they can be presented as clear solutions. Typical formulae are as follows:

No. 2192

Lead acetate	5
Sodium thiosulphate	15
Glycerin	80
Alcohol	100
Water (softened or distilled)	800
	1000

Water soluble perfume 0·5–1·0 per cent

No. 2193

Lead acetate	6
Sodium thiosulphate	12
Propylene glycol	100
Water (softened or distilled)	882
	1000

Water soluble perfume 0·5–1·0 per cent

If ammonium or monoethanolamine thioglycollate is used in place of sodium thiosulphate the rate of colouring is increased due to the stronger reducing action of the thioglycollate. To prepare the lotion dissolve the lead acetate in 100 parts of water. Mix the remaining ingredients and add the lead acetate solution.

Use of precipitated sulphur is shown in the following basic formula:

No. 2194

Precipitated sulphur	20
Lead acetate	10
Propylene glycol	100
Alcohol	100
Water (softened or distilled)	770
	1000
Water soluble perfume	0·5–1·0 per cent

This formula can be modified within the following range to suit particular requirements. It follows that the higher concentrations of lead and sulphur give darker shades.

No. 2195

Precipitated sulphur	20–	50
Lead acetate	5–	15
Glycerin (or propylene glycol)	50–	150
Water (softened or distilled)	925–	785
	1000	1000

The lotions are packed in amber bottles to avoid deposition of insoluble sulphur due to exposure to light.

Silver dyes were probably the most important group of preparations before the advent of the organic chemicals. They can be prepared to dye the hair a variety of shades from ash blonde to brown and black. These different tints are obtained by increasing the proportion of silver present. They are made in a variety of ways. For the two-solution type one bottle consists of silver nitrate solution with ammonium hydroxide and sometimes ammonium nitrate. The second lotion containing the 'developer' or reducing agent consists of a mixture of pyrogallic acid and sodium metabisulphite. Examples are appended:

No. 2196
No. 2 Solution

	A	B
Silver nitrate	20 grams	30 grams
Copper sulphate	–	0·2 grams
Ammonium hydroxide	*q.s.* to redissolve ppt.	120 cc
Distilled water	1000 cc	1000 cc

No. 2197
No. 1 Solution

	A	B
Pyrogallic acid	30 grams	–
Sodium metabisulphite	10 grams	–
Sodium sulphide crystals	–	40 grams
Distilled water	1000 cc	1000 cc

Use from 1 to 15 per cent of silver nitrate to prepare the No. 2 solution and from 1 to 5 per cent pyrogallic acid in the No. 1 solution according to the shade required. It is usual to pack No. 2 solution twice the size of the No. 1. Brush the hair with No. 2 solution immediately after No. 1 has been applied. Allow a few hours before shampooing.

A one-solution dye can be prepared as follows:

No. 2198

	A	B	C
Silver nitrate	30 grams	50 grams	40 grams
Copper sulphate	–	2·5 grams	–
Nickel sulphate	0·2 grams	–	–
Lead acetate	–	–	10 grams
Ammonium hydroxide	*q.s.*	*q.s*	*q.s.*
Distilled water	1000 cc	1000 cc	1000 cc

Sufficient ammonium hydroxide is added to redissolve the precipitate.

These dyes are much slower in their action and may require two or three applications.

The hair is first shampooed, and petroleum jelly or a firm cold cream is applied to the forehead, ears and neck before applying a silver dye. This prevents the skin being stained black by the silver compound.

Pyrogallic acid or Pyrogallol $C_6H_3(OH)_3$ is most important raw material in many hair dyes, and although when fresh and kept away from the light and air it occurs in fine white crystals, these readily oxidize and turn varying shades of brown according to the progressive degree of oxidation. A solution of this acid would appear therefore to make a good hair dye without any further additions, but in actual practice the development of colour is so slow that frequent applications are necessary to produce a tint of sufficient intensity. Oxidation of the acid is hastened if the solution contains alkali, and a non-metallic preparation can therefore be produced by the use of ammonia as follows:

No. 2199

Pyrogallic acid	100
Strong solution of ammonia	50
Alcohol	250
Water (softened or distilled)	700
	1100

Water soluble perfume 0·5–1·0 per cent

This hair dye will produce an effective brown tint. The use of free acid in combination with pyrogallic acid is not unusual, and both acetic and nitric have been used as follows:

No. 2200

Pyrogallic acid	50
Acetic acid glacial	10
Alcohol	290
Water (softened or distilled)	700
	1050

This dye will give the hair a pretty chestnut shade. Because of the reducing action of pyrogallic acid it is also used as the 'developer' in progressive hair dyes based on several metallic salts other than those of lead and silver. A few examples are given:

No. 2201

Cobalt dye

No. 1 Solution

Pyrogallic acid	30
Glacial acetic acid	2
Distilled water	1000
	1032

No. 2201 (*continued*)
No. 2 Solution

Cobalt sulphate	30
Copper sulphate	2
Distilled water	1000
	1032

Sufficient ammonium hydroxide is added to maintain a blue solution. Apply No. 1 solution and then brush in No. 2. Effective auburn shades are obtained.

No. 2202
Cadmium dyes
No. 1 Solution

	light golden	golden	ash blond
Pyrogallic acid	10 grams	15 grams	20 grams
30 per cent alcohol	1000 cc	1000 cc	1000 cc

No. 2 Solution

Cadmium sulphate	20
Distilled water	1000
	1020

Apply No. 1 solution and after 20 minutes brush in No. 2.

No. 2203
Copper dyes

Pyrogallic acid	30
Copper chloride	30
Nitric acid	1
Distilled water	1000
	1061

This dye is progressive and should be brushed in three or four times a day until the desired tint is obtained.

A two-solution copper dye is made as follows:

No. 2204
No. 1 Solution

Pyrogallic acid	20
Potassium carbonate	1
30 per cent alcohol	1000
	1021

No. 2 Solution

Copper sulphate	10
Ammonium chloride	3
Distilled water	1000
	1013

Mix equal quantities of the solutions immediately before applying to the hair. The tint will develop slowly, and when the desired degree of intensity has been obtained the hair may be shampooed.

It would seem likely that toxic or sensitizing effects would occur as a result of using materials such as heavy metal salts and pyrogallic acid. In use this seldom occurs, probably because the application is made with care to avoid staining of the scalp—an effect which is objectionable to users of hair dyes.

Oxidation hair colourants

Oxidation hair colourants are the most important group of permanent dyes, although their popularity has decreased to some extent in favour of the semi-permanent colourant which permits fashionable change of colour. On the other hand, permanent colouring remains popular to provide a lasting colour change or as a means of concealing grey hair. Although permanent dyes are not removed by a shampoo it is necessary to re-dye new hair growth.

Many detailed accounts of the chemistry of the oxidation hair dyes appear in the literature and the references mentioned in the text will be found useful to those who are particularly interested in this subject. These notes are intended to survey the subject and give a guide to formulators who should consider that many experiments and intensive trials must be carried out before embarking on the production of this type of product. This indeed, applies to all colourants and cannot be too strongly emphasized.

The oxidation dyes are based on synthetic organic chemicals or dye intermediates. Paraphenylenediamine and paratolylenediamine are the original favourite intermediates and are used either singly or in various blends to produce a range of shades. The principle of hair dyeing with these materials is to add the necessary oxidizing material to a solution of the dyestuff immediately before it is used. Various colour effects can thus be obtained by using different concentrations of the dyestuff and different strengths of oxidizing agent. The end product of the oxidation process is an azine dye which is considered to react with keratin to form insoluble azine derivatives and consequently give a permanent colouration. A disadvantage of treating hair with permanent colourants is the risk of subsequent skin trouble. Itching occurs as a result of a mild reaction but in severe cases dermatitis with erythema and swelling can occur. In the case of para dyes allergic reactions are considered to be due to soluble intermediate substances formed during the oxidation process.

When compared with the volume of sales it would appear that there are now very few cases of severe dermatitis. This may be due to the use of purer chemicals and also to the fact that the public have become more conditioned to read and follow specific instructions. With a view to ascertaining the sensitivity of a person's skin a cautionary statement is given on the package together with directions for carrying out a preliminary allergy test. This is effected by mixing a small quantity of the dye solution with an equal amount of the oxidizing agent and applying the mix to a clean small patch of skin just behind the ear. This is allowed to dry and covered with collodion, and if no irritation is felt after a few hours the process may be considered safe. Prominence given to the importance of this test together with a statement that the preparation may cause serious inflammation of the skin seems to have the necessary effect and deter casual use of the preparation. Risk of dermatitis can obviously be reduced if the user confines application to the hair and avoids staining of the scalp after use. Thorough washing and rinsing is necessary to remove soluble and irritant reaction products.

In addition to the original para-dyes already mentioned, several other chemicals are now available including direct dyestuffs and nitrated derivatives. All permanent colourants are often simply referred to as 'Para' dyes although obviously this can be incorrect if another type of material is used. Some of these are listed as follows:

ortho-aminophenol (ortho-hydroxyaniline)—brown shades.
para-aminophenol (para-hydroxyaniline)—red and chestnut tones,
meta-tolylenediamine (toluene-2,4-diamine)—all shades including black,
N-methyl-para-aminophenol—red and chestnut tones,
para-aminodiphenylamine (N-phenyl-para-phenylenediamine)—red tones.

In addition to the main dyestuff or mixture of dyestuffs, other materials are used in the dye solution as so-called modifiers. These additions are required to obtain certain colours or to improve the stability of the dye to the effect of light and wear, or permanent waving solutions. Others are used to give a bright and natural appearance to the hair as distinct from a dull, mat and lifeless effect. Modifiers are:

resorcinol: stabilizer for red or chestnut shades
pyrocatechol (catechol): stabilizer for brown shades

chlorohydroquinone (2-chloro, 1,4-dihydroxybenzene): for light and warm brown shades,
pyrogallic acid (pyrogallol): for light shades,
meta-dimethyl aminophenol:[1] stabilizer for use with para-phenylenediamine,
diamino-anisidine:[2] for warm blonde or red shades.

(1) Brit. Pat. 868325. Wella A.G. Darmstadt, Germany.
(2) Brit. Pat. 896167. Therachemie Chemisch Therapeutische G.m.b.H.

The vehicle used for the mixture of dyestuffs and modifier can vary. Distilled water can be used in some cases although an additional solvent is advisable since the dye intermediates have low solubility in water. An aqueous-alcoholic mixture containing from 20 to 50 per cent alcohol with 0·5 to 2·0 per cent glycerol is generally satisfactory as a vehicle for liquid type preparations. It is usual to add from 0·3 to 10 per cent of sodium sulphite to prevent oxidation and deterioration of the product during storage. The main purpose of the vehicle is to distribute the dye mixture evenly throughout the hair and to this end the product is often prepared as a cream or lotion using suitable surface-active agents. They can be either of a foaming or non-foaming type. The following cream base prepared with a non-ionic surface-active agent does not foam.

No. 2205

Mineral oil	20
Non-ionic emulsifier[1]	50
Cetyl alcohol	50
Water (softened or distilled)	880
	1000

Methyl parahydroxybenzoate 0·2 per cent

1 Abracol L.D.S. type–Bush, Boake, Allen.

Procedure: Heat the water to a temperature of 70°C and add to the melted fats and oil previously heated to the same temperature in a separate container. Stir during mixing and continue stirring until cold. The cream is of a suitable viscosity and can be used as a base for semi-permanent or permanent colourants. It is not affected by ammonia.

A cream with foaming properties can be prepared with monoethanolamine lauryl sulphate (27 per cent active) and 0·5 to 1·0 per cent ethylene glycol monostearate. A clear base with foaming properties is made using an alkyl sulphate or a sodium salt of sulphated lauryl alcohol ether. In this case solution of the dye intermediate is improved by the addition of 2 to 5 per cent of an alkanolamide (Brit. Patent 889327).

With few exceptions the oxidation-type colourant is effective as a result of a chemical oxidation process carried out under alkaline conditions using ammonium hydroxide as the alkali. This alkali is preferred since it appears to give greater colour fastness than other alkalis. Ammonia also contributes to the absorption of the oxidation products by reaction with keratin. The reaction is commenced by adding an oxidizing agent to the dye solution immediately before it is applied to the hair. The product is thus sold in a two-container package, one containing the dye solution and the other containing a suitable amount of oxidizing agent. Hydrogen peroxide (10 volume) solution is generally used for this purpose and is sometimes referred to in the trade as the activator or 'mordant'.

If the dye solution is presented as a cream or lotion the oxidizing agent may also be prepared in similar form. Sodium perborate or urea peroxide are also used as oxidizing agents. These materials or hydrogen peroxide solution can be prepared as creams or lotions based on the formula given for cream No. 2205. Conditioning effects can similarly be claimed when using a quarternary ammonium compound (see Chapter 6).

Para-dyes are also sold in solutions of anionic surface-active agents without an oxidizing agent as a colour shampoo, but these should be clearly distinguished from semi-permanent colourants. A shampoo base containing one or more dye intermediates is used to give slight colouring effects and highlights to medium and dark coloured hair, and as a brightener for light shades of hair. The lather is allowed to remain on the hair for a few minutes before final rinsing. More pronounced effects can be obtained if the lather is left on the hair for a longer period of time. An indication of the concentration of the main dye intermediates—(*para*-phenylenediamine or *para*-tolylenediamine)—required to produce a basic range of shades is given below:

	(per cent)
Black	1·0
Dark brown	0·9
Brown	0·5
Medium brown	0·45
Light brown	0·3
Dark ash blonde	0·2
Light blonde	0·1

Variations to these basic shades are made by blending paraphenylenediamine or paratolylenediamine with an additional dyestuff keeping about the same level total concentration of dyestuff in the final mix. Blends with paraminodiphenylamine for example used for reddish tones gives golden blonde to reddish blonde shades using a total concentration of from 0·15 per cent to 0·2 per cent total dyestuff depending upon the proportion of para amino diphenylamine in the mix. Similarly reddish toned dyestuffs used in total concentrations of between 0·2 to 0·3 per cent give light or mid velvet browns to light golden browns, and blends at total concentrations of 0·5–0·6 per cent give auburn to copper shades. These figures give a general guide towards preparation of a range of oxidative hair dyes.

The final performance of the dyestuff or dyestuff mixture also depends on the type and proportion of modifier which is used. A dye solution containing 20 per cent of modifiers on the total concentration of dye intermediates gives a much darker colour than a solution containing 50 per cent of modifiers using the same concentration of intermediates. Experimental work is thus carried out on the following lines:

1. select dye intermediate or blend of dye intermediates to be used.
2. select modifier.
3. determine concentration of dye intermediate or blend of dye intermediates to be studied with varying proportion of modifier.
4. determine quantity of alcohol necessary to obtain solubility of mixture.
5. determine quantity of dye solution and time required to obtain a particular effect.

All test work should be carried out using aqueous-alcoholic solutions before the dye solution is finally prepared in a suitable vehicle for use as the end product. The finished product is packed in coloured containers (photo-insensitive). These are kept full to prevent deterioration by light and air.

Probably the main disadvantage of oxidation-type hair dyes, particularly for the home user, is the necessity of a mixing procedure and the comparative difficulty of dyeing new hair growth uniformly. On the other hand, application of the dyes is more practical when they are presented as viscous liquids, creams, or in detergent bases. Suggestions for new methods of application include a solid stick form. This is applied after wetting the hair with hydrogen peroxide solution (R. Heilingotter Am. Pf. 75.5.19. May 1960). Dyes have also been prepared as powders and tablets. These contain a solid oxidizing agent such as sodium perborate, and an alkali, and are claimed to have excellent stability. The preparation is mixed to a paste with water when required for use (Brit. Pat. 917,840).

To prevent oxidation during storage 0·1 to 1·0 per cent of thioglycollic acid is used as an alternative to sodium sulphite, particularly in cream preparations. The ammonium thioglycollate so formed stabilizes the dye and also has some influence on the dyeing process probably on account of its reducing effect on keratin (*J. Soc. Cos. Chem.* **2**, 240. 1951). This effect led to consideration of preparations which combine a dyeing process with permanent waving (Brit. Pat. 721831. 1955; Brit. Pat. 876663. 1961). Use of thioglycollate and oxidation dyes is however, likely to increase skin sensitization at the concentrations required. Any dyestuffs used for such a purpose must therefore be stable to alkaline thioglycollate and not be affected by oxidation neutralizers.

Aerosol preparations

Aerosol sprays are used to apply temporary colours to the hair and a range of suitable permitted colours can be obtained from the cosmetic colour manufacturer. These are essentially alcohol-soluble, but selected colours which are water-soluble and others which are both oil and spirit soluble also give satisfactory results. All samples received from the manufacturer must be carefully tested using hair switches both individually and in combination blends to check their performance and behaviour in use. Some colours do, for example, colour the hair effectively but are difficult to remove by shampooing, whereas others rub off and are easily transferred to pillows and clothing.

Tests should be made in the completed aerosol because certain propellents can cause precipitation of colourant.

Perfumes also can affect the solubility of the colour in the aerosol system. Examples of the colour effects obtained with alcohol soluble colours are:

F D and C red No. 3 used at a concentration of 0·1 per cent gives a warm auburn shade to mid-brown hair; F D and C violet No. 1 used at a dosage of 0·02 per cent gives an ash blonde shade on bleached hair and an amethyst shade to grey hair; F D and C yellow No. 5 used at a dosage of 0·2 per cent gives golden highlights to mid-brown hair. This particular dyestuff is not completely soluble in alcohol (99·5%) and the formula of the spray base must be modified to include a suitable proportion of water to obtain complete solubility. In this case the container is charged with propellent 12/114.

Similar tests with individual certified alcohol-soluble dyestuffs are made to determine their effect using different shades of hair switches and are then blended to obtain a complete range of shades. It is important to use a lacquer base which is water-soluble and enables the colour to be removed easily by shampooing. The following formula makes a good base for this type of product:

No. 2206
(Temporary colour base)

Spray base:	
Polyvinyl pyrrolidone	30
Propylene glycol monolaurate	3
Alcohol 99·5% v/v	967
	1000

Alcohol soluble colour perfume	0·2–0·5 per cent

Procedure: Dissolve the colour in part of the alcohol and the Polyvinyl pyrrolidone in the remaining alcohol. Mix and add the perfume and propylene glycol monolaurate.

Container charge:	
Spray base	40
Propellent–11/12 (50 : 50)	60
	100

Container: internally lacquered tinplate or aluminium. Valve: standard with lacquered cup fitted with simple orifice button.

A variation of the coloured spray is the application of streaks to the hair. A dispersion of titanium dioxide is used for white streaks and metallic powders for gold and silver effects.

The lacquer base is modified to give a more rigid film as follows:

No. 2207
(Lacquer base)

	Spray base:	
	Polyvinyl pyrrolidone/vinyl acetate 60 : 40	60
	Alcohol 99·5% v/v	720
	iso-Propyl alcohol	200
	Propylene glycol	20
		1000
	Titanium dioxide	15 per cent
or	Metallic powder	10 per cent

Procedure: Disperse the titanium dioxide or metallic powder in the propylene glycol, and sufficient *iso*-Propyl alcohol to form a paste. Dissolve the PVP:VA resin in the remaining alcohol and add slowly to the colour mix with continuous stirring.

Container charge:	
Spray base	60
Propellent–11/12 (50 : 50)	40
	100

Container: if a metal container is used it must be internally lacquered after the can has been formed otherwise corrosion occurs. Valve: special–to allow fine particles to pass without clogging. The cup should also be lacquered after formation.

A glass marble is included in each can filled as an aid in dispersing the sediment which occurs during standing. Use metallic powders of a fine particle size and a valve of the type used for paints. The container must be selected carefully because metallic powders cause corrosion.

CHAPTER EIGHT

Lipsticks

Lipsticks have gained popularity during the past ten years to such an extent that they are now probably used more than any other single cosmetic product. This increase in popularity is undoubtedly influenced by the widening choice of colours available. Originally the function of the lipstick was to give a positive bright red colour to the lips and this was eventually modified to include blue-reds for brunettes and orange-reds for blondes. These early shades were all designed to give dramatic effects which at one time were considered an essential part of the allure and glamour attributed to make-up. Modern lipsticks still include several shades which are variations of true red, but no manufacturer would consider his range complete without including several pastel shades of rose, pink, or coral type. These soft toned colours appeal equally to the sophisticated and to the teenager making discreet experiments in the art of make-up. The modern range of colours is devised to meet various demands controlled by such factors as skin and hair colouring, fashion requirements, and the age of the user.

The lipstick is a product which illustrates a true cosmetic designed to beautify or improve natural beauty. Use of a brush is advocated to give a smooth outline, and its cosmetic application may be utilized in many ways. A thin mouth may be corrected by extending the lipstick above the upper lipline, or the mouth can be made to appear smaller by applying the lipstick well within the natural lipline. After application it is usual to remove any excessive grease or oil by blotting the lips with tissue. This practice favours the formulation of an emollient stick which is smooth and easy to apply rather than a stick which gives a matt appearance and drags on the lips when it is being applied.

In addition to ease of application the lipstick should also have the following characteristics:

(1) a good degree of indelibility,
(2) high retention of colour intensity,
(3) complete freedom from grittiness and be non-drying,
(4) a desirable degree of plasticity,
(5) a pleasant odour and flavour.

A lipstick should also have the following storage characteristics:

(1) a smooth and shiny appearance,
(2) freedom from 'bloom' or sweating,
(3) a suitable degree of firmness during reasonable variations of climatic temperature,
(4) retain plasticity without any tendency to dry-out or crumble.

The stick consists of colouring materials dispersed and suspended in a base prepared from a blend of oils and waxes adjusted to yield the desired melting point and viscosity. The temperature of the body varies between 36° and 38°C, the mouth exhibiting the higher figure. The lipstick must have a melting-point higher than this and appreciably higher if the product is expected to withstand exposure to hot climates. The most appreciated melting point is about 62°C, but from 55° to 75°C represent quite satisfactory products.

Composition

To consider the formulation of lipsticks the basic raw materials are conveniently classified as follows:

(1) wax mixture,
(2) oil mixture,
(3) bromo mixture,
(4) colours,
(5) preservatives and other additives.

Waxes and oils

The composition of the wax mixture is of prime importance. Best results are obtained by using a mixture of waxes of different melting points and adjusting the final melting point of the stick by adding a sufficient quantity of a high melting point wax. Sticks formulated on these lines do not have marked thixotropic properties but have a more cream-like or plastic consistency. Formulations which contain a high concentration of a single high melting point wax should be avoided. The following notes indicate the behaviour of some of the more important waxes used in lipsticks:

WHITE BEESWAX m.p. 62–64°C Use from 3–10 per cent.

A useful material to bind oils and higher melting point waxes. The molten wax shrinks slightly on cooling and thus helps preparation of moulded products. Higher concentrations give a dull waxy appearance and cause the stick to crumble during use.

CANDELILLA WAX m.p. 65–69°C Use from 5–10 per cent.

This material has a melting point of similar range to that of beeswax and gives a smooth and glossy appearance to a lipstick if used at a slightly higher dosage concentration than the proportion of beeswax. A mixture of candelilla wax and beeswax is suitable to use as the principle wax content of the product.

OZOKERITE WAX m.p. 60–80°C Use from 3–10 per cent.

The wax is useful to increase the melting point of the stick. Sticks containing more than 10 per cent ozokerite tend to crumble during application.

CERESIN WAX m.p. 60–75°C

This is sometimes referred to as mineral wax and is also used to denote a form of ozokerite purified with sulphuric acid. The name ceresin wax is now generally used to describe commercial products consisting of a mixture of purified ozokerite with other solid hydrocarbons to give waxes of various melting points. It is used to increase the melting point of the product in a similar manner to ozokerite.

CETYL ALCOHOL m.p. 45–50°C
CETOSTEARYL ALCOHOL m.p. about 43°C } Use from 2 to 3 per cent.

These materials have good emollient properties on the skin and are often included in lipstick formulations in an attempt to obtain similar effects. Only small proportions should be used, however, as concentrations higher than 5 per cent give the stick a dull appearance which develops to a 'bloom' on storage. High concentrations have also been known to be the cause of minute crystals forming on the surface of a stick after storage.

The oil mixture is required to blend intimately with the waxes to provide a suitable film when the stick is applied to the lips. It also functions as a solvent of eosin dyestuffs and as a dispersing agent for insoluble pigments. A suitable mixture is one which when incorporated in the formula enables the product to spread easily and at the same time give a thin film with good covering power. A thick or greasy film is not desirable.

Castor oil is used in many formulations on account of its unique properties. Suitable refined grades are of good colour and are odourless and tasteless. The oil is not as prone to rancidity as is olive oil or almond oil, but nevertheless it should not be used without the addition of a suitable antioxidant. Both olive oil and almond oil are unsuitable for use in lipsticks. The high viscosity of castor oil makes it an ideal vehicle to prepare dispersions of pigments, and these remain suspended during the mixing and moulding stages of manufacture. The oil was originally used as a solvent for bromo acids and although not more than 0·5 per cent bromo acid will dissolve in the oil, any undissolved dyestuff is finely dispersed in the finished product. Many solvents for bromo acids have been suggested for use in lipsticks, but with few exceptions these are unsuitable, either because of the volatility, odour or taste, or toxicity. A suitable product can still be prepared using castor oil as the only eosin solvent. The eosin should be dissolved by heating in the castor oil and any excess allowed to settle and removed by straining. This is a useful precautionary measure since the straining process removes gritty particles which are occasionally present and these can scratch and irritate the lips. As a general guide, about 40 to 50 per cent of castor oil is used in a stick. Higher proportions can be used but in these cases the waxes must be adjusted to give a high melting point otherwise the stick gives a thick greasy film.

Liquid paraffin or white mineral oils are used to give a glossy appearance after application. No more than 5 per cent should be used in the formula. Lipsticks containing a higher proportion than this come off the lips easily and also run or bleed around the lips. An equally effective gloss can be obtained with 2 or 3 per cent of isopropyl myristate or isopropyl palmitate, and this concentration does not affect the permanency of the film.

Oleyl alcohol is used in some formulations as a replacement for castor oil. Refined grades only are used since commercial grades have an unpleasant fatty odour and taste. The material is a better eosin solvent than castor oil and is useful to prepare sticks with high staining properties. A stick based on oleyl alcohol spreads smoothly and evenly, but gives an oily film unless the waxes are adjusted to give the product a high melting point. An antioxidant should always be used with oleyl alcohol.

A lipstick composition can be prepared by mixing suitable proportions of the materials already mentioned. It is more usual to add materials of a fatty nature, such as lanolin or one of the lanolin derivatives to act as blending agents of the liquid and solid ingredients. These materials are considered to have a plasticizing effect. They improve spreading properties and increase the thickness and stability of the film. They also prevent separation and migration of liquid ingredients from the solids, factors which cause sweating or bloom characteristics in the finished product. They also prevent brittleness and help to disperse the pigments and prevent them from settling out. In the composition of a formula these materials are considered as part of the oil mixture, being distinct from the solid ingredients.

Anhydrous lanolin (wool fat) is commonly used as a blender and the amount used can vary from 2 to 20 per cent. A high proportion of lanolin is useful to provide a particular emollient effect, or when fashion requires a thick unctuous film. Sticks containing a high concentration of lanolin may be greasy or sticky and the odour becomes noticeable particularly during storage. It is now more usual to have a stick which gives effective colour coverage with only a thin film of lipstick on the skin. In this type of formulation 5 to 7 per cent of lanolin is quite sufficient.

Lanolin absorption bases are often recommended as a replacement for lanolin to reduce stickiness. These are mainly proprietary mixtures of lanolin and hydrocarbons such as petroleum jelly or wax, and consequently have a corresponding lower plasticizing effect than an equal proportion of lanolin.

Lanolin derivatives are also used, but are often less effective than lanolin as a plasticizer.

iso-Propyl linoleate is a useful material to include in lipsticks as an additional emollient and protective against the drying effects and possible irritation caused by pigments and bromo acids.

Acetoglycerides

Acetoglycerides have certain unique properties whereby they act as blending agents and modify the rheological properties of mixture of oils, fats and waxes. Used in a lipstick they make the composition more plastic so that a suitable blend remains firm under warm conditions retaining its mobility and good spreading properties at low temperatures. They are non-drying and cause the lipstick to impart a non-greasy emollient film on the lips. These properties, together with their ability to give plasticity and prevent brittleness, make them suitable materials for stick products. The acetoglycerides may be prepared by acetylation of fats which have been previously modified by reaction with glycerin. Several commercial products are now available and these vary in composition depending upon the basic starting material and the method of manufacture. They are essentially mixture of mono and diglycerides which are either completely or partially acetylated. For purposes of formulation they are best considered in 2 main groups, the liquid and the solid forms. Their behaviour may be explained by considering the liquid acetoglycerides as having a plasticizing effect on the oily constituents, and the solid acetoglycerides as having a similar effect on the waxes. For this reason a mixture of 2 parts of a solid acetoglyceride with 1 part of a liquid acetoglyceride will give a better result than that obtained by using them singly. Theoretically such a mixture should have all the desirable features of lanolin, but in practice this is found not to be so. The ideal blend is a mixture of liquid and solid acetoglycerides, with some lanolin. With such a blend of acetoglycerides the proportions of lanolin and other ingredients are adjusted to obtain a lipstick with the desired physical characteristics.

Bromo mixture

There are a number of other fatty materials which are sometimes included in a lipstick blend. Some of these mentioned in the general list of lipstick raw materials which follow are best used only in small proportions to obtain slight modifications of texture or type of film required. In other cases, certain fatty materials have a specific function as a constituent of the bromo mixture. This term is used to denote that portion of the product which imparts an indelible stain as distinct from the opaque film of colour given by insoluble pigments. The modern trend often calls for a product with high staining qualities rather than one which covers the lips with a thick film of a vividly coloured material. The indelible type of stick has obvious advantages for, not only does the make-up appear more natural, but there is less pigment transfer during application and little need for constant touching up. Some degree of stain is always desirable and the composition of the bromo mixture is important whether it is used in a high stain product or in formulations containing low proportions of stain and lake colours. Essentially the bromo mixture is a solution of the staining dyestuff blended with suitable materials to enable the dyestuff to remain either wholly or partially in solution or intimately dispersed in the oil/wax base. The dyestuffs generally referred to as bromo acids consist of fluoresceins, halogenated fluoresceins and related water-insoluble dyes. When applied to the skin they form an indelible stain which is a slightly different shade than the acid. The bromo acids available as D and C colours fall broadly into two groups, those which are red and give a red or reddish-blue stain, and those which are orange-red and give a pink to yellowish-pink stain. Thus the stain used must be of the right colour for the finished product; blue staining dyestuffs, for example, should not be used in a pink or coral coloured stick. The degree of staining is a function of the amount of dissolved dyestuff, and it follows that this depends on the solvent used.

For general lipstick formulations which are not regarded as of the high staining type about 2 or 3 per cent of bromo acid is used. This amount is not completely dissolved if castor oil is the only solvent present in the formula and even if the undissolved dyestuff is finely dispersed it remains suspended in the oil/wax film. This does not penetrate or have the same affinity for the skin as the dissolved material and thus does not contribute much to the staining properties. It also happens that undissolved bromo acids often remain as small hard agglominates in the mass even after passing through a roller mill. If these are not detected and are present in the finished product they have a slight abrasive action and although this is not necessarily detected by the user they are considered to cause irritation with the possibility of allergenic effects. Because of the low solubility of bromo acids in castor oil and since a high degree of staining is desirable several materials have been examined for their potential use as solvents for bromo acids. Such a solvent should not be of too high a volatility. It should not have either a disagreeable odour or taste, or cause irritation, and should be compatible with the oil and wax mixture.

Butyl stearate is a useful material used in conjunction with castor oil. It acts as a partial solvent and has wetting properties which promote fine dispersion of undissolved dyestuff.

Tetrahydrofuryl alcohol and esters such as the acetate, stearate, and benzoate are effective bromo acid solvents, although some of the esters, particularly the acetate, have penetrating odours and require masking with suitable perfumes. Formulations containing solvents of this type tend to have a drying effect on the skin and can also cause dermatitis, but as they dissolve up to 25 per cent of bromo acids only small amounts are required to give good staining properties. To offset any drying effect the content of emollient ingredients of the lanolin type should be increased. Proprietary mixtures of esters of tetrahydrofuryl alcohol are available commercially as bromo acid solvents and it is claimed that these do not cause irritation.

When plastic containers are being considered for the end product tests should be carried out to determine whether the amount of solvent present is sufficient to attack the plastic.

Glyceryl derivatives, such as glyceryl monostearate, glyceryl monolaurate and diethylene glycol monostearate, are also useful to include with bromo acids. Of these, monostearate compounds are preferred since they have a better odour than monolaurates. Standard formulations can be modified to increase the staining properties by including a glyceryl derivative to partly replace waxes of similar melting point.

Propylene glycol, triethylene glycol and polyethylene glycol are also suitable eosin solvents. The latter, also known as macrogols, are mixtures of condensation polymers of ethylene oxide and water. Those with an average molecular weight of 200 to 700 are liquids, and those with a molecular weight of more than 1000 vary in consistency from a soft plastic texture to solid waxes resembling hard paraffin. A number is used in the name, for example, Macrogol 1500 indicates the average molecular weight of the polymer. If a high stain product is required the best effects are obtained by using a polyol. These materials are strongly hydrophilic and have some affinity for the moist surface of the lips. Dissolved dyestuff is transferred to the lips and becomes fixed as a stain. When incorporated in an existing formulation, however, they show tendency to bleed out if they are incompatible with the remaining ingredients of the base. This can be overcome by including one of the following materials:

propylene glycol monoricinoleate,
propylene glycol monolaurate,
propylene glycol monomyristate.

The myristate derivative has a better odour than the ricinoleate or laurate compounds. These materials act as co-solvents or coupling agents of the bromo acids solution and the oil/wax mixtures. The system can be demonstrated by preparing the following mixture:

Bromo acid	15
Propylene glycol	200
	215

A clear solution is obtained. To this solution add propylene glycol monomyristate as follows:

Bromo acid	15
Propylene glycol	200
Propylenc glycol monomyristate	100
	315

This blend gives a solution of bromo acid which mixes with mineral oil and castor oil and produces a homogenous mass with waxes. By regulating the amount of bromo acids and solvent the dyestuff solution is dispersed evenly throughout the mass. This is the basis of the high stain lipstick known as the two-phase system. In the so-called three-phase system bromo acid is used in excess of its solubility and undissolved dyestuff is finely dispersed in the blend. In this case the phases consist of:

(1) liquid phase containing dissolved bromo acid,
(2) dispersed phase containing bromo acid in co-solvent or coupling agent,
(3) solid wax phase.

Good staining is obtained in both types of system by transfer of bromo acid to the lips via the hydrophilic solvent.

Lipsticks containing a glycol or glycol derivative should be checked for their effect on plastics before being packaged in plastic containers.

Insoluble dyestuffs and lake colours

Insoluble dyestuffs and lake colours are the main colouring ingredients as distinct from the staining materials. They consist of calcium, barium, and aluminium lakes. Calcium and barium lakes prepared from azo type dyestuffs are also used. These are known as lake toners or simply toners. The amount used in a lipstick varies between 10 and 15 per cent depending upon the shade and opacity of the film. To accommodate the varying amounts of pigments used in a range of shades it is sometimes necessary to make adjustments to the composition of the base. Regulations for assuring the safety of colours used in lipsticks are under consideration and it seems likely that in due course of time all the colours used will require certification as a provision of the Food & Drugs Act. Manufacturers of dyestuffs are aware of impending legislation and can be relied on to supply 'safe' or 'permitted' colours at the present time, or such 'certified colours' as may result from new regulations. Thus although the cosmetic manufacturer assumes responsibility for the safety of the finished product in this particular instance he may rely on the supplier to a large extent.

In addition to the safety factor, there are a number of variables which occur with currently used permitted colours. These relate to certain physical characteristics and affect formulation and the behaviour of the mass during the manufacture and moulding procedures. When formulating to prepare a range of shades it is essential to examine all colour samples to determine whether they are suitable and with a view to fixing standards for subsequent supplies. Colours must be of a fine and even particle size to ensure smooth application. They must be of good opacity with adequate colouring and covering power. The behaviour characteristics of the colours with oil are also of importance since this affects the final consistency of the mass. The mass becomes thicker if colours with high oil absorption values are used, and when preparing a range of shades it becomes necessary in some cases to modify the formulation of the base owing to the varying behaviour of a particular blend of colours. Data obtained from these tests is useful in preparing a balanced formulation and also for correcting difficulties which may occur during manufacture. As a result of heating during processing, colours having a high oil absorption cause the mass to become gel-like and thixotropic, particularly if excessive heat is used, or when a batch is constantly being reheated for moulding. This problem also occurs if the colour contains an unduly high moisture content or when occluded air is present as a result of the grinding process. It is clear that preliminary tests are an important factor contributing to successful formulation and manufacture, and each supply of colours should be examined to ensure that they do not deviate from established standards.

Additives

It is essential to include a preservative in the product, and for this purpose 0·1 per cent of propyl-p-hydroxybenzoate is used. Concentrations of 0·2 per cent may cause a slightly burning sensation, and occasionally on sensitive skins this can trigger-off an eosin allergy reaction.

Titanium dioxide is frequently used as an additive to modify the shade of the basic pigments. The compound has a very high degree of brightness and has the greatest covering power of all other white pigments. In this respect it is superior to and is preferred to zinc oxide for use in lipsticks and other cosmetic products where covering power is required. It was used originally to obtain vivid effects with high proportions of colours, but is now frequently used with low proportions of colours to obtain delicate pastel shades, whilst at the same time retaining the necessary degree of opacity. Only grades of the Anatase form of titanium dioxide are suitable for cosmetic work. These are available with a low content of soluble salts and are specially milled to eliminate oversize particles. Material prepared to this high standard by the manufacturer is more readily dispersed to a paste during processing. For lipstick work the paste is prepared by first milling the material with a suitable quantity of castor oil. This is milled through a triple roll mill until smooth before adding to the colours or other ingredients of the mass. Titanium dioxide for use in oil/wax mixtures must also be dry, since moisture prevents adequate wetting with oil. This results in difficulties in handling and moulding owing to the formation of white streaks of the pigments which float persistently on the surface of the mass.

Several materials have been suggested for use as additives with the conventional lipstick to enhance the value of the make-up. These include an oil-soluble sunscreen to filter the sun's rays and prevent the lips from sun blisters. A silicone fluid can be used as a fixative and to prevent colours from bleeding on the lips. Polyvinyl pyrrolidone is considered to form a film on the lips and also reduce any tendency towards allergenic reactions. Used at a concentration of 0·5 to 1·0 per cent it is also used as a binding agent for the dyestuffs in high stain products. *iso*-Propyl linoleate is used to prevent drying effects and reduce the possibility of irritation by dyestuffs.

The perfume used in lipsticks must not contain any ingredients which are likely to be irritating or toxic. It should have a pleasant taste and cover any fatty odour of the base. Generally 1·0 per cent gives satisfactory results and at this concentration the risks of irritation are reduced to a minimum.

It is advisable to include antioxidants in all lipstick formulations to prevent rancidity developing on storage and a suitable synergistic mixture is prepared as follows:

No. 2208

Propyl gallate	60
Citric acid	40
Butylated hydroxy anisole	200
Propylene glycol	700
	1000

Use 0·1 per cent of this mixture calculated on the finished product.

In the following table a number of raw materials used in lipstick formulation are given, together with an indication of the quantities used. This varies between fairly wide limits as indicated:

	(per cent)
Acetylated monoglyceride (liquid type)	5–15
Acetylated monoglyceride (solid form)	5–15
Antioxidant	0·1–0·2
Candelilla wax	5–10
Carnauba wax	1–3

Castor oil	30–65
Ceresine	5–20
Cetyl alcohol	2–5
Cholesterine absorption base	5–20
Glyceryl monostearate	5–10
Hard paraffin	1–5
iso-Propyl myristate	2–5
iso-Propyl linoleate	1–5
Lanolin absorption bases	5–10
Lanolin, anhydrous	5–15
Lanolin derivatives	5–10
Liquid paraffin	1–5
Oleyl alcohol	1–10
Ozokerite	1–10
Propylene glycol monoricinoleate	5–15
Soft paraffin	1–5
Spermaceti	5–15
Stearic acid	1–5
Stearyl alcohol	5–10
White beeswax	5–20
Bromo acids	0·5–3
Solvent	5–10
Colours	5–15
Perfume	0·1–2
Preservative	0·1–0·2

Lipstick formulae

A selection of typical lipstick formulae follows. Each formula can be modified by slight alteration to the proportions of oils and waxes to adjust the melting point, spreadability, and amount of film on the lips according to the particular characteristics required. The first selection of formulae include typical proportions of lake colours and bromo acids. Titanium is not included, but all the formulae given are suitable for use with titanium dioxide. When this material is used it should be included as part of the total given as lake colour content. Perfume, preservative, and antioxidant should be added to all formulae as already indicated.

No. 2209

(Non-greasy type with good staining properties)

Castor oil	270
Beeswax	200

Mineral oil	30
Lanolin	50
Cetyl alcohol	20
iso-Propyl myristate	30
Ozokerite wax	100
Carnauba wax	25
Propylene glycol	110
Propylene glycol monoricinoleate	40
Lake colours	100
Bromo acids	25
	1000

No. 2210

Castor oil	390
Beeswax	50
Lanolin	30
iso-Propyl myristate	20
Ozokerite wax	50
Carnauba wax	40
Candelilla wax	70
Propylene glycol	60
Glyceryl monostearate (S.E.)	30
Acetylated monoglyceride (solid type)	70
Acetylated monoglyceride (liquid type)	50
Lake colours	120
Bromo acids	20
	1000

No. 2211
(Non-greasy type)

Castor oil	360
Beeswax	120
Mineral oil	20
Lanolin	30
iso-Propyl myristate	20
Ozokerite wax	50
Carnauba wax	20
Candelilla wax	20
Propylene glycol	100 G1
Glyceryl monostearate (self-emulsifying type)	50
Acetylated monoglyceride (solid type)	50

No. 2211 (*continued*)

Acetylated monoglyceride (liquid type)	20
Lake colours	120
Bromo acids	20
	1000

No. 2212

Castor oil	400
Acetylated monoglyceride (solid type)	50
Acetylated monoglyceride (liquid type)	25
Lanolin	20
iso-Propyl myristate	20
Propylene glycol monomyristate	100
Propylene glycol	20
Beeswax	75
Candelilla wax	70
Ozokerite	50
Carnauba wax	50
Lake colours	100
Bromo acids	20
	1000

The formulae which follow show the oil, wax, and bromo acid mixtures separately. The formulae are given as bases and lake colours can be added in the usual proportions as required. The proportions of bromo acids indicated are for normal or high stain sticks and are reduced for special sticks where low stain properties are required.

Oleyl alcohol is used in the following formula for an emollient stick with high stain properties. The high degree of staining is not obtained if the bromo acids are first dispersed in the castor oil instead of the mixture of oleyl alcohol and propylene glycol. Good staining is only obtained when the bromo acid mixture is prepared separately as shown in the formula.

No. 2213

Oil mixture	
Castor oil	390
Bromo acid mixture	
Bromo acids	30
Oleyl alcohol	200
Propylene glycol monomyristate	100

Wax mixture	
Candelilla wax	100
Ozokerite wax	80
Carnauba wax	100
	1000

A well-balanced base with medium staining and good storage properties in varying temperature conditions is as follows:

No. 2214

Oil mixture	
Acetylated monoglyceride (liquid type)	60
Acetylated monoglyceride (solid type)	25
Castor oil	450
iso-Propyl linoleate	10
Lanolin	20
iso-Propyl myristate	20
Bromo acid mixture	
Propylene glycol monomyristate	120
Propylene glycol	30
Bromo acids	20
Wax mixture	
Beeswax	75
Candelilla wax	70
Ozokerite wax	50
Carnauba wax	50
	1000

Successful preparation of lipstick shades depends to a large extent upon adequate dispersion of the lake colours in the lipstick mass. For convenience it is advisable to prepare dispersions of the lake colours in castor oil and use these for both experimental work and also during the manufacturing procedure. Dispersions are prepared by milling, generally as 25 per cent concentrations in castor oil and this can either be prepared as required or conveniently obtained from the manufacturer of cosmetic colours. It is also convenient to prepare or obtain similar dispersions of bromo acids and titanium dioxide. Use of these dispersions ensures consistency of the results obtained during laboratory experimental work and satisfactory reproduction of laboratory results when these are altered to manufacturing scale.

When using oil dispersions the castor oil is omitted from the basic formula and the balance of the amount required is added after the proportions of lake colours, bromo acids, and titanium dioxide dispersions have been determined by experiment.

A modification of the previous formula (No. 2214), convenient for use in experimental work, is made as follows:

No. 2215
(Lipstick base)

Oil mixture	
Acetylated monoglyceride (liquid type)	60
Acetylated monoglyceride (solid type)	25
Lanolin	30
iso-Propyl myristate	20
'Bromo acid mixture'	
Propylene glycol monomyristate	120
Propylene glycol	30
Wax mixture	
Beeswax	75
Candelilla wax	80
Ozokerite wax	50
Carnauba wax	60
	550

Details of lake colours, bromo acids and titanium dioxide required to obtain typical lipstick shades are given below using lipstick base formula No. 2215. The additives given are in 'paste' form consisting of 25 per cent dispersions of the material in castor oil. Modifications to the shades are easily made by varying the proportions given or by using alternative dispersions. In all examples the formulae are prepared to contain the correct proportion of castor oil (i.e. 450 parts per 1000) corresponding to formula No. 2214.

Details of colours and additives are given for strong positive coloured sticks, soft toned pastel shades, and iridescent or pearly types.

Basic shades

1 Basic red

Lipstick base No. 2215	550
Orange paste (bromo acid)[1]	40
Orange paste (bromo acid)[2]	40
Red paste (lake colour)[3]	120
Orange paste (lake colour)[4]	120
Titanium dioxide paste	130
	1000

2 Pink

Lipstick base No. 2215	550
Orange paste (bromo acid)[1]	40
Red paste (lake colour)[3]	40
Orange paste (lake colour)[4]	40
Titanium dioxide paste	320
Castor oil	10
	1000

3 Orange

Lipstick base No. 2215	550
Orange paste (bromo acid)[1]	40
Orange paste (bromo acid)[2]	40
Orange paste (lake colour)[4]	60
Titanium dioxide paste	300
Yellow paste (cosmetic oxide)[5]	10
	1000

For soft toned pastel shades the proportions of bromo acids and lake colours are reduced.

1 Pastel Rose

Lipstick base No. 2215	550
Orange paste (bromo acid)[1]	20
Red paste (lake colour)[3]	24
Orange paste (lake colour)[4]	16
Titanium dioxide paste	360
Castor oil	30
	1000

2 Pastel Peach

Lipstick base No. 2215	550
Orange paste (bromo acid)[2]	20
Orange paste (lake colour)[4]	25

2 Pastel Peach (*continued*)

Yellow paste (cosmetic oxide)[5]	10
Titanium dioxide paste	355
Castor oil	40
	1000

[1] Orange paste 25595
[2] Orange paste 25592
[3] Red paste 2561
[4] Orange paste 25119
[5] Cosmetic oxide paste 25717
} D. F. Anstead Ltd.

Iridescent or pearlescent lipsticks are made using natural pearl essence or a synthetic pearlescent pigment. The synthetic material can be of the crystalline bismuth oxychloride type or of platelets of mica coated with titanium dioxide or bismuth oxychloride. Crystalline bismuth oxychloride is available in several grades and they produce different effects of brilliance, sheen, and opacity depending on the crystal size. The coated micas also give effects of sheen and brilliance when used on their own and they are also used at low concentrations with bismuth oxychlorides for special sparkle and speckle effects.

The bismuth oxychlorides are available in powder form or as 70 per cent dispersions in castor oil which, like paste colours, are more convenient for use in lipstick work. These are used in the formulae which follow and the quantity of castor oil added is adjusted as with previous formulae. Titanium dioxide is not used because it masks the translucent brilliance of pigments based on bismuth oxychlorides and bromo acids are either used in small proportions or omitted from this type of lipstick.

1 Pearlescent rose

Lipstick base No. 2215	550
Red paste (lake colour)[1]	22
Orange paste (lake colour)[2]	18
Pearlescent pigment paste[3]	200
Castor oil	210
	1000

2 Pearlescent peach

Lipstick base No. 2215	550
Orange paste (lake colour)[2]	25

Pearlescent pigment paste[3]	200
Castor oil	225
	1000

[1] Red paste 2561—D. F. Anstead Ltd.
[2] Orange paste 25119—D. F. Anstead Ltd.
[3] Iriodin WR70—E. Merk, Darmstadt
or Bismuth Oxychloride NLY-L—Rona Pearl Co.

Coloured lustre pigments are prepared with titanium coated micas and used to obtain special effects. An attractive range of golden tones with brilliance and sparkle is made by adding varying proportions of from 5 to 10 per cent of a gold lustre pigment to a rose or peach type lipstick. Lustre pigments are generally supplied in powder form and should be made into a paste with sufficient castor oil before mixing with the lipstick mass.

No. 2216

Lipstick base No. 2215	550
Orange paste (lake colour)[1]	20
Yellow paste (cosmetic oxide)[2]	15
Lustre pigment powder[3]	50
Castor oil	365
	1000

[1] Orange paste 25119—D. F. Anstead Ltd.
[2] Cosmetic oxide paste 25717—D.F. Anstead Ltd.
[3] MP23 Sparkling gold pearl powder—Rona Pearl Co.

Attractive shades are also made by adding a pearlescent pigment to deeper and more intense colours. As an example a reddish blue or plum coloured lipstick is made as follows:

No. 2217

Lipstick base No. 2215	550
Red paste (lake colour)[1]	60
Blue paste (lake colour)[2]	10
Pearlescent pigment paste[3]	150
Castor oil	230
	1000

[1] Red paste 2561—D. F. Anstead Ltd.
[2] Blue paste 25670—D. F. Anstead Ltd.
[3] Iriodin WR70—E. Merk, Darmstadt
or Bismuth Oxychloride NLY-L-Rona Pearl Co.

A shimmer stick is made with a natural or synthetic pearl essence. It is applied to the lips as a second application over a more conventional lipstick colour to give highlights, gloss, and brilliance.

The sticks should spread easily and smoothly and are generally slightly softer than the normal lipstick. They are also used as part of the make-up of face and eyes to obtain special highlight effects.

There is increasing interest in the use of cosmetic products and make-up preparations by men. Many of the products used follow conventional lines and in certain cases a product is equally suitable for use by men and women. A typical example is a lipstick of low pigment content and covering properties which can be used both by men and women to enhance the appearance, lubricate the lips and prevent dryness. A suitable product is made using the basic lipstick formula already given by adding bromo acid and a suitable oil-soluble red dyestuff. In the following formula the proportion of oil soluble dyestuff can be increased to suit individual tastes.

No. 2218

Lipstick base (No. 2215)	860
Oil-soluble red dyestuff	2
Bromo acid[1]	20
	882

[1] Use a 25 per cent dispersion of D and C Red 21 in castor oil.

If an oil-soluble sunscreen agent is included in the oil phase of this stick, the product can be used by both men and women for sunbathing or skiing.

A formula is as follows:

No. 2219

Natural pearl essence[1]	200
Castor oil	338
iso-Propyl myristate	20
Coconut oil	160
Lanolin	55
Red paste (lake colour)[2]	2
Candelilla wax	70
Beeswax	50
	895

1 20 per cent dispersion in castor oil.

2 Red paste 2561–D. F. Anstead Ltd.

The proportion of beeswax and candelilla wax is increased if a firmer product is required, and the stick can also be made without a lake colour.

Manufacture

Manufacture of lipsticks varies only slightly according to the type of formulation, and a general procedure can be outlined. If a solvent is used for the bromo acid this solution is first prepared and set aside until required. If colour pastes are not being used the lake colours are first dispersed by mixing with a suitable quantity of oil, and the colour paste obtained is passed through a triple roll mill until smooth and free from agglomerates and gritty particles. Any titanium dioxide being used in the product is treated in a similar manner before it is added to the colour mix. The colour mixture is then mixed with the bromo acid mixture. The lower melting point fats and waxes are next melted together and mixed with the colours and bromo mixtures at the same temperature. This mixture is re-milled until perfectly smooth. The preservative and antioxidant is dissolved in any remaining oil and added to the mix. The high melting point waxes are now melted and added to the bulk at the same temperature. The perfume is finally added and the mass stirred thoroughly but gently to avoid incorporation of air. The mass should not be milled after the high melting point waxes have been added. Gentle stirring is continued until the mass is homogenous and it is then poured into moulds.

Use of an automatic ejection mould is preferred for large quantities and split moulds are satisfactory for small production runs. The mould is lubricated before pouring either with liquid paraffin or iso-propyl myristate. This should be carried out carefully and efficiently without leaving any surplus lubricant on the surface. The melted mass is poured at as low a temperature as is practicable to prevent the colours from settling out. For this reason it is usual to warm the moulds, otherwise the mass tends to set in ridges as it comes in contact with the cold surfaces. After moulding the sticks are chilled to encourage contraction of the waxes since this facilitates easy removal of the sticks.

Automatic ejection moulds are fitted with a water jacket which is used to warm or chill the mould as required.

Other lip products

Lip pencils are prepared in a similar manner to lipsticks but contain a higher proportion of high melting point waxes. The firmer consistency enables the product to be moulded in a long and slim form similar to a crayon. The lip pencil is generally sold as a companion product to a lipstick and is prepared in the same shade, the purpose being to use the harder pencil to give a firm outline on the skin before applying the softer lipstick.

Carnauba wax is used to increase the melting point of a lipstick, but in some cases use of this wax on its own makes the stick brittle, in which case the content of all the waxes is increased proportionally until the required degree of firmness is obtained.

The following formula which is suitable for the preparation of a lipstick can be modified to prepare a lip pencil by increasing the carnauba wax content from 30 parts to 70 parts:

No. 2220

Ingredient	Parts
Castor oil	400
Acetylated monoglyceride (liquid type)	150
iso-Propyl myristate	20
Lanolin	40
Acetylated monoglyceride (solid type)	50
Beeswax	70
Ozokerite wax	50
Carnauba wax	30
Candelilla wax	70
Lake colours	100
Bromo acids	20
	1000

Fashion trends vary concerning the appearance of the lips after make-up. A lipstick is sometimes required to give a natural appearance with a matt finish and at other times a high gloss is required either with or without a strong colour effect. A light application of an oil over the lipstick application is used to obtain a high gloss. A more adhesive preparation in the form of a lip jelly or lip gloss can be made using a mixture of fatty materials.

No. 2221

Petroleum jelly	500
Lanolin (anhydrous)	500
	1000

No. 2222

Castor oil	100
Lanolin (anhydrous)	900
	1000

The product is perfumed and tinted with an oil soluble dyestuff and applied either with the finger tip or by means of a lip brush.

Lip salve or lip pomade is used to prevent the lips from becoming dry or for treatment of cracked lips. Formulae are given:

No. 2223

Mineral oil (cosmetic quality)	450
Beeswax	250
Spermaceti	300
	1000

No. 2224

Lanolin (anhydrous)	150
Petroleum jelly (white)	500
Paraffin wax	350
	1000

The following formulae give softer products with good emollient properties:

No. 2225

Lanolin (anhydrous)	100
Petroleum jelly (white)	600
Paraffin wax	300
	1000

No. 2226

Lanolin (anhydrous)	50
Petroleum jelly (White)	550
Paraffin wax	300
iso-Propyl palmitate	100
	1000

No. 2227

Lanolin (anhydrous)	100
Mineral oil (cosmetic quality)	400
Beeswax	500
	1000

The salve is moulded similar to a lipstick and generally has a rounded end for application. The sticks are mainly prepared without colour, but an oil-soluble dyestuff can be added to prepare a rose or pink coloured product, if required.

Liquid lipstick consists of a suspension of lipstick pigments in a suitable liquid vehicle such as ethyl alcohol. A simple mixture such as this, however, suffers from uncontrolled spreading during application, lack of adhesion and too rapid an evaporation of the vehicle. These defects are overcome by the incorporation of a gelating compound such as ethyl cellulose to increase the viscosity and allow for greater control of application to the skin. After the evaporation of the alcohol, the ethyl cellulose remains on the lips as a varnish which also acts as the binder for the colourants. To prevent embrittlement of the varnish (and subsequent flaking on the lips) a suitable plasticizer such as castor oil is incorporated, which also reduces the evaporation rate of the alcohol.

Commercially available ethyl cellulose is graded according to the viscosity of a known concentration in a solvent. A grade must therefore be selected to give a suitable viscosity at the required concentration. A basic formula for a liquid lipstick is given as follows:

No. 2228

Ethyl cellulose (80–105 cps)[1]	90
Ethyl alcohol	768
Bromo acids	2
Castor oil	105
Lake colours	35
	1000

[1] The viscosity is that of a 5 per cent concentration by weight in 80 : 20 toluene: ethyl alcohol mixture, determined at 25° C.

It will be noted that the concentration of bromo acids and lake colours is less than those used in the normal type of solid lipsticks.

The advantage of this product is that a good indelible gloss is obtained without producing a greasy film on the lips.

Lip varnish is initially a clear varnish which is applied over the normal lipstick. Its value is mainly an aesthetic one to make the lipstick more permanent and thereby less likely to be inadvertently transferred to drinking utensils and clothing.

Such a product can be prepared by slight modification of the liquid lipstick formula and omission of the colourants.

No. 2229

Ethyl cellulose (8–11 cps)	100
Ethyl alcohol	900
	1000

CHAPTER NINE

Manicure Preparations

The nails should receive regular treatment as an essential part of grooming to maintain them in good condition. They should be cut regularly, particularly if they show any tendency towards brittleness and may be shaped preferably with an emery board. Modern manicure preparations are designed to promote nail hygiene and offer protection by external means against the effects of solvents and detergents, or other materials which are normally handled during the daily routine. The general condition of the finger nails is also dependent on certain general physical conditions which affect their growth such as nervous strain and alterations in glandular function, or dietary deficiencies involving amino acids, vitamins, and essential fatty acids. The nails are a hard keratin somewhat similar to that of hair. The protein structure which is arranged in fibres contains a high proportion of the sulphur rich amino-acid cysteine, a smaller proportion of methionone together with other amino acids such as tyrosine, lysine, and histidene. The nail is, however, not a homogenous structure but is composed of three layers, a soft layer known as the ventral nail, with hard keratin forming the intermediate layer, and the external layer known as the dorsal nail.

The nail also contains up to 12 or 14 per cent of water and fatty materials mainly in the form of cholesterol. Products designed as manicure preparations should wherever possible therefore avoid use of materials which either remove natural fat or water-soluble substances as these effects could interfere with the lattice like structure and contribute to splitting or breaking. It is for this reason that fatty materials are often included in manicure preparations.

Cuticle creams, oils, removers and nail bleach

The cuticle consists of a thin fold of skin which extends over the whitish area known as the lanula at the base of the nail. The growth of the cuticle is often irregular and consequently becomes unsightly and preparations are designed to improve its appearance.

Cuticle creams soften the cuticle and at the same time prevent the nails from becoming brittle and ribbed. A lanolin rich skin food used as a nightly application is suitable for this purpose:

No. 2230

(Cream)

Lanolin anhydrous	40
Beeswax (white)	10
Petroleum jelly	950
	1000

Perfume	0·3 per cent

Quaternary ammonium compounds are used as cuticle treatments either in solution or cream form and because of their substantivity to protein they also have a marked softening effect on the cuticle. Suitable materials are cetyl pyridinium chloride or stearyl dimethyl benzyl ammonium chloride.

No. 2231

(Lotion)

Stearyl dimethyl benzyl ammonium chloride	15
Alcohol	50
Diethyl phthalate	15
Water (softened or distilled)	920
	1000

Perfume (water soluble type)	1·0–2·0 per cent
Methyl parahydroxybenzoate	0·2 per cent

Procedure: Prepare a solution of the quaternary ammonium compound in water with the aid of heat. Allow to cool and add a suitable soluble colour if required. Add the mixture of alcohol, diethyl phthalate and preservative. Mix.

A product suitable for packing in a tube is made by including about 2·5 per cent of either methyl or hydroxymethyl cellulose. As an alternative prepare a cream with the addition of 5·0 per cent of ethylene glycol monostearate.

Cuticle oils are also used and these can be made with an oil soluble liquid lanolin diluted with a vegetable oil or a fatty acid ester.

No. 2232
(Oil)

Liquid lanolin[1]	750
Castor oil or iso-propyl myristate	250
	1000

1 Amerchol L-101 type—American Cholesterol Products, Inc.

iso-Propyl lanolate can be used as an alternative type of liquid lanolin. Regular use of a cuticle cream or oil loosens dead skin and cuticle, and helps to maintain it in a healthy condition. To improve the appearance it is usual to press the softened cuticle gently back and away from the nail using an orange stick tipped with cotton wool.

Cuticle removers are also applied to the softened cuticle in similar fashion. These are prepared either with the alkalis, sodium and potassium hydroxide, monoethanolamine, or triethanolamine or sodium carbonate which act by causing hydrolysis and swelling of the softened cuticle to facilitate its removal. It is emphasized that this practice should be carried out with care, otherwise regular application of an alkali may damage the nail bed and also cause dermatitis. The most effective type of solution is made with caustic potash and glycerine.

A suitable product is made using from 2 to 5 per cent of potassium hydroxide as follows:

No. 2233

Potassium hydroxide	50
Glycerin	200
Water (softened or distilled)	750
	1000

Methyl parahydroxybenzoate 0.2 per cent

A water-soluble type perfume is normally added.

Trisodium phosphate is used in modern type cuticle removers at a level of 8 to 10 per cent with 15 to 25 per cent of glycerin. A surface-active material, such as sodium or triethanolamine lauryl sulphate, can also be included and small proportions of a soluble lanoline derivative or potassium oleate have a softening effect without alkali harshness. A suitable formula is as follows:

No. 2234

Trisodium phosphate	100
Glycerin	200
Sodium lauryl ether sulphate[1]	30
Ethoxylated lanolin derivatives[2]	5
Water or Rose water	665
	1000

1 Empicol ESB. 30 type—Albright & Wilson Ltd.
2 Solulan 75 type—American Cholesterol Products, Inc.

Add a sufficient quantity of an alkali stable water-soluble dyestuff.

A variation of the above formula is given:

No. 2235

Potassium hydroxide	35
Glycerin	150
Propylene glycol	50
Alcohol	250
Water (softened or distilled)	515
	1000

The amines, monoethanolamine, and triethanolamine having a lower degree of alkalinity should be used at a concentration of 10 per cent, but even at this concentration they are not so effective as potassium hydroxide.

Nail bleaches are used to whiten the nails and remove discolourations, nicotine and ink stains. They consist of oxidizing agents such as hydrogen peroxide, or perborate, or citric, tartaric or hydrochloric acids. Formulae are given:

No. 2236

Hydrochloric acid (concentrated)	4
Glycerin	100
Water	896
	1000

No. 2237

Citric or tartaric acid	50
Water	950
	1000

No. 2238

Solution of hydrogen peroxide '20 volume'	500
Glycerin	100
Benzoic acid	1
Water	399
	1000

Nail Products

Powder nail polishes have largely been superseded by nail enamel or nail lacquer. A section of the public do, however, still use these preparations and they are also suitable for use by men. They are made principally from stannic oxide and are applied with a specially shaped chamois pad, which is used as a buffer to give the nails a burnished effect. Other possible constituents are:

kaolin,
kieselguhr,
zinc oxide,
talc,
powdered pumice (silica),
calcium carbonate (precipitated chalk).

These materials, however, are not as effective as stannic oxide mainly because they have less abrasive qualities. The powders are mixed together and sifted and are usually tinted with an aqueous solution of a bromo acid.

No. 2239

Stannic oxide	700
Talc	200
Zinc oxide	100
	1000

No. 2240

Stannic oxide	500
Kaolin	200
Calcium carbonate	300
	1000

Mixtures of these materials containing stannic oxide as the main abrasive can be prepared in paste form by mixing with glycerin or tragacanth or starch mucillage or they can be made into blocks by allowing such pastes to dry out at normal temperatures.

Nail enamel

Nail lacquer or nail enamel is based on nitrocellulose (cellulose nitrate) which is used as the film-forming ingredient. It is obtainable in varying grades of viscosity damped down for safety with industrial spirit or as a solution in a solvent such as butyl acetate. Special grades of nitrocellulose are available for use in nail lacquers and it is important that one of these stable forms is used as these avoid discolouration of the nails.

The selection of solvents used as the volatile portion of the lacquer is important and consists of a mixture suitably balanced so that the rate of evaporation prevents changes which cause precipitation of the nitrocellulose taking place in the residual film as it dries on the nail. For example, the low boiling point of acetone makes its use desirable for quick-drying enamels, but too rapid evaporation must be avoided otherwise even distribution on the nails becomes impossible. The rate of evaporation can be retarded by adding middle boilers such as alcohols and acetates. In order to assist the experimenter a list of the various solvents used, is appended with their boiling points given in degrees centigrade.

Nail enamels
Solvents

35	Ether	126	Diethyl carbonate
46	Carbon disulphide	134	Ethylene glycol monomethyl ether
50	Methyl acetone	135	Ethylene glycol monoethyl ether
55	Acetone	135	Ethyl lactate
56	Methyl acetate	138	Xylene
68	Ethyl acetate	140	Amyl propionate
70	Methyl ethyl ketone	140	*iso*-Amyl acetate
77	Carbon tetrachloride	145	Butyl propionate
78	Ethyl alcohol	154	Cyclohexanone
80	*iso*-Propyl alcohol	159	Ethylene glycol iso-propyl ether
80	Benzene	160	Hexalin
92	*iso*-Propyl acetate	160	Methyl cyclohexanone
96	Butyl formate	162	Furfural
99	Ethyl propionate	164	Diacetone alcohol
110	*n*-Butyl acetate	165	Methyl hexalin
110	Toluene	183	Hexalin acetate
110	Amyl formate	185	Ethyl oxalate
113	Butyl alcohol	185	Butyl lactate
120	Amyl alcohol	186	Glycol diacetate
125	Amyl acetate	190	Cyclohexanone phthalate

The following mixture is given as a suitable solvent blend:

No. 2241

Ethyl alcohol	150
Ethyl acetate	150
Butyl acetate	150
Toluene	550
	1000

A solution of 10 to 15 per cent of nitrocellulose in the above solvent blend provides the basis of a nail lacquer and such a mixture will dry out to produce a film. Such a film, however, would shrink on drying out and become brittle, and would not show much gloss; it is therefore necessary to add suitable plasticizers to give the film flexibility, and resins to provide hardness and gloss and reduce the tendency to shrinkage. The resin now accepted as being the best available for use in conjunction with nitrocellulose films for nail lacquers is a condensation product of para-toluene sulphonamide and formaldehyde (Santolites–Monsanto Chemical Co. Ltd.) and this is used at a concentration of between 5 and 10 per cent in the finished formulation, the exact proportion being dependent on the concentration and composition of the solvent and the other ingredients affecting film formation.

The choice of a suitable plasticizer is most important as this material has a great effect on the viscosity of the enamel, the volatility or rate of drying, and most important on the gloss and flexibility of the nitrocellulose film. Plasticizers used successfully are dibutyl phthalate, n-butyl stearate, and resorcinol diacetate. These materials are used either singly or together in the preparation using about 5 per cent of total plasticizer in the mix. A more complete list of plasticizers which can be used for experimental formulations is given below together with their boiling points in degrees centigrade.

203	Butyl tartrate
206	Camphor
209	Triacetine
212	Ethyl benzoate
220	Cyclohexanol oxalate
237	Butyl oxalate
245	Diethyl glycol
276	Triethylene glycol
278	Resorcinol diacetate
295	Ethyl phthalate
312	Dibutyl phyhalate
323	Benzyl Benzoate
336	Amyl phthalate
355	n-Butyl Stearate
410	Triphenyl phosphate
430	Tricresyl phosphate
–	Castor oil

When a satisfactory plasticizer mix has been evolved as being suitable for a particular formulation the finished product can often be improved by the addition of a small proportion of castor oil or a liquid acetylated monoglyceride. These materials have an additional plasticizing effect which make the film more flexible and consequently longer lasting with less tendency to chip. If too much is included, however, these additive materials will slow down the rate of drying. It will be seen, therefore, that innumerable combinations of nitrocellulose, resin, plasticizers, and solvents are theoretically possible in the formulation of nail enamel to give films which will vary in the degree of gloss, the thickness of the film and the rate of drying. It is, for example, possible to use a high content of nitrocellulose and plasticizer with a low resin content and a high proportion of solvent to give a fast-drying lacquer. On the other hand if the proportion of plasticizer and resin is increased a slower drying and thicker coating is obtained. The following is a typical formula for a basic clear nail enamel base:

No. 2242

Nitrocellulose	150
Resin	100
Plasticizer	30–50
Solvent blend	to 1000

The experimenter will soon discover the relationship of the composition of the solvent blend to the behaviour of the lacquer. Variations of volatile ingredients control both the viscosity and flow of the mix and the even drying out which must take place at a rate which gives a smooth even surface without precipitation of either the nitrocellulose or resin components.

A clear enamel base can be prepared on the lines given although in the trade it is now usual to purchase this as a 'mother lacquer', from manufacturers who specialize in this particular type of work. Pigments are added to the lacquer base to prepare the final product, and these should also be obtained from specialist manufacturers. The pigments are first dispersed by milling in a suitable vehicle and prepared into the form of coloured chips. The coloured chips are dissoved in the lacquer base and blended to prepare suitable shades. As an alternative concentrated tint bases, referred to as concentrated tinters, can be obtained consisting of colour chips dissolved in a lacquer base. These can also be blended to prepare suitable end products.

Thick or cream-type enamels are prepared by adding titanium dioxide as a filler, and this material is obtainable at different levels of concentration as a finely milled dispersion in a suitable vehicle for mixing with a lacquer base. Special additives are used to obtain the pearly or iridescent type of nail enamel. These are nacreous pigments and include natural pearl essence which is obtainable as a suspension or paste of guanine crystals (2-amino-6 oxypurine) —obtained from fish skin and fish scales. The crystals occur as a very thin transparent platelet and give a nacreous effect as light is reflected from the many separate parallel layers. Suspension of guanine crystals can be obtained at different concentrations and in different qualities which give varying effects from a pearly lustre, with exceptional covering properties, to a brilliant 'frosting' effect with moderate coverage. These variations are obtained by using both platelet and needle forms of guanine.

Synthetic nacreous pigments prepared from mica flakes or platelets coated with bismuth oxychloride (Bi-Lite—Mallinckrodt Chemical Works) or titanium dioxide (Timica—The Mearl Corporation) are also used. These materials are available in a range of attractive colours as concentrated dispersions in a nail enamel base and are less costly than those based on natural pearl essence. Coloured, synthetic, and natural materials are all available as dispersions in a common lacquer base and can be blended together to obtain the particular effect required. Guanine and Xanthine (dioxoporine) derivatives have also been prepared and are considered to confer improved solubility characteristics in many solvents. They do not give any pearly properties to the product but improve the strengthening effect and consequent durability of the lacquer film.

A base coat is sometimes applied to the nails before the coloured nail enamel. This consists of a clear concentrated lacquer base and forms a firm even surface on the nail and enables the subsequent application of coloured lacquer to be distributed evenly without showing streaks or variations in colour intensity in a relatively thin film, which is less liable to chip than several separate applications of a coloured lacquer.

Finally a top coat or hardener is used over the coloured lacquer. This consists of a clear lacquer base containing a lower proportion of plasticizer and a higher proportion of resin to give a hard surface film and high gloss. This complete manicure procedure is intended to enhance the final appearance of the nails and decrease wear characteristics such as the tendencies towards chipping from the edge of the nail, effects of abrasion and resistance to materials such as water, solvents, and detergents.

Use of a nail enamel dryer is a comparatively recent innovation designed to reduce the period of waiting time for the nail enamel to dry out and set after application. It is essentially an aerosol product, the purpose of which is to apply a thin film of highly volatile propellent. This results in more rapid evaporation of the total volatile solvents of the lacquer to effect more rapid drying. Consequently an aerosol can charged with propellent only gives a satisfactory 'quick dry' effect when sprayed on to freshly applied enamel. A small proportion of oily material and perfume is, however, generally included as indicated in the formula given below. The concentration of perfume should be not more than 0·2 per cent, as higher proportions may have a plasticizing effect on the lacquer. For the same reason fatty acid esters are not suitable for use in the product.

No. 2243
(Aerosol)

Oil base	
Almond oil	900
Perfume	100
	1000
Container charge—oil base	20
Propellent—114	80
	100

Container: internally lacquered aluminium or tinplate. Valve: standard as supplied by manufacturer.

As the spray deposits a thin film of vegetable oil over the surface of the nail it can be claimed that the dryer also softens or conditions the cuticle.

Enamel removers

Enamel removers may consist of any suitable solvent such as acetone or ethyl acetate, or a solvent blend based on toluene similar to that used as the solvent of the nail enamel. Most solvents particularly low boilers cause dehydration and remove natural greases from the nail. Such damaging effects are reduced to some extent by using solvent blends and also by including some fatty or oily material. Suitable fatty additives are castor-oil, butyl stearate, lanolin, fatty acid esters or glycols. Formulae are as follows:

No. 2244

Castor-oil	30
Diethylene glycol monoethyl ether	150
Acetone	820
	1000

No. 2245

Butyl stearate	30
Ethyl acetate	200
Butyl acetate	200
Acetone	250
Toluene	320
	1000

The products are made more attractive if they are perfumed, although up to 3·0 per cent of a heavy floral type perfume is often necessary to cover the objectionable solvent odour.

A nail enamel remover in 'cream' form can be prepared by dissolving waxes in the solvent to give a product of the desired consistency. A typical mixture can be prepared by dissolving 10 parts of beeswax and 10 parts of carnauba wax in 80 parts of the solvent. Although the mixture provides a convenient method of packaging and storing a highly volatile solvent, the solvent tends to 'break away' from the wax during use and gives a granular appearance. This feature, which is more or less inevitable, can be reduced to a minimum by the addition of a mixture of acetylated monoglycerides which acts as a plasticizer in the system. A small proportion of bentonite helps to keep the product in 'cream' form during use. The following formula is suggested;

No. 2246

Acetylated monoglycerides (equal parts solid and liquid type)	100
Beeswax	100
Carnauba wax	100
Bentonite	10
Methyl ethyl ketone	690
	1000

Procedure: Mix the monoglycerides and bentonite with the methyl ethyl ketone and dissolve the waxes in the mixture by heating gently, taking suitable precautions due to the inflammable nature of the solvent.

A formula for a cream enamel remover is as follows:

No. 2247
(Cream)

A	Diethylene glycol monoethylether	500
	Ethyl acetate	150
	Beeswax	25
	Stearic acid	160
	Microcrystalline wax (M.Pt. 140–150°F)	10
	Acetylated monoglyceride (solid type)	110
B	Triethanolamine	45
		1000

Procedure: Melt all the ingredients of part A. Stir the Triethanolamine into the mixture and allow to cool.

Nail white, strengtheners and elongators

Nail white is prepared as a thick paste using either zinc oxide or titanium dioxide in a glycerine tragacanth jelly or a base consisting of petroleum jelly and beeswax, as follows:

No. 2248

Titanium dioxide	250
Petroleum jelly	700
Beeswax	50
	1000

This type of product is used under the nail extremities to give a white edging, but has now largely been replaced by the nail white pencil made by pencil manufacturers of a soft crayon-like consistency. Such preparations are now rarely used because in modern manicure practice the nail enamel is applied to the edge of the nail and a white edging to the nail is neither required nor necessary.

Brittle nails may be hardened by the application of an aqueous solution of certain astringent salts. The products usually sold as *nail strengtheners* or *hardeners* are prepared from chlorides, sulphates, and acetates of aluminium, zirconium, and strontium, and also potassium, sodium, or ammonium alum (U.S.A. patent

No. 3034966). The lotion is usually applied with a small brush to the tips of the nails only, after washing and drying the hands to remove all oils and greases. After application the hand is held downwards for two or three minutes to allow drying to take place and a second application is made if the nails are particularly soft. When drying is complete a nail lacquer is applied to give a sealing effect. The treatment is continued two or three times each week.

The astringent salt is used at a concentration in the range from 1·0 to 5·0 per cent, hardening of the nail taking place more effectively at the higher concentration. To assist the application of an even coating and also to improve penetration of the solution 5·0 to 10·0 per cent of glycerine or propylene glycol is included. A typical formula is as follows:

No. 2249

Aluminium chloride	50
Glycerin	100
Solution of formaldehyde (40 per cent)	1
Soft soap	10
Alcohol	50
Water	789
	1000

A water-soluble perfume can be included in this solution.

Nail extenders or elongators (U.S. patent No. 2558139) consist essentially of a mixture of methyl methacrylate monomer and methyl methacrylate polymer which are either mixed immediately before use or during application on the nail. The low molecular weight monomer is presented in liquid form mixed with 1·0 per cent of a polymerization promoter such as *p*-phenyldiethanolamine or *m*-tolyl diethanolamine, and the polymethyl methacrylate in granular form is mixed with a peroxide catalyst generally benzoyl or lauroyl peroxide. The materials are mixed in the proportion of 2 parts of solid polymer with 1 part of the liquid monomer, to form a paste which is applied to the nail when polymerization takes place in about 10 minutes to form a hard plastic film. It should be recorded that the materials used for this type of product are all potential skin sensitizers.

CHAPTER TEN

Men's Toiletries

Increase in the sale of men's cosmetic products during World War II suggested that a new large potential market existed in this section of the industry. Before this time, cosmetic and toiletries for men were purchased mainly by women, but with interest growing, many new brands were produced and sales increased considerably and it is now accepted that men are buying these toiletries regularly. It is probably true to say that the real advances in this market have been made in the manner in which the products are presented and also by modern methods of marketing. A uniform range of products is offered and particular attention is given to packaging. Advertising is devised to associate the use of cosmetics as being an essential part of masculine behaviour, and special perfumes with so-called masculine appeal are used.

It is significant, however, that with few exceptions the range of products available remains virtually of the same type as the essential male requirements of many years ago. Thus it would appear that this new market has only developed to a limited extent. The fact that continued interest exists is shown by the appearance of the beauty parlour for men, and this may well indicate that there is a potential market for true cosmetic products as distinct from the present range of toiletries.

The two main groups of men's toiletries are those which are applied to the hair and scalp, and those used as part of the process of shaving.

Preparations for the hair and scalp are in the main concerned with fixing or tidying the hair. These include lotions, oils or brilliantines, and dressings or pomades. Also in this category are products used mainly after shampooing—the so-called stimulants, conditioners and friction lotions.

Hair lotions and tonics

Hair lotions and tonics or so-called tonic dressings with a light fixative effect are for use as a daily application and there is no doubt that many of these are expected to make hair grow on the baldest heads. While these cannot be regarded as specific for growing hair, their regular use undoubtedly helps to keep the scalp in good condition, and the friction produced either by the fingers or by a fairly stiff brush stimulates the blood flow to the hair follicle and thus contributes to their efficacy. Loss of hair is, however, a natural result of the cyclic activity of the hair follicle. After the growing phase, the hair has a resting phase when the germinal matrix becomes inactive. Eventually the hair becomes detached and moves up the follicle and falls out. During or after this period, cyclic activity develops a new germinal matrix which causes a new hair to grow.

This normal loss of hair is no doubt a useful factor in persuading men to use a hair lotion or tonic, because daily use appears to reduce loss of hair as noticed on the hairbrush or comb, merely because the hairs are removed regularly. When profuse loss of hair occurs at the temples and on top of the head with little indication of their being replaced, cyclic activity has stopped and baldness occurs. There are several theories as to the cause of baldness, but it is now generally accepted that the three main factors are heredity, presence of male sex hormone, and diseases of the scalp. It is considered that the presence of male sex hormone aids any hereditary tendency towards baldness; lack of the hormone in the male holds the balding process in check. According to this theory, male sex hormone activity which promotes sex characteristics, also affects the formulation of the subcutaneous fat of the scalp. This fat becomes thinner and eventually disappears, either prematurely in middle age, or naturally in old age. As a result, the scalp becomes more tightly attached to the skull and the hair follicles close up and baldness sets in. It follows that women are bald far more rarely than men because there is no similar hormonal activity and their layer of subcutaneous fat is thicker than a man's, and atrophies much later in life. Even so, male sex hormone does not appear to cause baldness if there is no hereditary tendency.

Several raw materials have been suggested and used for the formulation of hair tonics and lotions. Some of these are only intended to enhance the appearance of the dressed hair. Others are intended to function as active ingredients to maintain the scalp in good condition, since it can be argued that a healthy scalp is the first essential towards shining, healthy hair. Any effects on hair growth must be viewed in relation to the comments already made concerning hair loss. Unless an oily preparation is being offered as a tonic dressing, the vehicle frequently consists of a mixture of industrial methylated spirit diluted with water. In order to employ this raw material it is necessary to obtain a permit from the Excise authorities. At the time of application the formula of the product is disclosed and this must contain a sufficient proportion of a denaturant permitted for use by the authority. A permit is not required if iso-propyl alcohol is used in place of industrial methylated spirit, but this material is not very popular on account of its strong pungent odour which is difficult to cover. The concentration of alcohol in a lotion varies from as little as 10 per cent up to 90 or 95 per cent. It has been shown that alcohol acts as a solvent by removing a fatty acid protein complex from the hair, and high concentrations are not recommended for regular use since this solvent action can result in some breakdown of the protein structure (D.C.I. Vol. 92, No. 5, 1963, p. 652). Glycerin is a useful ingredient and is often included in a formula as a solvent. Since it remains on the scalp after evaporation of the more volatile constituents it tends to leave the hair sticky if too high a proportion is used. From 2 to 5 per cent is sufficient to give some emollient and lubricating effects and at this concentration it also acts as a light fixative. A selection of basic formulae follows. Several types are given and these are variously described for the hair or scalp, as lotions, tonics, or tonic dressings.

Bay rum originally contained rum which was used as a solvent for the bay oil, but nowadays a dilution of alcohol (industrial methylated spirit) and water is used.

No. 2250

Perfume	25
Alcohol	600
Glycerin	20
Water (softened or distilled)	355
	1000

Add sufficient water-soluble colour.

The bouquet of a good bay rum is suggestive of eugenol, but this natural isolate is seldom used. Typical perfumes can be prepared as follows:

No. 2251

Oil of bay	800
Oil of pimento	200
	1000

No. 2252

Oil of bay	600
Oil of cloves	200
Acetic ether	200
	1000

No. 2253

Oil of bay	750
Oil of pimento	150
Eugenol	100
	1000

Procedure: Dissolve the perfume oils in the alcohol. Add the glycerol, water and colouring. Mix well and filter bright using talc or kieselguhr.

A number of materials have been recommended for use in hair lotions as 'active ingredients'. These include many well known drugs which act as rubefacients and are considered to stimulate hair growth. A more modern approach is to apply a vasodilator to increase the blood flow in the region of the hair follicle. Whichever material is used as the active substance, it does however seem highly probable that much of the benefit of the lotion is due to the friction and massage employed during application. Massage is likely to increase localized temperature, improve circulation, and loosen the skin of the scalp, and as a result promote the supply of nutritive materials to the hair follicles.

Some of the materials used as active ingredients are as follows:

Cantharides	Capsicum
Pilocarpine	Resorcinol
Quinine	Salicylic acid
Ammonia	Sulphur
Rosemary oil	Cholesterol
Acetic acid	Mercuric chloride

Lotions containing cantharides are prepared as follows:

No. 2254

Tincture of cantharides	5
Solution of ammonia	50
Oil of rosemary	10
Alcohol	250
Glycerin	50
Water (softened or distilled)	635
	1000

Water-soluble perfume 0·5–1·0 per cent

No. 2255

Tincture of cantharides	10
Borax	40
Glycerin	50
Alcohol	50
Water (softened or distilled)	850
	1000

Water-soluble perfume 0·5–1·0 per cent

The alkaloid pilocarpine is also considered to stimulate hair growth. It is used in the following formula:

No. 2256

Pilocarpine nitrate	0·5
Quinine hydrochloride	2
Glycerin	50
Alcohol	50
Water	897·5
	1000·0

Water-soluble perfume 0·5–1·0 per cent

Cholesterol (cholesterin) is used in hair lotions as it is believed to have an effect on the activity of the sebaceous glands and also to be concerned with hair formation. It is known that about 5 per cent of cholesterol is present in sebum as secreted by the sebaceous glands leading into the hair follicle. Furthermore, the blood cholesterol varies according to the health of the individual. In illness there is a lowering of the blood cholesterol resulting in cholesterol deficiency of the sebaceous glands. During this condition the rate of hair growth is affected and loss of hair occurs. On the other hand excessive secretion of sebum is considered to contribute to hair loss. This effect may be due to the unsaturated groups present in sebum which affects the protein structure. It follows that any disturbances either in the rate of flow or in the composition of sebum can be intimately concerned with hair growth or keratinization. If, for example, a lowering of health results in a low blood cholesterol, whilst an excessive sebum secretion is maintained, severe irreversible hair loss could take place. In treating hair and scalp with a cholesterol lotion removal of fat is not attempted. Cholesterol is used as an anti-irritant fatty substance, with the object of retarding the excessive secretion of sebum. Other methods are based on removal of the excessive fatty secretion by means of solvents, alkalis or soaps. Shampooing should, however, be maintained at suitable intervals for regular removal of scales or to avoid build-up of grease.

Cholesterol is insoluble in water, but slightly soluble at 1 part in 100 parts of 90 per cent alcohol and to prepare lotions with a low alcohol content it is necessary to include some form of solubilizer. The following formula illustrates a lotion containing 30 per cent of alcohol using an ethylene oxide condensate of a fatty alcohol as the solubilizer.

No. 2257

Cholesterol	5
Alcohol	300
Ethylene oxide condensate[1]	100
Water (softened or distilled)	595
	1000

Methyl parahydroxy benzoate 0·15 per cent
Water-soluble perfume 0·5–1·0 per cent

[1] Empilan KB type—Albright & Wilson Ltd., Marchon Division.

Procedure: Dissolve the cholesterol in the ethylene oxide condensate. Add the alcohol, preservative and perfume, and finally the water.

The amount of solubilizer required varies according to the concentrations of cholesterol and alcohol present and also on the type of fatty alcohol and degree of ethoxylation of the solubilizer. In the above formula a suitable condensate is prepared from lauryl alcohol and contains 3 moles of ethylene oxide.

Cetomacrogol emulsifying wax can be used to increase the solubility factor of cholesterol. In the following system an opaque lotion is obtained.

No. 2258

Cholesterol	5
Alcohol	250
Cetomacrogol	10
Water (softened or distilled)	735
	1000

Methyl parahydroxybenzoate	0·15 per cent
Water-soluble perfume	0·5–1·0 per cent

This formula gives a pourable product which is thickened either by increasing the amount of cetomacrogol wax or the proportion of alcohol.

An attractive pearly lotion can be made to the following formula:

No. 2259

Cholesterol	5
Alcohol	250
Cetomacrogol	50
Water (softened or distilled)	695
	1000

Methyl parahydroxybenzoate	0·15 per cent
Water-soluble perfume	0·5–1·0 per cent

Polyoxyethylene condensation products of wool wax alcohols can also be used as solubilizers (Polychols—Croda Ltd.). A number of these materials are available and they differ from each other in the amount of ethylene oxide condensed with wool wax alcohols. To prepare clear aqueous lotions or lotions containing a low concentration of alcohol, condensation products having a high average number of ethylene oxide molecules are most suitable.

A lotion containing cholesterol can be prepared as follows:

No. 2260

Cholesterol	1
Wool wax alcohols condensation product	20
Alcohol	100
Water (softened or distilled)	879
	1000

Methyl parahydroxybenzoate	0·15 per cent
Water-soluble perfume	0·5–1·0 per cent

Procedure: Dissolve the cholesterol in the alcohol; add the condensation product, and finally the water and perfume.

The proportion of condensation product used in this formula is sufficient to solubilize the cholesterol in the presence of 10 per cent of alcohol. Clear lotions can be prepared containing a higher proportion of cholesterol, provided the amount of alcohol and solubilizer is adjusted.

No. 2261

Cholesterol	2·5
Wool wax alcohol condensation product	20
Alcohol	200
Water (softened or distilled)	777·5
	1000·0

Methyl parahydroxybenzoate	0·15 per cent
Water-soluble perfume	0·5-1·0 per cent

Water and alcohol soluble forms of cholesterol are also available and their use is illustrated as follows:

No. 2262

Soluble cholesterol[1]	5
Alcohol	100–200
Water	895–795
	1000

Methyl parahydroxybenzoate	0·15 per cent
Water-soluble perfume	0·5–1·0 per cent

[1] Solulan C-24 type—American Cholesterol Products Inc.

Some lotions and tonic dressings are prepared with high concentrations of alcohol, possibly on account of the refreshing or stimulating effect obtained during and after application. It is however advisable to include oily or fatty materials to reduce or avoid the solvent action of alcohol. A specimen type is given:

No. 2263

Cholesterol	2
Oleyl alcohol	250
Alcohol	748
	1000
Perfume	0·5 per cent

Castor-oil is also used as the oily material of lotions containing high proportions of alcohol, although in common with other vegetable oils it has a dulling effect on the hair. Lanolin condensation products or wool wax alcohols condensation products can also be used either with castor-oil or on their own as the oily material of a dressing. Two formulae of this type are given:

No. 2264

Castor-oil	150
Alcohol	850
	1000
Butylated hydoxyanisole	0·02 per cent
Perfume	0·5 per cent

No. 2265

Castor-oil	50
Wool wax alcohols condensation product	20
Alcohol	930
	1000
Perfume	0·5 per cent

Serious attempts to improve the condition of the hair, to promote its growth, and prevent hair loss receive constant attention in cosmetic research laboratories and from these studies new materials have been considered as active ingredients for use in hair tonics. Some of these materials are mentioned.

A mixture of unsaturated fatty acids including linoleic, linolenic and arachidonic acids occurs in some fats and oils and in certain seeds, e.g. linseed. This mixture is sometimes referred to as vitamin F. Deficiency of these materials results in liver disorders accompanied by tendencies toward skin irregularities of dryness, toughness, and lack of healing of broken skin. It is considered that the hair is also affected becoming brittle, fragile and thin. External application of esters of linoleic and linolenic acids are used successfully for treatment of skin irregularities and it would seem reasonable to consider their use in hair preparations.

The materials available commercially as Vitamin F are prepared from a mixture of unsaturated fatty acids and contain a quantity of units of activity per gramme, given as Shepherd-Linn units. The mixture is oil soluble and must be used in an oil type dressing or an emulsion system. The finished product should contain 50–100 Shepherd-Linn units per gramme. An antioxidant is necessary to prevent rancidity developing during storage. Available as an ester, isopropyl linoleate is stable under normal storage conditions and provides a convenient mixture of unsaturated fatty acid esters.

Pantothenic acid and biotin (Vitamin H) has been associated with symptoms of seborrheoic dermatitis, depigmentation and arrest of hair growth, and depilation of hair. The water-soluble calcium salt of pantothenic acid is suggested for use in hair lotions as a corrective treatment.

No. 2266

Calcium pantothenate	5
Glycerin	30
Alcohol	275
Water (softened or distilled)	690
	1000

Lactic acid to pH value	5·0–6·0
Methyl parahydroxybenzoate	0·15 per cent
Water-soluble perfume	0·5–1·0 per cent

The pH value of the lotion is adjusted to pH 5·0–6·0 with lactic acid and the lotion is packaged in amber coloured bottles to prevent deterioration by the effect of light.

Protein hydrolyzation products and nucleic acids are used in lotions on account of their close relationship to the protein structures of hair and skin. Hydrolysis of protein materials yields a mixture of amino acids and polypeptides similar to those of the complex protein structures present in living cells of the skin or scalp, and the keratinized structure of hair. Nucleic acids are found in cell nuclei (deoxyribonucleic acid or DNA) and in cellular material (ribonucleic acid or RNA).

Use of these materials on the hair and scalp is based on the notion that either the growth, or structure and rigidity of the hair will benefit by bringing them into close contact with the keratogenous zone, or near to the tissue surrounding the hair papilla or germinal matrix.

To prepare these hydrolysates, protein materials such as leather, skins, gelatin, wool and hair are treated with acids, alkalis or enzymes.

They are available commercially (Vericrest—Protean Chemical Corporation, New York; Cosmetic Polypeptides—Wilson & Co, Inc, Chicago) and their use in hair lotions is given in the following formula.

No. 2267

Protein hydrolysate	10
Glycerin	20
Calcium pantothenate	2
Lactic acid[1]	1
Alcohol	250
Water (softened or distilled)	717
	1000
Methyl parahydroxybenzoate	0·5–1·0 per cent
Water-soluble perfume	0·5–1·0 per cent

[1] Adjust to pH value 5·0–6·0.

Reference to the incidence of scurf or dandruff has already been made in the chapter dealing with shampoos. For treatment and control of both the dry and greasy forms of this condition, regular washing with a shampoo containing antiseptic materials is advocated. Irritation of the scalp occurs with both types of seborrhoea and this condition can be alleviated in between shampooings by the use of a lotion together with massage. Seborrhoea can also be accompanied by loss of hair, particularly when the greasy condition of seborrhoea oleasa is present. It is worth noting that in some cases the hair and scalp does not appear visibly greasy because the excess sebum is constantly being removed by the use of detergent type shampoos.

Lotions for dandruff are not often classified as hair tonics, although any preparation which corrects or alleviates either form of seborrhoea must contribute to a more healthy condition of the scalp, and consequently reduce the tendency for hair fall. On this account they can probably be more correctly described as hair tonics than those which are only used as dressings. Lotions for treatment are described as oily and non-oily according to the particular form of dandruff which is being treated. Even so, if an alcohol-based product is prepared to correct a greasy condition, it is advisable to include some oily or fatty material. A lotion should not be formulated to function as a degreasing agent.

Several antiseptics used for treatment are mentioned as active materials in the formulae of medicated shampoos and some of these are used in the following lotions:

No. 2268

Salicylic acid	2
Resorcinol	2
Oleyl alcohol	200
Alcohol	796
	1000

Perfume 0·5 per cent

Sulphur preparations have a fungicidal action when applied externally. In the following lotion sulphur is used together with salicylic acid.

No. 2269

Salicylic acid	2
Precipitated sulphur	30
Glycerin	3
Alcohol	100
Water (softened or distilled)	865
	1000

Methyl parahydroxy benzoate	0·15 per cent
Perfume	0·2–0·5 per cent

The sulphur remains as a suspension in this lotion. For a more cosmetically acceptable form of sulphur therapy solutions of polythionic acids and their salts are used. Quaternary ammonium polythionates are also suggested for treatment to combine the properties of sulphur and those of quaternary ammonium compounds (U.S. Patent: No. 2815344). The object of using soluble sulphur compounds is to promote penetration of the tissues and carry the antiseptic effect to the site of infection. These lotions tend to be drying and dosage should be controlled.

Shampoos based on quaternary ammonium compounds are used successfully for treatment of seborrhoea. Lotions can also be prepared with these compounds provided the concentration of the solution is controlled to prevent overdosage since this is likely to cause irritation. The lotion should be prepared to contain from 0·01 to 0·02 grammes of active material per application.

Quaternary ammonium compounds can also be used with fatty acids to prepare conditioning creams or dressings for men. Mixtures of these materials with fatty acids give complexes with certain desirable properties which vary according to the chain length of the fatty acid used. Complex compounds prepared with a low molecular weight acid give the most satisfactory type dressings because they are more soluble and do not have a dulling effect on the hair. The less soluble stearic acid forms a complex which has more fixative properties. This is permissible and often desirable when considering men's hair dressings. On the other hand if this product is used regularly it tends to build up on the hair giving a dulling effect. Conditioner dressings of this type have antiseptic properties to alleviate irritation and improve the appearance of the hair by giving body and lustre. They also give a much lower build-up than the more conventional oil-in-water or water-in-oil emulsion type dressings. In addition, by damping with a little water and combing, the hair can be controlled and re-styled. Formulation of these preparations is given in the following table using stearyl dimethyl benzyl ammonium chloride as the cationic-active ingredient. This is used with fatty acids of increasing chain length and a fatty acid ester. Semi-transparent compositions are obtained which vary in consistency as indicated.

	1	2	3
iso-Propyl palmitate	25	25	25
Lauric acid	25	–	–
Myristic acid	–	25	–
Stearic acid	–	–	25
Stearyl dimethyl benzyl ammonium chloride	100	100	100
Water (softened or distilled)	850	850	850

1 Thick, but pourable lotion.
2 Thick gel, of similar appearance to petroleum jelly.
3 Thick gel.

The mixtures should be allowed to stand for several days until stable complexes are formed.

Brilliantines and hair oils

Brilliantines or hair oils are formulated to give the hair a good shine and gloss, and to impart some slight controlling effect without the use of a heavy oil or grease. The modern trend is to give a natural appearance to the hair and this is achieved by applying a thin continuous film of an oily material on the hair surface without causing greasiness or stickiness. The dressings are generally based on white mineral oils of low viscosity or suitable mixtures of oily materials selected to give a product of the correct viscosity. Sometimes these preparations are sold as 'hair tonics' but any 'tonic' properties would only appear to be obtained during their application. When applied to the hair and scalp with gentle massage the dressing lubricates the scalp and loosens dandruff scale. Mineral oils of high viscosity, such as liquid paraffin B.P., are more difficult to spread and cause a build-up of grease after two or three applications. High viscosity oils also tend to make the hair dull and heavy. Vegetable oils have a similar effect. Low viscosity oils spread easily to give a thin coating and they achieve maximum gloss effect by causing reflection of light.

Mineral oils used in brilliantines should be pure white, show no sign of fluorescence, and be odourless. They should be as free as possible from unsaturated hydrocarbons which can cause irritation. Impurities removed during the refining of white oil consist of both unsaturated and aromatic hydrocarbons. If these materials are not removed, the oil rapidly deteriorates during storage, both in odour and colour. The deterioration is due to the formation of peroxides which develop when the oil is exposed to both light and heat. Even a highly refined white oil deteriorates when exposed to sunlight and for this reason these products always store more satisfactorily when they are protected by a carton. A simple test to determine the degree of refining can be made by carrying out the acid test described in The British Pharmacopeia: A measured quantity of the oil is treated with concentrated sulphuric acid, and if an abnormally high proportion of unsaturated and aromatic hydrocarbons is present the acid layer becomes yellow, brown, or black, according to the amount of unsaturated material present. With an oil of good quality the acid layer remains comparatively free from colour.

If a brilliantine is based solely on a low viscosity mineral oil, selection of a suitable perfume is often restricted because the perfume is insoluble in the oil. Many perfumery oils and resinous fixatives are not miscible in clear solution and in consequence require filtering, which is a slow and costly process, apart from the loss of perfume. For this reason, all perfumes should be tested for solubility. A favourite or popular odour type perfume can often be made oil soluble by using a suitable proportion of deodorized kerosene or *iso*-Propyl myristate in the formula. These materials also reduce the viscosity of the oil mixture and consequently increase the spreading properties and improve gloss effect.

The products are made by simple mixing of the ingredients, and they can be filtered to give a bright crystal-clear appearance, although in most cases this should not be necessary. If oil-soluble colours are used, these should be shelf tested for light stability. The following formula gives a product of suitable viscosity:

No. 2270

Mineral oil (cosmetic quality)	850
iso-Propyl myristate	150
	1000
Perfume	0·5–1·0 per cent

The proportion of *iso*-Propyl myristate can be increased up to 25 per cent if required.

Aerosol brilliantines are made with similar mixtures of mineral oil and low viscosity fatty acid esters. The spray forms a thin film on the hair with a good gloss. Oil-soluble lanolin derivatives can be included if required.

No. 2271

Brilliantine base	
iso-Propyl myristate	250
Mineral oil (cosmetic quality)	750
	1000
Perfume	0·5–2·0 per cent

Procedure: Dissolve the perfume in the *iso*-Propyl myristate and add the mineral oil. Container: plain aluminium or tin plate. Valve: Standard fitted with a button giving a fine spray.

An aerosol hair 'tonic' is made using an alcoholic solution of a fatty acid ester.

No. 2272

(Non-greasy tonic type)

Hair tonic lotion base:	
iso-Propyl myristate	250
Alcohol 99·5% V/V	750
	1000
Perfume	0·5–1·0 per cent

Procedure: Mix the perfume with the *iso*-Propyl myristate and add the alcohol.

Container charge:	
Hair tonic lotion or brilliantine base	70
Propellent–11/12 (50 : 50)	30
	100

Container: internally lacquered aluminium or tin plate. Valve: standard with lacquered cup fitted with a button giving a fine spray.

Blue brilliantine for use on ash blonde or grey hair is prepared by adding a suitable oil-soluble light fast blue dyestuff to the basic oil mixture.

Two-solution or separable brilliantines consist of an oily material and an aqueous-alcoholic portion. Colouring matter is not necessary, but a brighter appearance is obtained if both layers are coloured yellow. This type of product is also used as a more modern presentation of a lotion with oil. In this case 'active-materials' such as salicylic acid, cholesterol or lanolin derivatives are dissolved in either the aqueous or oily layer. Care should be taken to make sure each phase is crystal clear and the line of demarcation clearly defined.

No. 2273

Castor-oil	20
Alcohol	180
Mineral oil	800
	1000
Oil-soluble perfume	0·5 per cent

Use either oil-soluble or alcohol-soluble perfume, and colour each phase with oil-soluble, and alcohol soluble dyestuff.

No. 2274

Alcohol	150
Water	100
Mineral oil	750
	1000

Use perfume and colour as in previous formula.

Solid brilliantines now have restricted sales outlet in Europe, but are still popular in many parts of the world particularly in hot climates and also when high fixing characteristics are required. The cheap types largely consist of white or yellow petroleum jelly and more elegant preparations contain mineral oil and waxes. Some care is required in selecting the proportion of oil and wax required to maintain a clear transparent product. Oil-soluble dyestuffs are used to obtain the most popular yellow or green colours and oil-soluble perfumes must also be used. Sample mixtures are prepared as follows:

No. 2275

Yellow petroleum jelly	940
Beeswax	60
	1000

No. 2276

White petroleum jelly	900
Mineral oil	100
	1000

Solid brilliantines based mainly on petroleum jelly tend to drag and are dull and opaque in appearance. More elegant products can be prepared using the following:

No. 2277

White petroleum jelly	600
Microcrystalline wax (m.p. 140–150°F)	150
Paraffin wax	50
Mineral oil	200
	1000

No. 2278

White petroleum jelly	750
Paraffin wax	50
Mineral oil	200
	1000

Microcrystalline wax is preferred to increase the melting point, since this gives a more translucent product than that obtained with paraffin wax. When the product is required for use in hot climates, the melting point is adjusted by increasing the proportion of wax. Alternatively, a small quantity (about 0·5 per cent) of carnauba wax can be added. The following formula gives a firmer product:

No. 2279

White petroleum jelly	800
Microcrystalline wax (m.p. 140–150°F)	50
Mineral oil	150
	1000

An interesting modification can be made by adding a solvent to the petroleum jelly to increase the degree of transparency. The following formula gives a firm gel-like mix which liquefies and spreads readily:

No. 2280

Petroleum jelly	800
Methyl resinate	200
	1000

Hair pomades were originally prepared from the greases obtained from the enfleurage and maceration processes and were considered to have tonic properties for the hair. The modern pomade is a solid dressing or jelly used to fix the hair. It is more elegant than the original pomades but is required to be greasy and sticky similar to the solid brilliantines. A suitable product can be prepared as follows:

No. 2281

Acetylated lanolin	100
iso-Propyl lanolate	50
Mineral oil	300
Polyethylene glycol 400 monostearate	50

Petroleum jelly	450
Ethoxylated cetyl/oleyl alcohol[1]	50
	1000

[1] Empilan KL. 10 type—Albright & Wilson Ltd.

The opacity is controlled by the proportion of P.E.G. 400 M.S. used. The pomade has a good gloss and being miscible with water is less tacky on the hair and gives less build-up than the traditional solid brilliantine. The product liquefies to the touch and spreads readily. A non-greasy modern pomade can be prepared as follows:

No. 2282

Polychol 15[1]	250
Alkyl myristate	500
Glycerin	100
Water (softened or distilled)	150
	1000
Methyl parahydroxybenzoate	0·15 per cent
Perfume	0·5 per cent

[1] Croda Ltd.

The proportion of water given is sufficient to prepare a gel-like preparation.

A similar product in the form of a clear gel, and containing mineral oil, is given in the following formula:

No. 2283

Polychol 15[1]	200
iso-Propyl myristate	100
Super Hartolan[1]	30
Mineral oil	70
Water (softened or distilled)	140
	540
Methyl parahydroxybenzoate	0·15 per cent
Perfume	0·5 per cent

[1] Croda Ltd.

In some territories, particularly in Africa, solid brilliantines and pomades are referred to as pressing oils. These may be tinted with an oil-soluble brown or red dyestuff or can contain a black pigment to colour the hair. The preparations are used liberally, and as the name indicates, are used as hair straighteners or anti-curl pomades. A suitable product can be made as follows:

No. 2284

Mineral oil	130
Lanolin	50
Petroleum jelly	820
	1000

Include an oil-soluble brown or red dyestuff or black pigment and 0·5 per cent of an oil-soluble perfume.

Hair colour restorers

Hair restorers are made with sulphur and a lead salt, and are used ostensibly to restore the colour to grey hair. When the mixture is applied, the sulphur which is deposited on the hair shaft gradually darkens with the formation of lead sulphide. The product is applied daily to build up the deposit on the hair and until the required colour has been obtained. Application is then continued at intervals to maintain the colouration. A lotion can be made as follows:

No. 2285

Precipitated sulphur	25·0
Lead acetate	30·0
Glycerin	44·5
Sodium lauryl sulphate	0·5
Water (softened or distilled)	900·0
	1000·0

Methyl parahydroxybenzoate	0·15 per cent
Water-soluble perfume	0·5 per cent

Procedure: Rub down the sulphur to a smooth paste with the glycerin and sodium alkyl sulphate and add the perfume. Dissolve the lead acetate in 500 grams of the water and gradually add to the sulphur paste. Make up to volume and when bottling, make sure that the insoluble portion is evenly distributed.

The product can also be prepared in two separate solutions which are mixed together immediately before use. Sodium thiosulphate is used as the source of sulphur in the following solution product.

Solution No. 1

No. 2286

Lead acetate	30
Glycerin	150
Sodium lauryl sulphate	2
Water (softened or distilled)	818
	1000

Solution No. 2

No. 2287

Sodium thiosulphate	125
Propylene glycol	350
Citric acid	25
Water (softened or distilled)	500
	1000

A hair darkening cream or pomade can be made as follows:

No. 2288

Mineral oil (or petroleum jelly)	50
Cetomacrogol emulsifying wax	100
Glycerin	240
Precipitated sulphur	30
Lead acetate	30
Water (softened or distilled)	550
	1000
Methyl parahydroxybenzoate	0·15 per cent
Water-soluble perfume	0·5 per cent

Non-greasy hair creams

Non-greasy hair creams are mucilages generally made with powdered tragacanth and small quantities of glycerin. They have good fixing properties, but tend to have a dulling effect on the hair and the gum flakes off when the dried film is subsequently combed. This effect is reduced by using a polyol and also by including a small quantity of an oily material in the formula. As with all preparations based on tragacanth, it is important to select a good quality gum and check each delivery to maintain constant viscosity of the finished product. Special care should also be taken to check that the product contains an adequate preservative, since bacteria cause decomposition of the mucilage affecting both the viscosity and opacity of the product. A standard degree of opacity is obtained by including small quantities (about 0·5 per cent) of the tinctures of tolu or benzoin. Good results are obtained with the following basic formula:

No. 2289

Gum tragacanth	20
Alcohol	10
Glycerin	50
Water (softened or distilled 1st part)	400
Water (softened or distilled 2nd part)	520
	1000
Methyl parahydroxybenzoate	0·2 per cent
Perfume	0·5 per cent

Procedure: Mix the tragacanth gum with the alcohol and add the perfume. Dissolve the preservative in the glycerol. Add part of the water and gradually mix with gum solution. Add the remaining water with continual mixing, and allow to thicken. Strain through muslin before bottling.

From 2·0 to 5·0 per cent of castor-oil, mineral oil, or iso-propyl myristate can be included in this formula, if required. The oil is mixed with the gum and alcohol before adding water.

Attempts to replace gum tragacanth with materials such as alginates and cellulose derivatives are not successful. Aqueous gels having similar properties to tragacanth mucilages can be prepared using the synthetic carboxy vinyl polymer Carbopol 940 (B. F. Goodrich Chemical Co.). Aqueous solutions of the polymer increase in viscosity and become gel-like when they are neutralized by the addition of an alkali. Additives such as water-soluble lanolin derivatives, polyols, or protein derivatives can be added, but these materials must be selected to prevent clouding of the solution. A mobile mucilage is prepared as follows:

No. 2290

Carbopol 940	1·25
Glycerin	20·0
Triethanolamine	1·25
Soluble lanolin derivative[1]	5·0
Water	972·50
	1000·00
Methyl parahydroxybenzoate	0·15 per cent
Water-soluble perfume	0·5 per cent

[1] Solulan 97 type—American Cholesterol Prods. Inc.

Procedure: Dissolve the preservative in the glycerol, mix with the Carbopol 940 to a paste. Add 900 parts of water and stir until dissolved. Mix any water soluble additives in the mixture. Mix the triethanolamine with the remaining water and add slowly to the Carbopol solution, stirring gently. Finally add the perfume.

More concentrated gels are used either as hair-dressings for men, or as hair-setting preparations for women. Dyestuffs can be added to prepare a range suitable for use on different shades of hair.

No. 2291

Carbopol 940	1·70
Glycerin	15·0
Triethanolamine	1·70
Soluble lanolin derivatives	10·0
Water	971·60
	1000·00
Methyl parahydroxybenzoate	0·15 per cent
Water-soluble perfume	0·5 per cent

Emulsion-type dressings

Emulsion-type dressings can either be of the oil-in-water or water-in-oil type. Oil-in-water emulsions have the advantage of being miscible with water and can be thinned out with a wet comb to distribute throughout the hair. Since the oil is present in the inner phase of the emulsion the oil only comes into contact with the hair after the water has evaporated. They can be prepared with low proportions of oil and on this account are often preferred to the more greasy water-in-oil type emulsion. Waxes such as stearic acid and beeswax are used to prepare the emulsion, the latter also being useful to impart gloss to the hair. The viscosity of the product depends upon the gravity of the mineral oil used and its ratio of the water content. A thin pourable dressing with low oil content is prepared to the following formula.

No. 2292

Mineral oil	250
Beeswax	50
Stearic acid	60
Triethanolamine	10
Water (softened or distilled)	630
	1000
Methyl parahydroxybenzoate	0·15 per cent
Propyl parahydroxybenzoate	0·02 per cent

Procedure: Melt the solids in the mineral oil and add the emulsifying agent to the water. Add the aqueous solution with continuous gentle stirring to the oil/wax mixture. Cool to about 35–40°C and add the perfume.

A similar emulsion with a higher oil content is prepared as follows:

No. 2293

Mineral oil	440
Beeswax	10
Stearic acid	44
Triethanolamine	6
Water (softened or distilled)	500
	1000

Methyl parahydroxybenzoate	0·15 per cent
Propyl parahydroxybenzoate	0·02 per cent
Perfume	0·5 per cent

Oil-in-water emulsions can also be made using glyceryl monostearate (self-emulsifying type). In the following formula it is used as an auxiliary emulsifier to triethanolamine stearate.

No. 2294

Mineral oil	355
Stearic acid	25
Glyceryl monostearate (S.E.)	20
Propylene glycol	45
Triethanolamine	10
Water (softened or distilled)	545
	1000

Methyl parahydroxybenzoate	0·15 per cent
Propyl parahydroxybenzoate	0·02 per cent
Perfume	0·5 per cent

A variation of this emulsion using propylene glycol monostearate as the emulsifier is given in the next formula:

No. 2295

Mineral oil	300
iso-Propyl myristate	90
Propylene glycol monostearate	60
Non-ionic emulsifier[1]	90
Water (softened or distilled)	460
	1000

Methyl parahydroxybenzoate	0·15 per cent
Propyl parahydroxybenzoate	0·02 per cent
Perfume	0·3–0·5 per cent

[1] Abracol L.D.S. type—Bush Boake Allen

In the following formula polyvinyl pyrolidone is used as an additional stabilizer. This emulsion is suitable for export.

No. 2296

Mineral oil	375
iso-Propyl myristate	80
Cetomacrogol emulsifying wax	60
Sulphated lauryl alcohol ethoxylate	30
Polyvinyl pyrolidone	2
Water (softened or distilled)	453
	1000

Methyl parahydroxybenzoate	0·15 per cent
Propyl parahydroxybenzoate	0·02 per cent
Perfume	0·3–0·5 per cent

Comparisons have been made between the relative merits of using oil-in-water emulsions with those of the water-in-oil type. The popularity and success of the latter, which at one time were referred to as brilliantine cream, is probably due to the fact that hair is more readily wetted with oil than water. As the oil film surrounding each particle of a water-in-oil emulsion wets the hair quickly after application, it makes the hair easy to comb and dress, and gives an immediate gloss effect. In the case of an oil-in-water emulsion, the hair becomes dull with the outer film of water. Fixative properties are good, and these can be enhanced with a wetted comb, but this effect is often accompanied by a white film of emulsifier, particularly when the emulsion contains a soap. The gloss effect of the oil only becomes apparent after the water has evaporated.

Water-in-oil emulsions invariably contain a high proportion of oil with the result that they soil clothing and are greasy in use, whereas oil-in-water emulsions can be prepared to contain less oil. Hair-dressings based on water-in-oil emulsions are difficult to prepare and after storage often show slight separation of oil on the surface of the emulsion. The stability of the product is influenced by:

(1) the specific gravity of the oil used,
(2) the ratio of oil to water,
(3) the manufacturing procedure.

Correct selection of emulsifying agents is obviously essential, but good results are generally obtained using calcium hydroxide and beeswax. Other suitable emulsion systems are based on fatty acid soaps, wool alcohols, lanolin, and sorbitan derivatives. Emulsion stability is increased in some cases by the addition of 0·5 per cent of zinc stearate, or by using small amounts of oil-in-water emulsifiers. With all formulae, it is emphasized that the manufacturing procedure of mixing technique, temperature of mix, homogenization, and rate of cooling have an important bearing on the appearance and stability of the final product. During experimental work, records of all these conditions should be made, and the effect of small variations noted, so that a process can be repeated accurately. By its nature, a water-in-oil type of hair dressing is required to break down rapidly and for this reason successful products are often based on a simple formula, indicated as follows:

No. 2297

A	Beeswax	35
	Mineral oil	375
	Paraffin wax	10
B	Lime water (freshly prepared)	580
		1000
	Methyl parahydroxybenzoate	0·1 per cent
	Propyl parahydroxybenzoate	0·05 per cent
	Perfume	0·3 per cent

Procedure: Melt together the beeswax, mineral oil and paraffin wax to a temperature of 65–70°C. Heat 20 parts of the lime water to a temperature of 25°C and add slowly to the oil phase using medium to high speed stirring or whisking. Add the perfume when the temperature of the mix is about 35–40°C. The remaining quantity of lime water is added cold. If a whisk type mixing appliance is not used, an equally effective mixing device must be used to achieve a 'beater' effect. A mixing device fitted with large blades is suitable. The emulsion is finally homogenized and filled into jars. This type of emulsion depends upon the formation of calcium soaps of fatty acids and consequently only freshly prepared lime water must be used. This is prepared from a one per cent mixture of calcium hydroxide and distilled water by mixing thoroughly and repeatedly. The mix is then set aside until clear and the clear solution siphoned off as required for use. Freshly prepared lime water of the British Pharmacopoeia is required to contain not less than 0·15 per cent w/v of $Ca(OH)_2$.

Saccharated solutions of calcium hydroxide are also used and are prepared in a similar manner. Sucrose added to the mixture increases the solubility of the calcium hydroxide. Modifications of the basic formulae already given are designed to increase the stability of the emulsion, although a stronger interfacial film is not always considered desirable. Use of auxiliary emulsifiers is indicated in the following formulae. Preservatives and perfume are used as for the previous formula.

No. 2298

Beeswax	35
Mineral oil	375
Petroleum jelly	25
Lanolin	20
Lime water (freshly prepared)	545
	1000

No. 2299

Beeswax	35
Mineral oil	350
Petroleum jelly	50
Wool wax alcohols	8
Lime water (freshly prepared)	557
	1000

No. 2300

Beeswax	40
Mineral oil	375
Petroleum jelly	50
Zinc stearate	20
Lime water (freshly prepared)	515
	1000

No. 2301

Beeswax	35
Mineral oil	300
Petroleum jelly	25
Sorbitan sesquioleate	10
Stearic acid	10
Lime water (freshly prepared)	620
	1000

In the following formula borax and sorbitan sesquioleate are the emulsifiers:

No. 2302

Beeswax	40
Mineral oil	325
Petroleum jelly	25
Sorbitan sesquioleate	20
Borax	5
Water (softened or distilled)	585
	1000

Cetyl alcohol is used in the following three formulae, to increase the viscosity. The emulsion containing the lowest proportion of cetyl alcohol is the most elegant, but the higher proportions of cetyl alcohol give firmer products which are suitable for use in hot climates.

	No. 2303	No. 2304	No. 2305
Mineral oil	580	580	580
Hartolan	8	8	8
Sorbitan monostearate[1]	30	30	30
Cetyl alcohol	15	10	5
Polyoxyethylene sorbitan monostearate[2]	50	50	50
Water (softened or distilled)	317	322	327
	1000	1000	1000

[1] Span 60–Honeywill & Stein Ltd.
[2] Tween 60–Honeywill & Stein Ltd.

The following formula gives a less viscous water-in-oil emulsion which is just pourable from the container.

No. 2306

Mineral oil	450
Lanolin	15
Sorbitan sesquioleate	15
Beeswax	15
Zinc stearate	5
Borax	3
Water (softened or distilled)	497
	1000

The base of an aerosol hair cream is made of pourable viscosity which becomes more firm as it is dispensed from the aerosol. When applied from the hands it is required to break and spread easily on the hair. The following formula gives a satisfactory product.

No. 2307

Hair cream base:

A	Ethylene oxide condensate of cetyl oleyl alcohol[1]	80
	Polyethylene glycol 400 monostearate	80
	Mineral oil (cosmetic quality)	200
	Silicone oil[2]	10
	Paraffin wax	40
B	Water (softened or distilled)	590
		1000

Perfume	0·3–0·7 per cent
Phenyl mercuric nitrate	0·01 per cent

[1] Empilan KL 10 type–Albright & Wilson Ltd., Marchon Division.
[2] Silicone MS 200/100 type–Midland Silicones Ltd.

Procedure: Heat A and B independently to 75°C and Add B to A slowly and with continuous stirring. Cool with stirring adding the perfume at about 35°C.

Container charge:

Hair cream base	86
Propellent 12/114 (50 : 50)	14
	100

Container: internally lacquered aluminium on tinplate. Valve: standard with lacquered cup fitted with a foam button.

Shaving creams

During recent years the elegance and efficiency of these products have made them more popular than the shaving stick. Their manufacture is, however, more difficult, and considerable experience is necessary for successful production on a large scale, together with close analytical control at all stages of the process. A good shaving cream should have the following characteristics:

1. A small quantity must give an abundant lather.
2. When applied to the face there must be no smarting or astringent effect on the skin.
3. The lather must be creamy and close without any apparent condensation after application.
4. The cream must remain soft in thc tube and not go lumpy if overheated. It must be sufficiently tacky to adhere to both brush and face and yet be easily washed off the razor.
5. It must not corrode either the nozzle or the close end of the tube.
6. The perfume must be fresh and not too lasting.

To obtain the above qualities it is absolutely essential that all raw materials be of the best and purest obtainable. At one time it was customary to saponify lard with potash lye, but this did not yield all the essential characteristics of a good cream. Lard was replaced either entirely or in part by best white neutral tallow, coconut oil, olive oil, peanut oil, or even sesame oil, and by replacing part of the potash lye with soda lye, many good products were obtained. Purified stearic acid is now used as the fatty acid. Triple pressed has the highest melting-point, is the purest, and yields the whitest product. It is comparatively easy to calculate by analytical methods the exact amount of alkali required to produce an absolutely neutral cream. Stearic acid alone, however, does in fact possess the disadvantage of producing a cream which does not have the necessary lathering properties. This drawback is overcome by adding some coconut oil, or if desired, some tallow as well. In general, four to eight times as much fatty acid is employed as free fat. When coconut oil alone is added to the stearic acid the ratio of about 1 to 7 should meet all requirements as to creaminess of lather. Exceptional foaming properties may be obtained by replacing part of the stearic acid with myristic acid.

A mixture of potash and soda is also preferable but the latter is kept in low proportions and seldom exceeds 15 per cent of the total alkali. The use of alkaline carbonate is not recommended since the CO_2 evolved remains occluded in the soap and shrinkage may result. The total fatty acid in a cream varies between 35 and 50 per cent, softness being enhanced by the addition of about 5 per cent of glycerin. The latter also has the advantages of improving the wetting properties, preventing too rapid drying of the lather, making the skin more elastic, softening the beard and facilitating the transit of the razor over the face. Other wetting agents of interest are alkyl sulphates and triethanolamine stearate which also enhance creaminess of the lather.

The real secret of plasticity, however, is not as is frequently supposed, in absolute neutrality but in the presence of a few per cent free acid–added after the saponification is completed. Alternatively this may be affected by the addition of the necessary quantity of boric acid to the nearly finished cream. Pearliness may be induced by running in a small quantity of concentrated solution of soap and thoroughly mixing.

The manufacturing process is commenced by melting and filtering the free fats and saponifying them with the soda lye and part of the potash lye in the presence of the glycerine. The remainder of the potash lye together with some water is now run in, the heat being maintained all the time. The melted stearic acid is then added in a continuous stream the whole being stirred throughout (slow stirring is preferable). Samples are tested, and the necessary additions of acid or alkali made. The cream is now superfatted by means of boric acid, additional stearic acid, lanolin or occasionally fixed oils, and boiling is continued until a perfectly even cream is obtained. With all formulae further examination is necessary so that the total fatty acid in the cream can be adjusted to the standard percentage. The steam is then turned off and gentle agitation continued while the cream cools. The perfume is then added and after manufacture, the cream is allowed to stand for a period of 3-4 weeks before packaging. During this time, which is sometimes referred to as the 'curing' process, changes occur in the crystal formation of the cream. The cream is finally turned out by machine filling. A formula for shaving cream is as follows:

No. 2208

Stearic acid	280
Coconut oil	40
Tallow	30
Potash lye at 20° Baume (approx.)	300
Soda lye at 20° Baume (approx.)	50
Glycerin	50
Boric acid	20
Sodium alkyl sulphate	20
Water (softened or distilled)	210
	1000
Perfume	0·3–0·5 per cent

Proceed as above, making adjustments necessary after saponification.

Some other examples are given:

No. 2309

Stearic acid	300
Myristic acid	100
Coconut oil	50
Glycerin	60
Sodium hydroxide (approx.)	2
Potassium hydroxide (approx.)	18
Triethanolamine stearate	10
Water (softened or distilled–approx.)	460
	1000
Perfume 0·3	0·5 per cent

Proceed as above adjusting the cream to 3 per cent free fatty acid.

No. 2310

Stearic acid	380
Olive oil	20
Coconut oil	60
Glycerin	40
Lecithin	20
Potassium hydroxide 80 per cent (approx.)	16

Sodium hydroxide 90 per cent (approx.)	3
Distilled water (approx.)	460
	1000

Perfume 0·3–0·5 per cent

Proceed as above, adjusting the cream to 2 per cent free fatty acid.

Another successful product can be prepared as follows:

No. 2311

Stearic acid	350
Coconut oil fatty acids	100
Potassium hydroxide	60–80
Sodium hydroxide	20
Glycerin	100
Water (softened or distilled)	370
	1000

Perfume 0·3–0·5 per cent

Adjust the proportion of potassium hydroxide to a 2–3 per cent free fatty acid finish. In this formula the potassium coconut fatty acid soaps give a fine abundant lather to which the sodium soaps give suitable 'body'.

Several materials have been used as additives to shaving creams. A small proportion of menthol is used to give a cooling effect on the skin. Chlorhexidine is used as a bactericide. Silicones are used to increase lubricity and prevent razor drag. They also give the skin a smooth soft feel which is an aid to closer shaving.

Changes in temperature conditions during storage cause variations to occur in the texture of shaving creams. The soaps increase in viscosity or become gel-like when subjected to increases in temperature. In low temperature conditions the cream becomes hard and is difficult to apply. These variations can be reduced by adding 0·5 per cent of borax to the product. Corrosion of tubes on storage is prevented by addition of 1.0 to 1.5 per cent of sodium silicate. Use methyl parahydroxybenzoate 0·15 per cent and propyl parahydroxybenzoate 0·02 per cent as a preservative in all formulae.

Brushless shaving creams

Brushless shaving creams are used to keep the beard soft for the duration of the shave after it has previously been softened by washing the face with soap and warm water. The creams are applied to the wet face and must therefore mix with water to spread evenly on the skin. They consist of suitably modified stearate soaps. A small proportion of oil is desirable in the cream to act as a lubricant, and a humectant is required to increase adherence of the cream on the skin, although if the proportion of humectant is too high the cream becomes sticky, and causes the razor to drag. The viscosity of the cream is adjusted by addition of waxes. It must be of suitable consistency to spread easily and be sufficiently firm to support the hairs during shaving. Modern formulations may include surface-active agents, emollients and antiseptic materials. In the formulae which follow include as preservative methyl parahydroxybenzoate 0·15 per cent and propyl parahydroxybenzoate 0·02 per cent. Satisfactory products can be made as follows:

No. 2312

Stearic acid	160
Mineral oil	140
Spermaceti	20
Glycerin	60
Dilute solution of ammonia (10 per cent NH_3)	20
Water (softened or distilled)	600
	1000
Perfume	0·5 per cent

No. 2313

Glyceryl monostearate (S.E.)	120
Stearic acid	50
Glycerin	30
Mineral oil (cosmetic quality)	10
Water (softened or distilled)	790
	1000
Perfume	0·5 per cent

A softer product with a pearly appearance is obtained with the following formula.

No. 2314

Stearic acid	150
Myristic acid	60
Cetyl alcohol	30
Diethyl glycol monoethyl ether	50
Sodium lauryl sulphate	30
Potassium hydroxide (80 per cent)	10
Borax	10
Water (softened or distilled)	660
	1000
Perfume	0·5 per cent

The standard method of procedure is used for manufacture, namely the water soluble components are heated together and added to the melted constituents of the oil/fat phase at a temperature of 65–70°C. The cream is stirred gently during emulsification and cooling. The stirring procedure often requires careful control, and should not be continued until the cream is cold, since this produces a soft product. Too rapid stirring causes entrainment of air which also affects the final appearance of the cream. As a general guide, continue gently stirring to a temperature of from 45–47·5°C and then allow the product to cool without further stirring. If the stirrer is removed above this temperature the cream will become gritty and have less pearl. After manufacture the batch is allowed to stand in cool conditions to mature. This period of time for maturation can vary from overnight to as long as 7 days, but it is during this time that pearliness develops and the texture of the product becomes stable. After the maturing period the cream is stirred gently before being filled cold into tubes or jars.

The following formulae contain a small proportion of a surface-active material. This is mixed with a small quantity of water and added to the emulsion as indicated in the formulae.

No. 2315

Stearic acid	250
Cetyl alcohol	10
Mineral oil (cosmetic quality)	20

No. 2315 (*continued*)

iso-Propyl myristate	20
Triethanolamine	15
Water	580
Triethanolamine lauryl sulphate (50%)	5
Water (softened or distilled)	100
	1000

Perfume 0·5 per cent

No. 2316

Stearic acid	300
iso-Propyl myristate	20
Glycerin	50
Triethanolamine	7·5
Potassium hydroxide	3
Water(softened or distilled)	509·5
Triethanolamine lauryl sulphate	10
Water (softened or distilled)	100
	1000·0

Perfume 0·5 per cent

Aerosol shaving lathers

Formulation of an aerosol shaving lather requires particular care and several experimental preparations must be formulated to obtain a lather which has the desired characteristics. The body and texture of the lather depend upon the amount of fat, water and propellent present. The fat content should be sufficient to give a product which is not soft and sloppy, but farily rigid and stable without being a dry type of lather. Can corrosion is a problem and storage trials should always be carried out before marketing a product. Choice of perfume which is stable under these conditions and will not induce corrosion is important. The following formula gives good results:

No. 2317

Shave cream base:

A	Myristic acid	20
	Stearic acid	60

Alcohol 96% v/v	5
Lanolin	10
Glyceryl monostearate (S.E.)	25
Mineral oil (cosmetic quality)	20
Triethanolamine lauryl sulphate[1] (40 per cent active)	10
Glycerin	50
Triethanolamine	40
Water (softened or distilled)	760
	1000

Perfume	0·5–1·0 per cent
Methyl parahydroxybenzoate	0·15 per cent
Propyl parahydroxybenzoate	0·02 per cent

[1] Empical TL 40 type–Albright & Wilson Ltd., Marchon Division.

Procedure: Heat A and B with half the water only to 70°C. Pour B into A with continuous stirring. Add the remainder of the water, and mix. Cool with slow stirring, adding the perfume at about 35°C.

Container charge:	
Shave cream base	87
Propellent–12/114 (10 : 90)	13
	100

Container: internally lacquered aluminium or tin plate; plastic coated glass or polypropylene. Valve: standard with a lacquered cup and fitted with a foam button.

The container charge with this type of product is a question of personal choice. If too much propellent or a propellent of too high a pressure is used then the foam will be too rigid and will not 'pick-up' on the skin.

Pre-electric shave preparations

Pre-electric shave preparations are mainly designed to dry the skin. Consequently they are used more frequently and sell more readily during summer weather, or in hot climates when the skin is often covered with perspiration. To this end, the products are either based on alcoholic lotions, or on powders. Electric shave lotions contain a relatively high percentage of alcohol, which helps to dry up the film of perspiration on the skin and is also alleged to cause a contraction of the erector pili muscles. To supplement this action, the lotions may contain an astringent which tightens the skin causing the hair to stand out more to facilitate cutting. Antiseptic materials can also be added in order to prevent any infection which can occur as a result of damage to the epidermis by the cutting action. At the same time, the lotion should contain some material of an oily nature to lubricate the hairs with a thin film of oil and this acts as a cutting lubricant. The concentration of oily material is equally as important as the concentration of alcohol used, since if there is too much oil present, a pasty mass is formed with the cut-off whiskers, and this clogs the cutting edges of the razor.

In the following formula using *iso*-Propyl myristate as the oily material excellent results are obtained at the concentration given:

No. 2318

iso-Propyl myristate	120
Alcohol	880
	1000

Perfume 0·5–1·0 per cent

No. 2319

iso-Propyl myristate	120
Menthol	0·5
Alcohol	879·5
	1000·0

Perfume 0·5–1·0 per cent

Procedure: Dissolve the menthol in half of the total volume of alcohol and dissolve the perfume in the remainder. Mix the two solutions and add the *iso*-Propyl myristate.

No. 2320

iso-Propyl myristate	120
Menthol	0·5
Zinc phenolsulphonate	1
Alcohol	878·5
	1000·0

Perfume 0·5–1·0 per cent

Procedure: Liquid acetoglycerides are also used as the oily material of electric shave lotions. They have an emollient effect on the skin and provide adequate lubrication of the beard.

No. 2321

Liquid acetoglyceride[1]	120
Menthol	0·5
Boric acid	10
Alcohol	869·5
	1000·0

Perfume 0·5–1·0 per cent

[1] Acetoglyceride LC.—Bush Boake Allen

Procedure: Dissolve all the ingredients in the alcohol. Cool to 0°–5°C and filter.

Use of a vasoconstrictor such as adrenalin, or epinephrine (pseudo-ephedrine), has been suggested for use in electric shave lotions. Local application of these materials has a pilomotor action which causes contraction of the skin, making the hair fibre stand further out of the follicle. The hair is cut whilst it is in this extended position, and after cutting, as the effect of the adrenalin wears off, the hair appears to retract below the skin surface. A concentration of 0·5 per cent adrenalin hydrochloride is used and the lotion is adjusted to an acid pH value (German Patent No. 1032482), since adrenalin is unstable in neutral or alkaline solutions.

An aerosol form of pre-electric shaving lotion is made as follows:

No. 2322

Pre-shave base:	
iso-Propyl myristate	120
Menthol	0·5
Boric acid	2
Alcohol 99·5% v/v	877·5
	1000·0

Perfume 0·5–1·5 per cent

Procedure: Dissolve the menthol and boric acid in the alcohol. Add the perfume and *iso*-Propyl myristate. Stir well.

Container charge:	
Pre-shave base	45
Propellent—11/12 (50 : 50)	55
	100

Container: internally lacquered aluminium, tinplate, or plastic coated glass. Valve: standard with lacquered cup.

Electric shave powders

Electric shave powders are applied immediately before shaving to dry the skin and provide slight lubricating action. They consist mainly of talc with suitable proportions of metallic stearate to promote adhesion of the powder to the skin, and provide the slip required for ease of shaving. An absorbent powder such as kaolin or magnesium carbonate can also be included. To avoid damaging the cutting edges of the razor, it is essential that the ingredients used should contain a minimum of abrasive materials.

No. 2323

Ingredient	
Zinc or magnesium stearate	70
Kaolin	50
Talc	880
	1000
Perfume	0·5 per cent

The powder can be made to stick form by mixing to a paste with a dispersion of magnesium aluminium stearate in water. The mass so formed is milled until smooth and then extruded to the desired shape, and finally oven dried to expel excess moisture. An alcoholic stick based on a solution of sodium stearate in alcohol is also used as a pre-shaving aid. The proportion of stearate is kept to a minimum, to prevent clogging of the razor. A lubricant is again included in the product as indicated in this formula.

No. 2324

Ingredient	
Stearic acid	30
Sodium hydroxide solution (10 normal)	10·5
iso-Propyl myristate	100
Alcohol	859·5
	1000·0
Perfume	0·5 per cent

After-shave preparations

After-shaving preparations are intended to cool and refresh the skin, allay irritation, be mildly astringent and in some instances when soap has been used, to neutralize any alkali left on the skin after shaving. They are also used to relieve the feeling of tautness and soreness and to disinfect and promote healing of skin which has become damaged during the shaving operation. They are used mainly in the form of clear lotions containing between 25 and 50 per cent of alcohol, and antiseptic, emollient, or haemostyptic materials can be included. Concentrations of alcohol above 50 per cent cause excessive stinging and are not generally used. Many of them contain extract of witch-hazel together with such substances as menthol, glycerin, boric acid, alum, potassium oxyquinoline sulphate and chloroform. On account of the relatively low concentrations of alcohol used, attention must be given to the solubilization of the perfume. Select a perfume which is water soluble or one specially made to be soluble in low concentrations of alcohol. If necessary include a surface-active material or one of the commerical preparations available as a perfume solubilizer. A simple preparation based on bay rum mixed with peppermint oil and glycerol is made as follows:

No. 2325

Peppermint oil	10
Glycerin	50
Bay rum	940
	1000

Another formulation based on witch-hazel also includes boric acid, glycerin and menthol dissolved in alcohol.

No. 2326

Boric acid	20
Glycerin	30
Menthol	1
Witch-hazel extract	849
	1000

Water-soluble perfume 1–3 per cent

No. 2327

Alum	20
Glycerin	30
Menthol	1
Witch-hazel extract	250
Alcohol	250
Water (softened or distilled)	449
	1000

Water-soluble perfume 1-3 per cent

In the following formula for a clear lotion, glycerin is used for emolliency and potassium oxyquinoline sulphate as an antiseptic.

No. 2328

Glycerin	30
Potassium oxyquinoline sulphate	1
Perfume solubilizer[1]	40
Alcohol	400
Water (softened or distilled)	539
	1000

Perfume 1-2 per cent

[1] Polyoxythylene Sorbitan monolaurate type.

Procedure: Mix the perfume with the perfume solubilizer and add the alcohol. Dissolve the oxyquinoline sulphate in the water; add the glycerin and mix with the alcohol solution.

Further formulae are given:

No. 2329

Glycerin	30
Witch-hazel extract	225
Alcohol	125
Menthol	1
Water (softened or distilled)	619
	1000

Water-soluble perfume 1-3 per cent

Procedure: Mix the glycerin with the witch-hazel extract, and dissolve the menthol and perfume in the alcohol. Mix the two solutions and filter make up to volume.

Suitable antiseptic or antibacterial agents for use in after-shave lotions include quaternary ammonium compounds, selected for low irritation potential, and chlorhexidine diacetate. Quaternary ammonium compounds also leave the skin feeling soft and supple after application.

No. 2330

Glycerin	30
Cetrimide	1
Menthol	1
Alcohol	400
Water (softened or distilled)	568
	1000

Water-soluble perfume 1–3 per cent

No. 2331

Glycerin	20
Chlorhexidine diacete	2
Menthol	1
Alcohol	400
Water (softened or distilled)	577
	1000

Water-soluble perfume 1–3 per cent

An after-shave preparation in the form of a gel is of interest as a product suitable for travelling. There is no leakage problem and a small quantity can be conveniently dispensed from a tube pack as required for use.

A suitable formula based on a neutralized carboxy vinyl polymer is made as follows:

No. 2332

Carbopol 934[1]	10
Menthol	1
Alcohol	450
Di-isopropanolamine	8
Water (softened or distilled)	531
	1000

Water-soluble perfume 1–3 per cent

[1] B. F. Goodrich Chemical Company.

Procedure: Dissolve the menthol in alcohol, then dispense the Carbopol 934 in the aqueous alcoholic mixture. The gel is then formed by the addition of the di-isopropanolamine.

A thick gel can be prepared by increasing the proportion of Carbopol 934, and an antiseptic can be included if required. A quick breaking foam after-shave preparation is formulated so that the foam stays long enough to put onto the face without running off the hand. In the following formula allantoin is included as a healing agent.

No. 2333

Quick breaking foam base:

A	Polawax A. 31[1]	20
	Alcohol 96% v/v	621
	Menthol	0·5
	Camphor	0·5
B	Allantoin	1
	Water (softened or distilled)	357
		1000·0

Perfume	0·5–1·5 per cent
Methyl parahydroxybenzoate	0·15 per cent
Propyl parahydroxybenzoate	0·02 per cent

[1] Croda Ltd.

Procedure: Warm the Polawax in the alcohol gently and add the rest of A together with the perfume, stirring all the time. Dissolve the allantoin in the water warming gently. Whilst still warm add B to A with stirring. Fill out whilst still warm.

Container charge:

Q.B.F. base	91
Propellent–12/114 (10 : 90)	9
	100

Container: plastic coated glass or polypropylene. Valve: tilt action fitted with a simple orifice button—no metal parts should be in contact with the product.

If the base is not filled into the containers whilst still warm, it solidifies and becomes difficult to handle.

After-shave creams are sometimes used especially for sensitive skins when an alcoholic lotion can be more irritating than refreshing. In such cases pourable emulsified lotions are more acceptable, although these do not provide the refreshing effect of a quick-drying lotion. On the other hand, a cream imparts emolliency, relieves tenderness, dryness and roughness, and provides opportunity to include either oil-soluble or water-soluble additives in the emulsion. Use 0·15 per cent methyl parahydroxybenzoate and 0.02 per cent propyl parahydroxybenzoate as preservatives in the formulae which follow.

No. 2334

Liquid acetoglyceride[1]	25
iso-Propyl myristate	50
Ethylene glycol monostearate	5
	25
Water (softened or distilled)	895
	1000

Perfume	0·5–1·0 per cent

1 Acetoglyceride L/C type—Bush Boake Allen.

No. 2335

iso-Propyl palmitate	50
Propylene glycol monostearate	50
Cetyl alcohol	25
Glycerin	50
Water (softened or distilled)	825
	1000

Perfume	0·5–1·0 per cent

No. 2336

Stearic acid	30
Liquid acetoglyceride[1]	35
iso-Propyl myristate	35
Emulsifying wax[2]	2
Triethanolamine	10
Glycerin	30
Water (softened or distilled)	858
	1000

Perfume	0·5–1·0 per cent

[1] Acetoglyceride L/C type—Bush Boake Allen.
[2] Emulsifying wax B.P.

A firm cream can be made as follows:

No. 2337

Glyceryl monostearate (self-emulsifying type)	120
Stearic acid	50

No. 2337 *(continued)*

iso-Propyl myristate	20
Glycerin	50
Water (softened or distilled)	760
	1000

Perfume 0·5–1·0 per cent

An interesting after-shave cream which also provides the refreshing sensation given by the use of alcohol is made to the following formula:

No. 2338

Glyceryl monostearate (self-emulsifying type)	150
Mineral oil (cosmetic quality)	40
Glycerin	60
Boric acid	10
Water (softened or distilled)	689
Menthol	1
Alcohol	50
	1000

Perfume 0·5–1·0 per cent

Procedure: Dissolve the menthol in the alcohol and add this to the emulsion at a temperature of about 30°C.

An after-shave stick based on an alcoholic solution of soap provides another convenient method of relieving any feelings of discomfort due to shaving. This product suffers from the slight drawback that it leaves a somewhat thicker film on the skin and tends to have a slightly alkaline reaction.

No. 2339

Sodium stearate	60
Glycerin	50
iso-Propyl myristate	10
Menthol	1
Alcohol	839
Water	40
	1000

Perfume 0·5–1·0 per cent

Add a small proportion of a water-soluble dyestuff.

Procedure: Heat all the ingredients under reflux until a clear solution is obtained. Pour into moulds. Allow to set and pack in air-tight containers.

After-shave powders are still popular for use after shaving and give a smooth matt appearance to the skin. They are based on a high proportion of talc with additives such as stearates to increase adhesion, boric acid to counteract alkalinity, and a suitable bactericid, such as cetrimide or chlorhexidine diacetate. A smooth invisible powder is desirable and to this end they are tinted to a pale sun-bronze with a red or yellow cosmetic oxide. Examples follow:

No. 2340

Calcium carbonate	50
Zinc stearate	100
Kaolin	50
Boric acid	20
Chlorhexidine diacetate	5
Talc	775
	1000
Perfume	0·5 per cent

No. 2341

Boric acid	30
Magnesium stearate	40
Talc	930
	1000
Perfume	0·5 per cent

When selecting perfumes for shaving preparations, particularly white creams, these should always be tested to ensure that they do not discolour the product. Many desirable perfumery notes obtained with materials such as vetivert, patchouli and the eugenols can cause troubles with discolouration. The question of irritation of perfumery materials must also be considered. Perfumes for shaving preparations follow current masculine trends. Lavender, lavender-colognes and fougere types remain popular, though preference is increasing for modifications of these basic types with floral, spicy and bay top notes. The selection of a suitable perfume often becomes more important in after-shaving preparations when these can be purchased just as much for the lingering effect of the perfume as for their functional performance. This also applies to other toiletries which are now prepared exclusively for use by the male. Products such as deodorants, colognes, bath preparations, and talc are essentially female cosmetics modified only by introducing a suitable fragrance. Indeed, it is probably true to say that only a few years ago men's toiletries were selected and purchased purely as functional products, whereas to-day the same product may be purchased on account of a particular fragrance. There are also signs that the tastes and fashion of fragrances for men are changing and sophisticated floral bouquets with aldehydic top notes may well be acceptable in the near future. It is sufficient to note that the unobtrusive fragrance which was customary has now been replaced by a stronger nuance which is required to persist and is accepted both at home and in the office.

CHAPTER ELEVEN

Rouges and Eye Cosmetics

Rouges can be applied either before or after powdering, depending upon the type of product being used. All types of rouge are made in numerous shades varying from the palest of pinks to the deep blue reds and the conventional rouge contains a high proportion of red or reddish brown pigments. Products containing lower proportions of pigments are used as toners or blenders for special highlighting effects. The pigments used must be carefully chosen and checked for resistance to bleeding. Bleeding can occur as a result of the moisture and sebum present on the skin surface or as a result of the ingredients of the base used for the product. For this reason the pigments used should be checked for resistance to bleeding in the presence of water, oil, alcohol, and the perfume. A small proportion of a suitable bromo acid is often included (red or orange tone) as this gives a natural blushing effect which is particularly attractive when used in products containing low proportions of pigment.

Liquid rouges

Liquid rouges do not have as great an appeal as the cream or powder types, but they are nevertheless attractive products and are very easily applied, if correctly formulated. A suitable product is prepared as follows:

No. 2342

A	Cetyl alcohol ethoxylate[1]	30
	iso-Propyl myristate	20
	Cetyl alcohol	5
	Diethylene glycol monostearate	45
B	Glycerin	200
	Water (softened or distilled)	700
		1000

Perfume	0·2 per cent
Pigment	2·0 per cent
Titanium dioxide	2–3 per cent
Methyl parahydroxybenzoate	0·15 per cent
Bromo acid	0·02 per cent

[1] Lubrol type—I.C.I. Ltd.

Procedure: First prepare the emulsion base by heating together the ingredients of (A) to a temperature of 70–75°C. In a separate vessel dissolve the preservative in the glycerine by heat. Add the water and heat to 75°C. Add B to A with slow continuous stirring. Continue stirring and allow to cool adding the perfume at about 35°C. Finally add the pigment and titanium, and mix well.

The following liquid rouge contains a higher proportion of oil which contributes to even application:

No. 2343

A	Mineral oil	120
	Acetylated monoglyceride (liquid type)	120
	iso-Propyl myristate	120
	Oleic acid (redistilled)	80
B	Triethanolamine	40
	Water (softened or distilled)	520
		1000

Perfume	0·3 per cent
Pigments	2–3 per cent
Methyl parahydroxybenzoate	0·15 per cent
Butylated hydroxyanisole	0·01 per cent
Bromo acid	0·02 per cent

Procedure: Disperse the pigments and eosin in A and mill. Heat A and B independently in separate vessels to 70–75°C. and add part B to A with stirring. Cool with stirring, adding the perfume at about 35°C.

In this cream the degree of oiliness can readily be controlled by varying the ratios of mineral oil, acetylated monoglyceride, and *iso*-Propyl myristate. As the proportion of acetylated monoglyceride is increased the product becomes less oily and more emollient. It is advisable to include an antioxidant such as propyl gallate or butylated hydroxyanisole to prevent rancidity developing on storage.

Powder rouges

Powder rouges are prepared as compressed powders or compacts. They consist of pigmented powders generally formulated with a high proportion of talc and containing a binding agent to enable the mix to be compressed in tablet form. Typical formulae for the powder base are as follows:

No. 2344

Zinc stearate	170
Rice starch	170
Talc	660
	1000

Perfume	2·0 per cent
Pigments	12–15 per cent

No. 2345

Light magnesium carbonate	100
Prepared chalk	200
Zinc oxide	100
Rice starch	150
Talc	450
	1000

Perfume	2·0 per cent
Pigments	12–15 per cent

No. 2346

Zinc oxide	100
Lanolin	300
Light Magnesium carbonate	200

No. 2346 (*continued*)

Talc	400
	1000

Perfume	2·0 per cent
Pigments	12–15 per cent

When the compact is to be made by the dry process the binding material is included in the mix. This can consist of 1·0 per cent of tragacanth or acacia powder. As an alternative binding material small amounts of a lanolin derivative and *iso*-Propyl myristate can be used, as given in the following formula:

No. 2347

Kaolin	50
Calcium carbonate (precipitated)	50
Magnesium carbonate	50
Zinc stearate	50
Talc	750
Pigments	50
	1000

Perfume	2·0 per cent
Binder: *iso*-Propyl myristate, Lanolin absorption base	equal parts

Procedure: Mix the binding materials together. Add the perfume and mix with the magnesium carbonate. Add to the remaining ingredients and mix well. Grind and compress into godets.

For manufacture by the damp or wet process the binding agent is prepared separately and is based on materials such as tragacanth gum, a mixture of soap and tragacanth gum, and polyvinyl pyrrolidone. Formulae are given as follows:

No. 2348

Tragacanth	10
Alcohol	20
Water (softened or distilled)	970
	1000

Methyl parahydroxybenzoate	0·2 per cent

Procedure: First wet the tragacanth with alcohol, and then add the water in a continuous stream and stir well. Allow to stand for 24 hours stirring occasionally, and pass through muslin. This is then ready for use. Dissolve the preservative in the water, with the aid of heat before adding to the tragacanth.

No. 2349

Tragacanth	12
Soap chips	40
Water (softened or distilled)	948
	1000

Methyl parahydroxybenzoate 0·2 per cent

Procedure: Prepare a concentrated solution of the soap chips with sufficient hot water and use this to prepare a tragacanth paste. Add the remaining water in a continuous stream and mix well. Allow to stand for 24 hours stirring occasionally, and pass through muslin before use. Dissolve the preservative in the water with the aid of heat before adding to the soap tragacanth mix.

About 12 to 15 per cent of the binding agent solution is mixed with the previously prepared coloured powder mix. For the wet type process sufficient is added to prepare a fairly damp paste which is then moulded in shallow capsules of about 1¼ inches in diameter. The blocks are then dried at normal temperature in a current of air. If undue heat is applied during the drying time they will split, and apart from this the perfume will be damaged. During the drying time the compacts are stood on blotting paper to absorb any excess moisture. When dry they are trimmed with a special rotating knife to give a perfectly smooth finish to the surface. They are then stuck with a suitable adhesive to the base of the container. For the so-called damp process the coloured powder base is mixed thoroughly with about 5 to 10 per cent of binding agent solution. The amount of moisture required in the mix can be judged by pressing a quantity between finger and thumb when it should be sufficiently damp to hold together without crumbling. If the mix is too wet at this stage it should be broken into suitable sized pieces and allowed to dry. When the correct degree of moisture has been obtained the mix is compressed either by automatic or hand operated machine direct into a godet. The compacts are then allowed to stand in a current of air at normal temperature to dry off before being fixed into containers.

Rouge creams

Rouge creams are often preferred to powder rouges as they are more easily applied over a foundation cream and give a suitable base for face powder. The powder type tend to leave sharply defined edges to the make-up, which look unnatural unless they are carefully applied. Rouge creams are prepared as anhydrous preparations based on thixotropic blends of waxes, oils and fatty acid esters, or as emulsions of the cold cream type. Vanishing cream type emulsions can be used also but should include up to 8 or 10 per cent of glycerin or propylene glycol. This is used as a pigment dispersing agent and also promotes even application of the product. Products made using a vanishing cream base should preferably be packaged in tubes. Formulae are given.

No. 2350
(Anhydrous type)

Petroleum jelly	400
Acetylated lanolin[1]	50
Acetylated monoglyceride (solid type)	100
iso-Propyl myristate	390
Ozokerite	50
Carnauba wax	10
	1000

Perfume	2·0 per cent
Pigments	5·0 per cent

[1] Modulan type—American Cholesterol Products Inc.

Procedure: Mix the ingredients together and add the pigments and mill.

Pigments used for cream rouges should be obtained as a semi-processed colourant in one of the basic ingredients used in the product and the formula adjusted accordingly.

No. 2351
(Cold cream type)

A	Mineral oil	460
	iso-Propyl myristate	100
	Spermaceti	100
	Beeswax	120
B	Borax	5
	Water (softened or distilled)	215
		1000

Perfume	1·0 per cent
Pigments	5·0 per cent

Methyl parahydroxybenzoate	0·15 per cent
Propyl parahydroxybenzoate	0·02 per cent

Procedure: Disperse the pigment to a smooth paste with the *iso*-Propyl myristate. Heat together the remaining ingredients of A to a temperature of 70–75°C. Add the pigment dispersion. Heat ingredients of B in a separate vessel to 70–75°C. Add B to A with slow continuous stirring and allow to cool adding the perfume at about 35°C. Fill warm.

No. 2352

(Vanishing cream type)

A	Stearic acid	150
	Hartolan[1]	10
	iso-Propyl myristate	25
	Lanolin	25
	Beeswax	50
	Glyceryl monostearate (self-emulsifying type)	70
B	Propylene glycol	50
	Glycerin	30
	Water (softened or distilled)	590
		1000

Perfume	1·0 per cent
Pigments	5·0 per cent
Methyl parahydroxybenzoate	0·15 per cent
Propyl parahydroxybenzoate	0·01 per cent

[1] Croda Ltd.

Procedure: Melt together the ingredients of A and heat to 70°C. In a separate vessel heat B to 70–75°C and add B to A with stirring. Allow to cool and add the perfume at 35°C. Add the pigments in a semi-processed form, preferably as dispersed colourant in propylene glycol, making the necessary adjustment to the formula.

Eye make-up

Cosmetic preparations for the eyes are now an essential feature of the facial make-up, assuming as much importance as the use of lipsticks. The basic steps of the eye make-up involve use of an eyebrow pencil to enhance the natural eyebrow or to modify the outline after plucking; use of eye shadows for the eyelid and for shadowing towards the eyebrow; use of crayons or liners to contour the eyes, and mascara to enhance the appearance of the lashes. As with all types of coloured preparations, correct choice of pigments is most important and these should only be obtained from a manufacturer able to supply the special blends of permitted cosmetic oxides and ultramarines required. Reference has already been made to the range of colours available. For eye make-up a limited number of specialized colours are also available. To facilitate dispersion wherever possible, reduced colourants should be obtained on bases similar to one of the ingredients in a particular formulation.

Eyebrow pencils

Eyebrow pencils are prepared in a range of colours from black through brownish-black and brown to blue. The true pencil type made by pencil manufacturers are fairly hard and generally used to lightly accentuate the natural line and natural hair. Separate leads are also made for use in an automatic case. A softer version also made by the pencil manufacturer is of the wax or crayon type, and is more often used to improve the browline after plucking. The true cosmetic eyebrow pencil is made of similar composition to a lip-pencil with a higher proportion of waxes than a lipstick to increase the melting point so that the pencil can be moulded as a thin stick and sharpened to a point. A suitable formula is as follows:

No. 2353

Beeswax	24
Ozokerite	24
Butyl stearate	9
Lanolin	3
Castor-oil	25
Mineral oil	15
	100
Perfume	1–2 per cent
Butylated hydroxyanisole	0·002 per cent

Procedure: Melt the ingredients together and add the dispersed colour. Stir well and strain through muslin. Add the perfume and pour into moulds.

Eyeshadow

Eyeshadow is used to give a background of colour to the eyes and there is virtually no limit to the number of shades which are used, from the more conventional blue, green, and mauve, to very light or transluscent shades of beige, grey, and white. Gold and silver colourings are used for evening wear. Eyeshadows are prepared in wax bases as creams, in stick form, as loose powders, or as compressed powders.

Wax or cream bases and sticks

Wax or cream bases and sticks must be firm enough to suspend the pigments, but sufficiently thixotropic to spread easily by gentle stroking over the eyelids. After application the residual film should not be tacky or form into 'lines'. A suitable anhydrous cream base can be made as follows:

No. 2354

Petroleum jelly	480
Liquid lanolin[1]	40
Microcrystalline wax (m.p. 65–70°C)	80
Beeswax	40
iso-Propyl myristate	360
	1000
Perfume	0·5 per cent

[1] Fluilan type—Croda Ltd.

Procedure: The ingredients are melted together and mixed. To prepare the finished eyeshadow up to 25 per cent of pigment is included, i.e. 25 parts of pigment mixed with 75 parts of the base. As already indicated an adjustment to the formula is necessary when dispersed pigments are used.

The pigments are included in the following formula:

No. 2355

Petroleum jelly	200
Liquid lanolin[1]	50
Ozokerite	150
iso-Propyl myristate	400
Pigments	200
	1000
Perfume	0·5 per cent

1 Fluilan type—Croda Ltd.

Procedure: Mix dispersed pigments with the petroleum jelly, liquid lanolin, and *iso*-Propyl myristate adjusting the formula to allow for dispersing agent. Heat to about 70–75°C and mill. Add the ozokerite and stir until uniform. Allow to cool and add the perfume at about 30–35°C. Mix and pour into suitable containers.

To prepare a 'pearlized' or 'translucent' eye shadow include either a natural or synthetic pearl essence or a lustre pigment. The following base spreads easily and gives an even non-greasy film:

No. 2356
(Base only)

iso-Propyl myristate	80
Beeswax	120
Petroleum jelly	800
	1000

Translucent eye shadows are prepared from this base as follows:

No. 2357

Base No. 2356	700
Pigment	50
Lustre pigment[1]	200
Talc	50
	1000

[1] Bi-Lite—Mallinckrodt Chemical Works; or Shinju TM White—The Mearl Corporation.

The quantities given in this formula for pigment and lustre pigments refer to the dry weight and when used in this form these materials should be sieved before incorporating in the base. Adjustments to the formula are made when pigment dispersions and pearl or lustre suspensions are used. Use of talc is not essential, but gives a little more cover on the skin and does not suppress the pearlescent effect.

Pastel shade effects and increased covering properties are obtained by replacing the talc with an equal proportion of titanium dioxide. This should also be obtained as a ready milled dispersion in a suitable vehicle. 'Pearl' or 'translucent' effects are also enhanced by using a suitable proportion of magnesium myristate, as indicated in the following formula:

No. 2358

Base No. 2356	700
Pigment	50
Lustre pigment	200

Magnesium myristate[1]	50
	1000

[1] Satinex–Bush Boake Allen.

Cream eye shadows are made using a non-fatty or cream type vehicle and are suitable for tube packing. A base formula is prepared as follows:

No. 2359
(Cream base)

A	Lanolin	10
	Mineral oil	100
	Stearic acid	25
	Glyceryl monostearate (self-emulsifying type)	30
B	Glycerin	50
	Triethanolamine	5
	Water (softened or distilled)	780
		1000
	Perfume	0·5 per cent
	Methyl parahydroxybenzoate	0·2 per cent
	Propyl parahydroxybenzoate	0·05 per cent

Procedure: Heat A and B independently to 75°C and add B to A slowly with continuous stirring. Cool with stirring, adding the perfume at about 35°C.

The finished product is prepared by adding 10 parts of pigment to 100 parts of the above base. A shimmer or pearlescent effect is made as follows:

No. 2360

Cream base No. 2359	1000
Pigment	100
Pearl suspension in mineral oil[1]	300
	1400

[1] Mearlmaid MO–The Mearl Corporation.

A suspension of a pearl essence in mineral oil can also be used as the additive for an anhydrous cream base, when the proportion of oily material in the formula should be correspondingly reduced.

Eyeshadow stick performs the same function as the anhydrous

cream and is prepared by moulding in similar manner to the preparation of lipsticks. The base is requied to be thixotropic to give smooth and even application without being greasy or sticky. In the formulae which follow the proportions of pigment, lustre pigment, and talc refer to the dry material.

No. 2361
(Stick base)

Beeswax	50
iso-Propyl palmitate	160
Petroleum jelly	250
Oleyl alcohol	50
Paraffin wax	220
Cetyl alcohol	30
Lanolin	40
	800

Perfume	0·5 per cent
Propyl parahydroxybenzoate	0·01 per cent
Butylated hydroxyanisole	0·02 per cent

Procedure: Heat the *iso*-Propyl palmitate, petroleum jelly, oleyl alcohol, and lanolin together. Add the preservative and antioxidant.

The waxes are added during manufacture of the completed stick as follows:

No. 2362

Base No. 2361	820
Pigment	80
Lustre pigment	100
	1000

Procedure: Mix the pigment with the melted oils and fats of the base and mill. Add the melted waxes and lustre pigment and mix, and allow to cool before adding the perfume. Pour into moulds and allow to set.

The above formula gives a stick of good colour intensity with a 'glitter' or 'pearl' effect. Pastel shades with glitter and translucent effect are made by modifying the proportions of colour and lustre additive as follows:

No. 2363

Base No. 2361	720
Pigment	30

Talc	50
Lustre pigment	200
	1000

Powder eyeshadow

Powder eyeshadow is prepared either as loose or compressed powder as an alternative to the cream or stick form. It is comparatively easy to apply and blends on to the eyebrow and does not readily form into 'lines', and for this reason is often preferred to other forms of eyeshadow. A suitable base is as follows:

No. 2364

Talc	400
Magnesium myristate	440
Titanium dioxide	160
	1000

Pigments—use about 7 per cent depending on the tinctoral properties of the pigment.

Procedure: Mix the pigments with a proportion of the talc and sieve. Sieve the titanium dioxide separately, and add the pigment blend followed by the remainder of the talc and magnesium myristate. Mix and sieve.

To prepare pearlized or translucent eyeshadow powders the basic formula is modified, and opaque materials such as titanium dioxide are not used as these prevent or mask the light reflecting properties of pearlescent materials. A formula for a pearlescent eye shadow powder is as follows:

No. 2365

Talc	350
Pigment	150
Pearlescent pigment[1]	500
	1000
Perfume	0·5 per cent

[1] Bi-Lite 20—Mallinckrodt Chemical Works.

Procedure: Sieve together the talc and pigment and mix. Add the pearlizing agent, sieve and mix. The powder should not be milled, as this procedure reduces the light reflecting properties and pearlescent characteristics.

The following powder spreads smoothly and evenly and gives a transparent appearance with a good glitter effect.

No. 2366

Magnesium myristate	350
Pigment	150
Pearlescent pigment[1]	500
	1000
Perfume	0·5 per cent

[1] Shinju TM White–The Mearl Corporation.

Pearlized or pearlescent eye shadow powders are made with standard conventional colours and in a wide range of pastel shades. Pearlescent effects are obtained by using natural pearl essences, prepared from crystalline fish guanine, synthetic 'crystalline' bismuth oxychlorides, and pearlescent pigments prepared by deposition of bismuth oxychloride on platelets of white mica. The materials have good gloss and slip properties and also give some degree of opacity to the products. In this respect they replace titanium dioxide as an opacifier particularly as titanium dioxide tends to mask the brilliance and pearlescent effect. Zinc stearate and kaolin can be used to increase the opacity if this property is required.

Highlight effects with additional sparkle and brilliance are obtained by using lustre pigments based on a titanium dioxide coated mica. Coloured lustre pigments generally referred to as iridescent colours based on titanium coated mica reflect and transmit colour and are used for special effects. An attractive range of pearlized and translucent eye shadows, is made without using an opacifier as follows:

No. 2367

Pigment	150
Pearlescent pigment[1]	250
Talc	600
	1000

[1] Bi-Lite 20–Mallinckrodt Chemical Works.

No. 2368

Pigment[1]	150
Lustre pigment	250
Talc	600
	1000

[1] MP-23—Rona Pearl Company.

More opaque pearlescent powders are prepared as follows:

No. 2369

Pigment	150
Kaolin	200
Zinc stearate	75
Lustre pigment[1]	250
Talc	325
	1000

[1] MP 30—Rona Pearl Company.

No. 2370

Pigment	150
Kaolin	150
Zinc stearate	75
Lustre pigment[1]	250
Pearlescent pigment[2]	50
Talc	325
	1000

[1] MP 30—Rona Pearl Company.
[2] Bismuth oxychloride NLZ—Rona Pearl Company; Bital—Rona Pearl Company; Bi-Lite 20—Mallinckrodt Chemical Works; Shinju TM White—The Mearl Corporation.

The formulae given for opaque powders are suitable for processing as compressed eye shadow powders to use neat, or with water and brush applicators, as shadows or eye-liners.

Compressed powder eye shadow is made following the procedure described under compressed powders using 5 to 6 per cent of a lanolin ester (*iso*-Propyl lanolate—Amerlate P; or acetylated lanolin alcohols—Acetulan, American Cholesterol Products Inc.) as the binding material. The lanolin ester is also used to disperse the perfume and is sprayed on the powder/pigment mix after the first sieving or milling process. The mix is re-milled or re-sieved and the pearlizing agent added so that the lustre is not affected by milling. Finally, the powder is compressed into godets adjusting the pressure to ensure ease of take-off, and application with a sponge-tipped applicator.

Eyeliners

Eyeliners are used after the eyeshadow has been applied to give the eyes a more striking appearance. The desired effect is obtained by using either a soft pencil or crayon or more usually with a product in liquid or cake form applied with a brush. The products are made in a wide range of shades from white through pastel shades to blues, greens, and black.

Liquid eye-liners consist of a suspension of pigment in a base containing a film-forming material which helps to fix the product after application and prevent smudging. The liner should also spread easily and smoothly and dry out fairly rapidly.

A small proportion of an emulsifying wax such as glyceryl monostearate or diethylene glycol monostearate is useful to help smooth application, and to increase the viscosity and suspending properties of the vehicle. A suitable proportion of a surface-active agent also assists spreading properties and even application. Glycerin is used to give a smooth film and prevent drying out and caking, but if used in too high a proportion slows down the rate of drying and causes smudging on the eye after application. The final consistency of the product depends to some extent on the specific volume or density of the pigment so that a basic formula is not necessarily suitable for all the shades required. When this occurs the volume of pigment used should be adjusted with a suitable amount of light magnesium carbonate, or a magnesium aluminium silicate (Veegum—R. R. Vanderbilt Company Inc.). Suspensions of these materials tend to thicken on standing, so that final evaluation of a product should not be made until samples have been allowed to stand, to allow time for this adjustment of viscosity to take place.

Colloidal silica* and magnesium aluminium silicate are also used on their own account as pigment suspending agents. Consequently formulations for liquid eye liners can vary from shade to shade according to any variations in the physical characteristics of the

* Aerosil—Degussa.

pigment used. Typical formulae are as follows:

No. 2371

Magnesium aluminium silicate[1]	25
Film-former (polyvinyl pyrrolidone)	20
Glycerin	50
Lanolin	5
Diethylene glycol monostearate	30
Sodium lauryl sulphate (needles)	10
Pigment	100
Water	760
	1000

Methyl parahydroxybenzoate	0·15 per cent
Propyl parahydroxybenzoate	0·01 per cent

[1] Veegum-type HV—R. T. Vanderbilt Company, Inc.

If a perfume is required in this product use only 0·1 to 0·15 per cent of a tested non-sensitizing and non-irritant type.

Procedure: Disperse the silicate with sufficient water to a liquid consistency. Dissolve the P.V.P. in water. Warm gently and add to the melted stearate and lanolin. Stir to form a cream. Dissolve the preservative in the glycerin and disperse the pigment in the solution. Mill if necessary and add to the mix. Finally add the sodium lauryl sulphate, dissolve in a little water and mix.

The following formula illustrates the slight variations which are required due to variations in the characteristics of the pigment:

No. 2372

Magnesium aluminium silicate[1]	28
Film former (polyvinyl pyrrolidone)	20
Glycerin	24
Lanolin	5
Diethylene glycol monostearate	30
Sodium lauryl sulphate	10
Colloidal silicon dioxide[2]	5
Pigment	120
Light magnesium carbonate	30
Water	708
	1000

Methyl parahydroxybenzoate	0·15 per cent
Propyl parahydroxybenzoate	0·01 per cent

[1] Veegum-type HV—R. T. Vanderbilt Company, Inc.
[2] Aerosil—Degussa.

Procedure: Disperse the pigment with the colloidal silica and light magnesium carbonate and carry out the remaining procedure as already given.

Use of a carboxy vinyl polymer as a suspending and dispersing agent is illustrated in the following formula:

No. 2373

Carboxy vinyl polymer[1]	5
Film former (polyvinyl pyrrolidone)	15
Glycerin	50
Sodium lauryl sulphate	5
Pigment	100
Triethanolamine	5
Water	720
	900

Methyl parahydroxybenzoate	0·15 per cent
Propyl parahydroxybenzoate	0·01 per cent

[1] Carbopol 934—B. F. Goodrich Chemical Company.

Procedure: Disperse the carboxy vinyl polymer with part of the water and add the triethanolamine with stirring. Add the polyvinyl pyrrolidone dissolved in an adequate portion of the water. Dissolve the preservative in the glycerin and disperse the colours in the solution, and add to the solution and mix well.

Adjustments to the pigment dispersion using light magnesium carbonate or colloidal silica must likewise be made to this formula if necessary.

Cake eye-liners resemble cake eyeshadows but generally contain a higher proportion of pigment and sufficient binding agent to form a cream or paste, for application with a wet brush. They can be made according to the wet mix and drying procedure described for cake make-up, using a cream base as the binding material. Alternatively they can be made by following the dry-mix method described for compressed powders, using an acetylated lanolin alcohol as the binding agent. A suitable base is made as follows:

No. 2374

Zinc stearate	100
Precipitated calcium carbonate	60
Kaolin	50
Pigments	100

Talc	690
	1000

Titanium dioxide is only included in the base if it is required to prepare specific shades. It is not required for distinctive colours such as blacks, browns, blues, and greens.

Procedure: The pigments are first milled with the calcium carbonate and the mix is added to the remaining ingredients in a sifter mixer. Add about 5·0 per cent of acetylated lanolin alcohol by spraying during the final mixing. Slow spraying and thorough mixing is essential. Use of the lanolin derivative as the binding agent gives the compressed cake a creamy consistency with good adhesive properties when it is subsequently wetted.

Mascara

Mascara or eyelash cosmetic is one of the most popular of eye make-up preparations. It is used either to darken or colour the lashes and to give an illusion of greater density and length. This practice has indeed been followed since the time of the early dynasties. Kohol or Kohl was used by the ladies of ancient Egypt for darkening their lashes when materials used included antimony sulphide, lead sulphide, copper oxide, and magnesium oxide. These were mixed with oil and applied with a stick along the eyebrows and under the eyelids. More recently, mascaras were prepared in 'cake' form and consisted of a mixture of lamp black and soap bound together with a little water, the mixture then being stamped out in moulds.

Cake mascara is still a convenient and popular product and is now prepared using ethanolamine soaps as an emulsifier and binding material. Solutions of ethanolamine soaps are only mildly alkaline and consequently less likely to cause irritation than the more alkaline sodium soaps used originally. The mascara is prepared in moulded bars and applied with a little water and rubbed on to the cake. The moisture forms an emulsion with the soap and the resultant paste or cream is thus transferred to the lashes. The use of ethanolamine soaps enables oils and waxes to be included in the base and the proportion of these materials can be varied to control the required degree of hardness and water repellency. The most satisfactory results are obtained by using natural waxes such as beeswax, carnauba wax, and spermaceti, since these are more compatible with a soap base than hydrocarbon based oils or waxes. The balance between the proportions of waxes and soaps is important to maintain a moisture repellent film which is free from smudging. Of the waxes used, carnauba wax and beeswax increase the glossy appearance of the cake, whereas the soap gives it a dull appearance, and if an excessive proportion of carnauba wax is used the resultant film becomes brittle. Correct balance of materials is, therefore, essential. Hard, brittle, or flaky films can be corrected by including a small proportion of lanolin, or a fatty acid ester.

Silicone oils can also be used to increase resistance to moisture. Glyceryl monostearate is useful as a binding material which also helps to form a smooth paste when the cake is wetted before application. Hydrocolloid gums, and thickeners such as hydroxy methyl cellulose and alginates can also be used in small proportions. These increase the viscosity of the aqueous phase and make the proportions of soap and oils or waxes a little less critical in a formulation. They also help to bind the pigments onto the lashes as the emulsion dries out. Mascara is prepared in several shades with black remaining a favourite colour, followed by brownish-black and blue-black. A range of cosmetic grade carbon blacks is available for use in eye cosmetics. These are manufactured by the 'channel' process and are available in a range of particle sizes. For cosmetic uses the most suitable grades are those of particle size in the 20 micron region. For brownish-black carbon black is blended with a suitable cosmetic brown oxide, and cosmetic ultramarine blue is used with carbon black for a blue-black shade. For greens and yellow-greens use chromium oxide and hydrated chromium oxide for more brilliant blue-greens. Cosmetic grades should always be used, and with all shades particular care should be taken during manufacture to ensure that the pigment is thoroughly 'wetted out' with a suitable liquid component part of the base, before being incorporated in the final mix. Titanium dioxide can be used to reduce the intensity of the colour and this mateiral should also be thoroughly wetted in similar fashion during manufacture. Formulae for cake-mascara are as follows:

No. 2375

A	Carnauba wax	50
	Beeswax	150
	iso-Propyl myristate	30
	Stearic acid	350

	Glyceryl monostearate	50
B	Triethanolamine	150
C	Sodium alignate	70
	Ultramarine blue	120
	Carbon black	20
	Titanium dioxide	10
		1000

Use of perfume is not essential for mascaras, but if required use 0·1–0·15 per cent of a rose type perfume of a non-sensitizing and non-irritant type. As a preservative use:

Methyl parahydroxybenzoate	0·15 per cent
Propyl parahydroxybenzoate	0·01 per cent

It is also advisable to include 0·002 per cent of butylated hydroxyanisole as a precautionary measure to prevent oxidation and rancidity.

Procedure: Melt together the ingredients of A with the exception of the carnauba wax. Use this hot oil/wax mixture to wet-out the pigments already mentioned, and maintain at a temperature of 85°C. Heat the triethalamine to 80°C and add with slow and continuous stirring to the melted oils/wax mixture. Mill until smooth. Finally add the carnauba wax. Remelt and mix well and pour into moulds.

In the above formula the blend of waxes and fatty acid ester provide suitable hardness, and glyceryl monostearate and triethanolamine soap form an emulsion in situ when a wet brush is rubbed on the cake. A disadvantage which is occasionally associated with the use of a cake-type mascara is the variation in the amount of water which is applied by individuals during use. In this respect, a cream-type product which is used with a dry brush is considered by some to be an improvement. The consistency can be varied from a lotion to a thick cream. A suitable formula is as follows:

No. 2376
(Cream)

A	Stearic acid	120
	iso-Propyl myristate	70
	Glyceryl monostearate (self-emulsifying type)	50
B	Glycerin	50
	Triethanolamine	35
	Water (softened or distilled)	575

No. 2376 (*continued*)

C	Pigments	100
		1000

Use the preservatives and antioxidant system already given for cake-mascara.

Procedure: Wet-out the pigments as described for the previous formula using the melted waxes and add B with slow stirring. Mill until smooth.

A cream lotion type can be prepared as follows:

No. 2377
(Cream lotion)

A	Stearic acid	105
	Microcrystalline wax	20
	Carnauba wax	65
	Paraffin wax	55
	Cosmetic mineral oil	15
B	Water (softened or distilled)	492
	Triethanolamine	48
C	Pigments	200
		1000

Use the preservative and antioxidant system given above and carry out the procedure as already described.

A lotion-type cream with good lasting properties can be prepared from an emulsion based on triethanolamine stearate, containing about 5 per cent of polyvinyl alcohol, or a film-forming resin, such as polyvinyl acetate, or a film-forming co-polymer. The resin acts as a pigment dispersant, and also binds the pigments onto the lashes.

No. 2378
(Lotion)

A	Stearic acid	25
	iso-Propyl myristate	40
B	Triethanolamine	10
	Polyvinyl acetate	50
	Water (softened or distilled)	775
C	Pigments	100
		1000

Use the perfume, preservative, and antioxidant system as previously given (No. 2391). The procedure is similar to the previous formula. If there is any tendency for the pigments to settle in this lotion include about 1·0 per cent of hydrated aluminium silicate (bentonite)—prepared as a suspension using a suitable proportion of the water given in B.

Automatic mascara or mascaramatic is conveniently packaged in a tubular type container fitted with an applicator. This consists of a rod fitted either with a grooved end, or a suitably shaped spiral brush, so that the mascara can be applied in a rotating manner to the lashes. The tubular container is designed with a restriction at the opening which effectively removes excess liquid as the applicator is withdrawn for use. The distribution and depth of the grooving of the rod type applicator, and the length and arrangement of the bristles of the spiral brush applicator are most important, and play an important part in the success of the product. Attempts to design applicators using moulded plastic have been unsuccessful, because these do not distribute the mascara evenly.

The liquid mascara used is designed to have a viscosity which can be readily and evenly applied by the specifically designed applicator. It should also be quick drying and moisture proof. The product can be based on a liquid cream mascara of suitable viscosity, prepared with ethanolamine soaps or on a film-forming resin of similar composition to an eye-liner. These products, do, however, tend to dry out over a period of time, particularly when the container is being constantly opened and closed in use. The best results are obtained using a suitable mixture of waxes in a non-aqueous system based on a medium to heavy grade of an isoparaffinic solvent. A low molecular weight polyethylene is used to form a gel with the solvent to act as a colour dispersant and suspending agent, and also to leave a pigmented film on the lashes after evaporation of the solvent. The success of the product, therefore, depends upon the strength of the polyethylene film, the evaporation rate of the solvent, and the proportion and type of waxes used to determine the final deposition on the lashes. A formula is given as follows:

No. 2379

Carnauba wax	17
Beeswax	45
Ozokerite wax	45
Microcrystalline wax	10
Polyethylene	35

No. 2379 (*continued*)

Isoparaffin[1]	580
Pigments	100
	832

[1] Isopar H—Esso petroleum Company Ltd.; Shellsol T—Shell Chemical Company Ltd.

Procedure: Mix the polyethylene with about half the amount of solvent using a propeller type mixer with moderate agitation. Heat to about 100°C and stir. Remove the heat and add to remainder of the solvent and cool rapidly. The purpose of gelling in half the solvent is to enable 'shock' cooling to be obtained by the addition of the remaining portion. Thorough dispersion and rapid cooling are essential for the preparation of a stable gel. Wet-out the pigments in the gel, add the melted waxes and mix thoroughly.

A variation of this formula is given using a firmer gel and a lower content of waxes. First prepare the gel by the method described:

No. 2380

Polyethylene	120
Isoparaffin	880
	1000

The mascara is then made as to the following formula:

No. 2381

Pigment	300
Colloidal silicon dioxide[1]	18
Carnauba wax	160
iso-Propyl myristate	5
Gel	517
	1000

[1] Aerosil—Degussa.

Procedure: Heat the gel to 70–75°C. and wet out the pigment and colloidal silicon dioxide. Add the wax and *iso*-Propyl myristate. As already indicated the properties of the product depend on the proportion of waxes in relation to the strength of gel used.

In addition the choice of pigment is important and samples should be evaluated to assess their functional properties in this type of product. Some samples do not wet out satisfactorily, or do not stay in suspension and as a result may cause smudging after application. Use of colloidal silicon dioxide as indicated in the above formula will help to suspend the pigment and overcome these difficulties. The exact quantity required often varies from shade to shade and can only be determined by experiment.

Eyelash lengtheners are used to make the lashes appear longer and thicker. They are made by including up to 5 per cent of nylon or rayon fibres in a mascara or mascaramatic preparation. The formulations given should be adjusted by a slight reduction in the proportion of gel or wax, to compensate for the effect on the viscosity caused by the addition of the fibres.

Eye make-up removers and eye lotions

Eye make-up removers consist of suitable low viscosity oils such as light mineral oil used either alone or mixed with a fatty acid ester. *iso*-Propyl myristate is the most suitable ester for this purpose because of its low viscosity, but *iso*-Propyl palmitate is also satisfactory.

No. 2382

iso-Propyl myristate	250
Light mineral oil	750
	1000

The product is usually sold in the form of cotton pads which are saturated with the oil. If plastic containers are used these should be carefully checked to make sure that they are not affected by the fatty acid ester. More viscous blends of oils containing emollients such as lecithin, lanolin, or lanolin derivatives, are used as tissue oils to apply around the eye or to the throat, as a treatment for lines and wrinkles. Suitable formulae are as follows:

No. 2383

Olive-oil	100
iso-Propyl myristate (or *iso*-Propyl palmitate)	100
Castor-oil	800
	1000

No. 2384

iso-Propyl lanolate	50
Lanolin	50
Olive-oil	100
iso-Propyl palmitate	200
Castor-oil	600
	1000

In both these products use up to 0·2 per cent of a non-irritant perfume, and include 0·002 per cent of a butylated hydroxyanisole as an antioxidant.

Eye lotions are recommended as a beauty aid to keep the eyes clean and bright, and to relieve strain and irritation. Solutions should be *iso*-tonic with lachrymal secretion and the product should be prepared using an aseptic technique and sterilized either by autoclaving or by filtration and packed in previously sterilized containers. Boric acid and sodium borate are used as mild bactericides and fungicides, and witch hazel water has a mild astringent and cooling effect. Zinc sulphate has a stronger astringent effect, and is also used to relieve inflammation in conjunctivitis. Formulae are as follows:

No. 2385

Boric acid	10
Zinc sulphate	2
Distilled water	988
	1000

No. 2386

Sodium chloride	5
Sodium borate	5
Glycerin	20
Witch hazel	50
Distilled water	920
	1000

No. 2387

Boric acid	10
Zinc sulphate	2
Glycerin	20
Distilled witch hazel	100
Distilled water	868
	1000

Chlorbutol can be included in any of the above lotions as a mild analgesic at a dosage of 0·1 to 0·2 per cent, and as a preservative at 0·5 per cent. Where chlorbutol is not included use 0·1 per cent of methyl parahydroxy benzoate as a preservative.

Cosmetic eye drops based on a solution of methylene blue are antiseptic and impart an attractive blue colour to the whites of the eyes.

No. 2388

Methylene blue	0·02
Sodium chloride	9·00
Distilled water	990·98
	1000·00
Methyl parahydroxybenzoate	0·1 per cent

CHAPTER TWELVE

The Skin–Preparations for the Face and Hands

The use of salves and unguents for preserving and beautifying the skin dates back to earliest antiquity, when they were generally prepared by digesting aromatic gum resins, roots, flowers, etc. with fats and oils. The first notable change in the constitution of these cosmetics appears to have been made about the second century A.D., when the Greek physician Galen (who practised in Rome) added water to his salves, and by so doing laid the foundation of our modern cold cream and cleansing cream. This is probably the only skin application which has stood the test of time successfully, and although there is little doubt that it has been much modified, the principle remains the same. As recently as twenty years ago formulation of cosmetic products was largely an empirical business because very little was known about the functions of the products or of the raw materials used. Nowadays large sums of money are spent every year on research dealing with the effect of various materials and preparations on the skin, with the result that the cosmetic chemist is in a much better position, not only to assess the usefulness of new and existing raw materials, but he is also able to avoid those materials which are likely to cause skin damage. Although all cosmetic preparations are intended for application to the body, it is those which are applied to the facial skin which often demand greater attention. These can be classified into two main groups–make-up preparations which provide decoration; and preparations used for the care of the skin. The latter are in the main intended to fulfil a particularly useful purpose; to supplement the natural functions of the skin, improve its health and condition and help to prolong its youthful appearance. It is useful, therefore, to have a clear conçeption of the skin, its functions and the conditions which exist at and near its surface.

Structure of the skin

The layers of the epidermis are shown in Figs. 12.1 and 12.2. These are the stratum germanativum, or growing cells, the

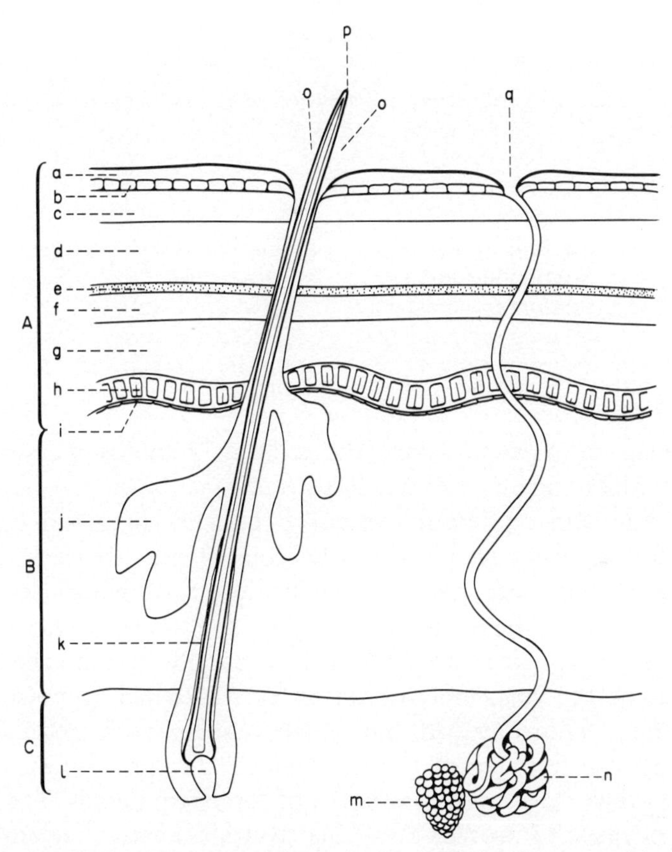

Fig. 12.1 Schematic presentation of cross-section of human skin key (A) epidermis; (B) dermis; (C) subdermis; (*a*) layer of skin-surface fat; (*b*) cornified cells; (*c*) corneum; (*d*) stratum lucidum; (*e*) barrier of skin absorption; (*f*) stratum granulosum; (*g*) malpighian layer or stratum mucosum; (*h*) stratum germanativum; (*i*) basal layer; (*j*) sebaceous gland; (*k*) hair sheath (internal); (*l*) hair papilla; (*m*) fat; (*n*) eccrine sweat gland; (*o*) space filled with grease and air; (*p*) hair; (*q*) sweat-duct pore.

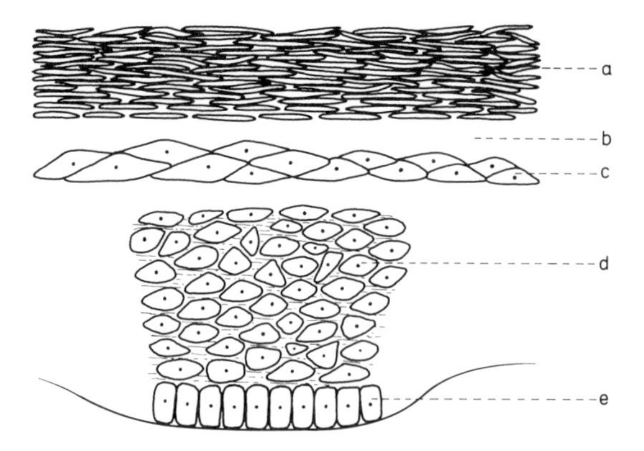

Fig. 12.2 Diagram to illustrate distribution of layers of epidermis showing horny zone A–C, and germinative zone D, E.

Stratum corneum (a)	(horny layer)
Stratum lucidum (b)	(clear layer)
Stratum granulosum (c)	(granular layer)
Stratum spinosum (d)	(prickle cell layer)
Stratum basale (e)	(basal cell layer)

Malpighian or pigment layer, the stratum granulosum, stratum lucidum, the corneum, and finally the cornified layer.

The true skin or dermis contains a network of blood vessels, hair follicles, sweat glands and sebaceous glands. Beneath this is the subcutaneous fatty tissue into which project the lower ends of the hair bulbs. The cells which form the lower layers travel outwards during their life cycle and eventually become the dead flat cells of the corneum, the outer layer of which is constantly being shed. This outward movement, ending with exfoliation, underlines the function of the skin, which is to provide a natural barrier against the inward movement of foreign materials. Thus the skin also serves to protect the living tissues beneath it against the elements, the absorption of toxic substances and the invasion of micro-organisms.

The sebaceous glands situated in the dermis are most abundant wherever there are hairs, and consist of small saccules communicating with a common duct opening into the neck of the hair follicle or sometimes directly onto the skin surface. They secrete sebum, the oily fluid which lubricates the hair shaft and skin. Sebum is of fatty composition with some hydrophilic properties, and is normally present on the skin and hair in a semi-solid state. It contains free and combined fatty acids, free cholesterol, and cholesterol esters, waxes, triglycerides, and the hydrocarbon squalene. In addition, there are also present small amounts of dihydrocholesterol and several other sterol-like substances, including pro-vitamin D.

There are two types of sweat glands; the eccrine glands, which are present over most of the body and secrete the clear aqueous liquid which helps to regulate body temperature, and the apocrine glands which are present only in certain regions (notably under the arms) and secrete a whitish turbid fluid, the function of which is unknown. The eccrine glands extend downwards from the surface, finally in the form of a coil, each gland being a separate unit; and they may be stimulated either by direct heat, by axon reflex, or reflexes through the central nervous system. Eccrine sweat is composed of 98–99 per cent of water containing common salt, together with small quantities of urea, glucose, lactic acid, ammonium salts, amino-acids, and other minor ingredients. It generally has a pH value between 3·8 and 5·6.

The apocrine glands, which are larger than the eccrine glands, generally open into a hair follicle and are only stimulated by the central nervous system. The composition of apocrine sweat is still undetermined, but it has been reported to contain iron, proteins, reducing sugars, ammonia and cholesterol. Even when no visible sweating is taking place, there is a continuous loss of water from the skin surface in the form of vapour. This is sometimes called insensible perspiration, as distinct from the visible or sensible type. This invisible moisture loss originates from the eccrine glands, and is partly a continuous loss of water through the skin itself, in which vaporization occurs before it reaches the surface.

The normal skin surface, then, is covered with the secretion products of the sebaceous, eccrine, and apocrine glands, which help to maintain the moisture balance of the underlying tissues. The cells and flakes of the cornified epithelium are also present. These are the conditions which concern the cosmetic chemist and influence the creation of functional cosmetic products. In addition, special preparations are designed to suit particular skin conditions.

Emulsion systems

It is true to say that all cosmetic creams and lotions are derived from the study of cleansing creams and vanishing creams. These are essentially emulsion systems, in one case based on beeswax acid soaps and in the other on stearate soaps. The soaps thus formed act as an emulsifying agent which is used to disperse a given material as globules in another material, that is, when the two materials are themselves immiscible such as oil and water. The emulsifying agent is thus distributed between the oil and water at the so-called interface. The dispersed liquid which is present in the form of globules will be the internal phase, and the liquid which is preferentially wetted by the emulsifying agent will form the external or continuous phase.

Water is an important ingredient of skin creams and lotions because to a large extent it is the moisture content of the skin which controls its healthy appearance and undue loss of moisture which takes place when the skin is exposed during severe weather conditions causes dehydration accompanied by shrinking. The absorption or retention of moisture by the epidermal tissue is to some extent desirable in order to maintain the skin surface in a soft and pliable condition. It is for this reason that many cosmetic products take the form of emulsified preparations containing some fatty or oily material.

Ideally emulsions should be formulated to give preparations which are neither too dry nor too greasy during application. They should spread easily and evenly and in some cases they should also have adhesive properties and provide coverage to the skin. They should also be attractive to the eye, and to this end it is desirable that the emulsion should have a glossy appearance. In addition many emulsions are tinted to colour the skin and the selection of colours, preparation of desirable shades, matching and control of coloured preparations are important factors which must be taken into account.

In the formulation of modern cosmetic creams and lotions the ingredients of the two phases are selected to perform an adequate function and the correct choice of emulsifying agent is equally important in order to obtain a stable system of the consistency and texture required.

Emulsifying materials

Emulsifying materials belong to the group known as surface-active agents because they concentrate at the interface. Their behaviour is due to the fact that they possess both hydrophilic and lipophilic properties and can be held at an oil-water interface in a particular way. The lipophilic portion will become orientated towards the oil and the hydrophilic portion towards the water. The attraction for the emulsifying agent of one liquid over that of another, determines the type of emulsion. Thus an 'oil-in-water' emulsion sometimes referred to as an 'obverse' emulsion denotes that the oil is dispersed as the internal phase in water; and a 'water-in-oil' emulsion also referred to as a 'reverse' emulsion denotes that water is dispersed in the oil which is the external phase.

The conception of molecular orientation and the formation of a monolayer at the emulsion interface has resulted in the development of a large range of surface-active materials, many of which are useful emulsifying agents and are used extensively in cosmetic emulsions. Surface-active agents may be classified into the following groups:

(*a*) *Anionic emulgents*

These are so termed because they are 'anion-active', the surface active portion of the molecule being contained in the negatively charged anion. This group includes soaps of long chain fatty acids, soaps of certain ammonium derivatives (e.g. triethanolamine), and organic sulphates and sulphonates. Sodium stearate, for example, ionizes in the following way:

$$C_{17}H_{35}COONa \rightleftharpoons C_{17}H_{35}COO^{-}-Na^{+}$$

If the hydrocarbon chain is represented by a tail and the polar radical by a head, the anionic emulsifier may be represented as in Fig. 12.3. The anion active radical is oil-soluble, and the polar portion water-soluble. Therefore the molecule will become oriented at an oil/water interface as in Fig. 12.4.

(*b*) *Cationic emulgents*

These are so termed because the surface active portion of the molecular is contained in the positively charged cation. A typical

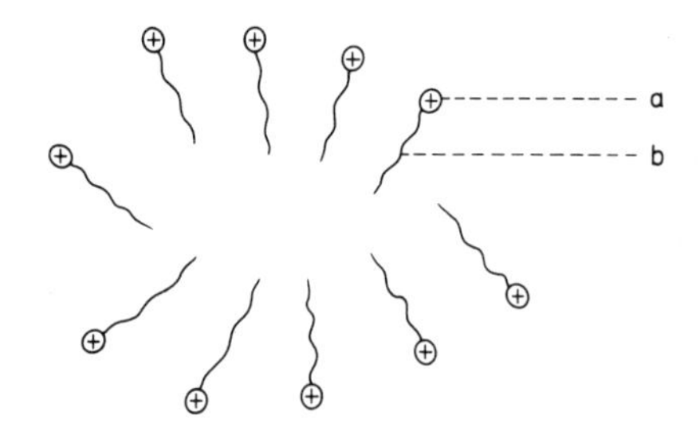

Fig. 12.3 (*a*) Positively charged polar heads; (*b*) negatively charged hydrocarbon tails.

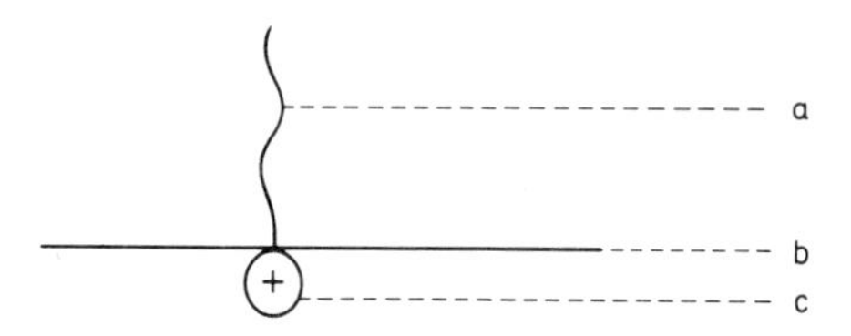

Fig. 12.4 Distribution of anionic emulsifier at emulsion interface (*b*), showing water-soluble anion (*c*) and oil soluble hydrocarbon tail (*a*).

example of this group of compounds is cetyl trimethyl ammonium bromide, which ionizes as in Fig. 12.5.

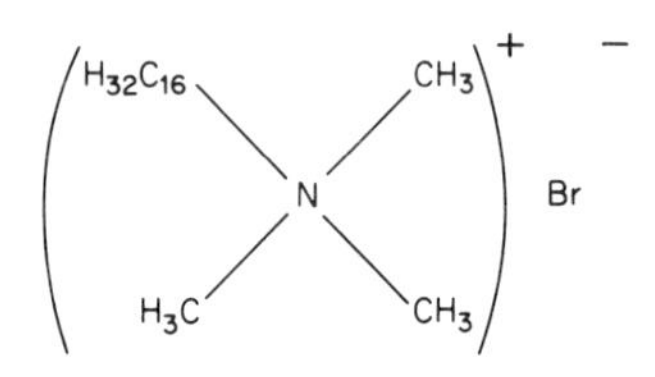

The cationic activity of these compounds is attributed to the basic nitrogen atom; thus amines and their salts, quaternary ammonium compounds, and hecterocyclic nitrogen bases and their salts will fall into this category. As in the case of the anionic emulgents, the lipophilic portion or cation active radical becomes orientated towards the oil, and the hydrophilic portion towards the water.

(*c*) *Non-ionic emulgents*

These compounds do not ionize in solution. They are available as groups of compounds which fall within certain definable categories as follows:

1. Products marketed as 'spans' which are based on esters of fatty acids and sorbitol. These are lipophilic or oil loving; they are soluble in oils and organic solvents and generally insoluble in water.
2. The polyoxyethylene derivatives of the sorbitol esters, which are hydrophilic or water loving, are marketed as 'tweens'. These are generally soluble or dispersible in water.
3. Polyoxyethylene derivatives of fatty acids, marketed as 'myrj' emulsifiers, are also hydrophilic and generally soluble or dispersible in water.
4. Polyoxyethylene lauryl ethers, known commercially as 'brij' emulsifiers, also have some hydrophilic properties.

Further comprehensive groups of emulgent materials are available as esters and polyoxyethylene reaction products of lanolin and beeswax. The reaction products modify the characteristic properties of lanolin and beeswax, altering the appearance and texture and producing materials with interesting solubility and surface-active characteristics.

A considerable number of raw materials which can be classified within these groupings are now available from manufacturers and this has led to new opportunities in formulation of cosmetic products based on emulsion technology. Because of the large number available it is however often difficult to select the most suitable materials for a particular application. The most useful method of classification was worked out by William C. Griffin of the Atlas Powder Company, who devised the H.L.B. System, a title derived from the term 'Hydrophilic-Lipophilic Balance'. The system is used to indicate the behaviour of an emulsifying material according to the proportion of hydrophilic and lipophilic groups present in the molecule. A material with a low H.L.B. value tends to be predominantly lipophilic in character and forms water-in-oil emulsions, and a substance of high H.L.B. value, with predominantly hydrophilic properties, tends to produce oil-in-water emulsions. These values, however, only give an indication of the behaviour characteristics of the materials and are not necessarily a measure of the emulsion stability.

In certain emulsion systems a hydrophilic emulsifier of high H.L.B. value is combined with a lipophilic emulsifier of low H.L.B. value, to give a balanced distribution at the interface of the emulsion. In this case the final H.L.B. value of the system indicates the effect of the total of the two emulsifiers, for example:

1 part of emulsifier of H.L.B. value	15 = 15
4 parts of emulsifier of H.L.B. value	5 = 20
5 parts of total emulsifier of H.L.B. value	35
Indicating an effective H.L.B. value of	35 ÷ 5 = 7·0

To use this system to advantage it is necessary to know the relationship of an H.L.B. value and its function in relation to the raw materials used as ingredients of an emulsion system. This is achieved by preparing emulsions with individual raw materials to determine the H.L.B. value necessary to produce a stable emulsion and the result indicates the 'required H.L.B.' of the particular raw material. Because this is a rather laborious procedure, several other techniques have been suggested for determination of required H.L.B. For example, the relationship of an oil to its spreading coefficient or the dielectric constant. It is also accepted that the method used to prepare an emulsion also has a relationship to the required H.L.B.

At the present time work in this area of investigation is by no means complete, and consequently only a comparatively small amount of data is available regarding the required H.L.B. values. On the other hand a large amount of data is available of the H.L.B. values of surfactant materials because these values can be determined by calculations based on either their analytical or composition data.

In the case of surfactants based on a fatty acid ester, for example, the H.L.B. value is calculated from the saponification number of the ester and the acid number of the fatty acid according to the following formula:

$$\text{H.L.B.} = 20\,(1 - S/A)$$

where S = saponification number of the ester
and A = acid number of the acid.

A different formula is used for calculation of H.L.B. values of materials which do not give clear saponification.

Further information giving details of these methods of calculation is given in Griffin's original reports (*Journal of the Society of Cosmetic Chemists*, **1954**, 5, p. 249). Tables of calculated and determined H.L.B. values are also given in this work and details of

H.L.B. values of specific emulsifiers can always be obtained from the manufacturers of the materials if the reader wishes to formulate on this basis. Formulation using the H.L.B. systems is therefore a method of balancing H.L.B. values of surfactants used as the emulsifier, with the required H.L.B. of the material to be emulsified. If, for example, an oil-in-water emulsion is to be made the oil being held within the water phase, the hydrophilic-lipophilic balance is predominantly hydrophilic and the emulsifiers will require to be hydrophilic in behaviour rather than lipophilic. This is represented by high H.L.B. values within the range of 10 to 16 or 17. The required H.L.B. of the oil or oil phase will also be of the same order. Conversely to prepare a water-in-oil emulsion when the water is held within the oil phase, the hydrophilic-lipophilic balance will be predominantly lipophilic.

The emulsifiers are not required to have sufficient surface activity towards water and those with low H.L.B. values are used. Under these conditions the required H.L.B. of the oil phase is also lower than that required for an oil-in-water system. For this reason the oils, fats and waxes used in an emulsion have two required H.L.B. values, that of a water-in-oil emulsion being lower than that required for an oil-in-water system. Data concerning required H.L.B. values is also published by Griffin, these being based on the H.L.B. values of surface-active materials which are required to give optimum results in the particular type of emulsion required.

The H.L.B. system of formulation is at the present time a useful method towards the preparation of stable emulsion systems and because the number of emulsifiers available is so large and apparently still increasing, the system provides a means of selecting particular emulsifiers likely to be most suitable to perform a particular function. Further work is to be expected, however, regarding the determination of H.L.B. values and particularly required H.L.B. values, where there may be a relationship between the required H.L.B. value and chemical composition. As an example it is required to prepare the following oil-in-water emulsion:

Oil	25	(required H.L.B. = 10)
Petroleum jelly	5	(required H.L.B. = 10)
Beeswax	12	(required H.L.B. = 10)
Emulsifier	5	
Water	53	

Calculate the required H.L.B. of each material in the oil phase as a percentage of the total oil phase i.e. 42 parts.

								Required H.L.B.		
Oil	=	25/42	=	62%	x	10	=	6·2
Petroleum jelly	=	5/42	=	12%	x	10	=	1·2
Beeswax	=	12/42	=	27%	x	10	=	2·7
Total required H.L.B.								10·1

Since the required H.L.B. is 10·1 then emulsifiers must be used which have H.L.B. values near to this value i.e. from about 9 to 11. As the emulsifiers use a blend of a predominantly hydrophilic emulsifier, e.g. polyoxyethylene sorbitan monostearate of H.L.B. value 15 with a smaller proportion of a lipophilic emulsifier—sorbitan monostearate of H.L.B. value 4·7. If these are used in the ratio 40:60 the combined H.L.B. value equals

	40/100	x	4·7	=	1·8
plus	60/100	x	15	=	9·0
					10·8

The H.L.B. value 10·8 is near to the required H.L.B. indicating that a suitable choice of emulsifier for the first experiment is as follows:

Sorbitan monostearate	2 parts
Polyoxyethylene sorbitan monostearate	3 parts
total	5 parts

Experiments are then continued using blends of the two emulsifiers over a narrow range of H.L.B. values to determine the most effective blend which will obtain a stable emulsion.

One of the most useful applications of the H.L.B. system is in the preparation of transparent emulsions utilizing surfactants of high H.L.B. value to obtain so-called solubilization of oils. When a surfactant is added to an oil the interfacial tension is gradually reduced as the surface activity is increased until the oil is completely dispersed in an aqueous medium forming a transparent emulsion. This is achieved by using surface active materials with great affinity to water indicated by the high H.L.B. value of 16. In the same way as oils and waxes have varying required H.L.B. values, it follows that essential oils or mixed blends used in perfumery compounds have differing required H.L.B.s and different proportions of oil to surfactant are required to obtain solubilization. Clear solutions are obtained when the proportion of oil to surfactant provides sufficient surface-active effect.

Polyoxyethylene sorbitan monolaurate (Tween 20) H.L.B. value 16·7, and polyoxyethylene sorbitan monooleate (Tween 80) H.L.B. value 15 are commonly used for this purpose and solubilization is generally obtained by using 1 part of oil to 5 to 6 parts of the solubilizer in 100 parts of total solution. Higher proportions of solubilizer may be required according to the solubility characteristics of the oil(s). These two materials (Tween 20 and Tween 80) are unstable in conditions of extreme acidity or alkalinity. In these cases polyoxyethylene lauryl ethers are more suitable (Brij emulsifiers).

Certain other materials are available which have suitable characteristics for solubilization for example a range of polyoxyethylene oleyl ethers (Croda Ltd.). Some manufacturers offer materials as specific 'solubilizers', which may be either single chemicals or blends of surface-active materials (Bush Boake Allen). As these are all surface-active it sometimes happens that if a high proportion is used to obtain clarity of a particular oil or perfume blend the resulting solution will show undesirable frothing. It is, therefore, advisable to experiment with several materials before finalizing a formulation. The author has found that blends of surface-active materials are more effective as a result of combined surface-activity and consequently can be used in a lower proportion with consequent minimum frothing of the resultant solution.

In addition to their use to prepare transparent emulsions in water, solubilizing materials have a useful function when formulating perfumery products where a low alcohol content is desirable such as after-shave lotions containing 40-50 per cent of alcohol or colognes containing 70-80 per cent of alcohol. The solubility characteristics of a perfume compound in alcohol can be increased by the addition of the solubilizer and the amount required, which should be kept to a minimum, is determined by experiment.

After-shave lotion

No. 2389

Perfume compound	40

No. 2389 (*continued*)

Alcohol	400
Solubilizer	a sufficient quantity
Water (distilled)	to 1000

Mix the perfume compound with the alcohol. Mix well and add the water with continuous stirring, when a cloudy dispersion will be obtained. Add the solubilizer in small separate additions mixing well after each addition and continue the additions until clarity of solution is obtained. This method is used to determine the minimum amount of solubilizer required for the particular perfume compound used in a formulation. It is necessary to repeat the experimental procedure for each individual perfume compound as it is used in a particular concentration of alcohol. By using this method higher concentrations of perfume compound can be used in products which normally contain comparatively low proportions of alcohol.

Formulation

From the foregoing it will be seen that the preparation of cosmetic creams, milks and lotions depends firstly upon the selection of ingredients which will provide the oil phase and aqueous phase of an emulsion system, these ingredients being selected for their particular properties, and their effect during and after application to the skin and secondly a suitable emulsifier must be selected to determine the type of emulsion system to be used. When these requirements have been decided the final composition of the product can be modified for consistency and effect on the skin by adjusting the separate ingredients of the two phases of the emulsion.

Many satisfactory products are made with anionic emulsifiers, but the versatile non-ionic emulsifiers are used particularly to prepare lotions and creams of pourable viscosity. The H.L.B. system can be applied when formulating as an aid to providing emulsion stability.

The four main types of cosmetic preparations for the facial skin are as follows:

(*a*) cleansing preparations,
(*b*) those which are intended to protect or nourish the skin,
(*c*) preparations to tone and stimulate the skin including those which are designed as corrective or treatment preparations,

(*d*) make-up preparations designed to embellish or enhance the appearance.

Liquefying cleansing creams

Cleansing of the skin normally infers removal of make-up and other forms of skin soil. Soil normally found on the skin has already been discussed and consists of a mixture of the excretions of the sebaceous glands and the sweat glands. This is readily removed either by oils on their own or combined in a suitable emulsion system. Such an emulsion is provided when the skin is washed with soap and water, but this method often removes all the oils from the skin surface and does not leave any protective film. In hard water areas particularly, calcium soaps remain as a film on the surface of the skin. Surface-active agents which do not form insoluble lime salts can be used as cleansing materials, but they also degrease the skin and leave it in a dry and sensitive condition. Preparations for cleansing are in the main based on a mineral oil of good quality and suitable viscosity. A liquefying cleansing cream is made by mixing such an oil with petrolatum and suitable waxes. The products which do not contain water are white and translucent solids, have a thixotropic effect when applied to the skin, and have excellent cleansing properties. It is essential that the creams should liquefy readily in use, and they can be formulated to contain small proportions of other fatty materials such as vegetable oils, fatty acid esters, or lanolin. Although not as popular as emulsified products, their effectiveness is indicated by their long acceptance as theatrical cleansing creams. The following is a typical formula for a cleansing cream of this type:

No. 2390

Mineral oil (cosmetic quality)	800
Petroleum jelly	150
Ozokerite	50
	1000
Perfume	0·3–0·5 per cent
Propyl parahydroxybenzoate	0·05 per cent

Procedure: Heat together until uniform. Add the perfume and fill whilst still warm.

This product melts easily to the touch and has an attractive transparency. The proportion of ozokerite given in the formula is just sufficient to solidify the mixture of mineral oil and petroleum jelly. If preferred it can be increased by 1 or 2 per cent. A small quantity of a suitable oil-soluble colour is often added. In these preparations the choice of oily material to be used as the cleansing vehicle is all-important. An oil of high viscosity gives a greasy, sticky film on the skin which is difficult to remove. A similar result can be caused by an incorrectly proportioned mixture of petroleum jelly and a light mineral oil. The correct balance of oils and waxes is also essential so that the product remains firm during various temperature conditions, but liquefies readily when it is applied to the skin. The best results are obtained when ozokerite is used to give the required setting point. Ozokerite has a melting point in the range 60–75°C. and this is sufficiently high to give the mixture a thixotropic effect.

Microcrystalline waxes are also suitable for these products. These differ from paraffin waxes in their structure and are more flexible in use. They are also available in a range of melting points, which is usually higher than that of ordinary paraffin wax. A suitable product can be prepared as follows:

No. 2391

Microcrystalline wax (m.p. 140–150°C)	250
iso-Propyl palmitate	650
Mineral oil (cosmetic quality)	100
	1000
Perfume	0·3–0·5 per cent
Propyl parahydroxybenzoate	0·05 per cent

Procedure: Heat the ingredients together and mix until uniform. Add the preservative and perfume and mix. Fill warm.

Cold cream

Cold cream is an emulsion in which the proportion of fatty and oily material predominates, although when it is applied to the skin a cooling effect is produced due to slow evaporation of the water contained in the emulsion. At one time the bases in general use were almond oil, lanolin, and white wax, but the oil has now been largely replaced by mineral oil, which does not become rancid on keeping like the vegetable oils. Spermaceti is sometimes used instead of, or in combination with, white beeswax, and borax is used to aid emulsification. As a pharmaceutical ingredient, cold cream appeared in the 1949 issue of The British Pharmaceutical Codex, where it was known as Ointment of Rose Water. The formula of this traditional preparation is as follows:

No. 2392

Almond oil	610
White beeswax	180
Borax	10
Rose water	200
	1000
Oil of rose	1

Procedure: Melt the beeswax in the almond oil, and add, with constant stirring, the borax previously dissolved in the rose water; add the oil of rose, and continue to stir until cold.

The method of manufacture of cold creams is similar to that given in The Pharmaceutical Codex and follows the general procedure used for the preparation of emulsified products. The waxes are first melted in the oil with other oil soluble ingredients and heated to a temperature of 75–80°C. The aqueous phase containing borax and any other water-soluble ingredients is heated to a similar temperature and added to the oil/fat mixture with slow stirring. The perfume is added when cool, and the cream filled warm. For large scale production, filling can be carried out in two stages. The jars are first filled almost to the top of the jar when the cream is hot, and then allowed to set. The final surface of the cream is obtained by 'finishing' with a sufficient quantity of cream which is just pourable. In this way a brilliant white, smooth and shiny surface is obtained.

The traditional cold cream of the beeswax–borax type is represented by the following formula:

No. 2393

Mineral oil (cosmetic quality)	450
Beeswax	160
Borax	10

No. 2393 (*continued*)

Water (softened or distilled)	380
	1000
Perfume	0·3–0·5 per cent

As a preservative for all cold creams and cleansing creams use:

Methyl parahydroxybenzoate	0·12 per cent
Propyl parahydroxybenzoate	0·02 per cent

Procedure: heat together the mineral oil and beeswax to a temperature of 75°C. In a separate container dissolve the borax in the water and heat to 75°C, and add slowly, with continuous stirring, to the oil/wax mixture. Cool with stirring, adding the perfume when the temperature has fallen to about 35°C.

The consistency and texture of the product can be varied by replacing part of the mineral oil with white petroleum jelly. Up to 15 per cent of the petroleum jelly can be used in the product. The basic formula can also be modified by replacing a proportion of the mineral oil with a suitable fatty acid ester. Use of *iso*-Propyl myristate in the oil phase increases the thixotropic properties and allows the cream to spread easily, and on account of the lower viscosity of the oil phase, improves the cleansing action of the product. A typical modern cold cream is prepared as follows:

No. 2394

Mineral oil (cosmetic quality)	400
Beeswax	160
iso-Propyl myristate	50
Petroleum jelly	50
Borax	10
Water (softened or distilled)	330
	1000
Perfume	0·3–0·5 per cent
Methyl parahydroxybenzoate	0·1 per cent
Propyl parahydroxybenzoate	0·02 per cent

Cleansing creams

The function of a cold cream remains primarily as a cleansing agent, the emulsified product being used to assist penetration of oil and any fatty material to make the finished product more effective and easier to use than those which are not emulsified. It follows that cleansing creams are basically cold creams slightly modified in some cases to be somewhat softer, to melt quickly and thereby facilitate ease of application and spreading. The creams are generally applied liberally night and morning to the neck, cheeks, chin, nose, and across the forehead, to remove make-up or normal skin surface soil. After gently massaging over the skin, the surplus cream is removed with tissues. When the skin is dry or flaky the residual film of oil is sometimes allowed to remain on the skin as a moisturizing base for make-up. In other cases the oil is removed either with a skin freshener or toning lotion, or with soap and water. Effective cleansing creams can be made by modifying the previous formulae for cold creams as follows:

No. 2395

Mineral oil (cosmetic quality)	400
Beeswax	160
Alkyl myristate[1]	50
Borax	10
Water (softened or distilled)	380
	1000

Perfume 0·3–0·5 per cent

[1] Bush Boake Allen.

No. 2396

Mineral oil (cosmetic quality)	450
iso-Propyl myristate	110
Spermaceti	100
Beeswax	120
Borax	5
Water (softened or distilled)	215
	1000

Perfume 0·3–0·5 per cent

The formulae mentioned will all give satisfactory cleansing preparations, but modifications of these basic formulae can be prepared by varying the proportions of oil and fatty materials and using alternative emulsion systems.

The following formula contains a high proportion of oil together with lanolin. The cream is an effective cleansing agent and a thin residual film can be left on the skin when extra emolliency is required.

No. 2397

Mineral oil (cosmetic quality)	500
Stearic acid	100
Lanolin	50
Triethanolamine	12·5
Water (softened or distilled)	337·5
	1000·0

Perfume 0·3–0·5 per cent

Satisfactory cleansing creams can be prepared to the following formulae:

No. 2398

Mineral oil (cosmetic quality)	250
Beeswax	120
Paraffin wax	50
Petroleum jelly	100
Lanolin	20
iso-Propyl myristate	100
Borax	10
Water (softened or distilled)	350
	1000

Perfume 0·3–0·5 per cent

In the following formula self-emulsifying glyceryl monostearate is used as the primary emulsifier and sorbitan sesquioleate as an emulsion stabilizer. A lower proportion of beeswax is used in this system with a comparatively high proportion of petroleum jelly. A lighter cream is obtained by adjusting the proportions of mineral oil and petroleum jelly.

No. 2399

Mineral oil (cosmetic quality)	100
Lanolin	20
Petroleum jelly	120
iso-Propyl myristate	150

Beeswax	50
Glyceryl monostearate (S.E.)	100
Sorbitan sesquioleate	30
Glycerin	25
Water (softened or distilled)	405
	1000
Perfume	0·3 per cent

The following formulae give effective cleansing preparations and are of interest:

	No. 2400	No. 2401
Mineral oil (cosmetic quality)	300	220
Beeswax	120	120
Spermaceti	100	100
iso-Propyl myristate	20	–
Petroleum jelly	50	100
Oil-soluble lanolin alcohols	50	50
Paraffin wax	–	20
iso-Propyl palmitate	50	50
Borax	10	10
Water (softened or distilled)	300	330
	1000	1000
Perfume	0·3–0·5 per cent	

Some manufacturers continue to use vegetable oils in cleansing preparations. This is either on account of aesthetic reasons, since they are undoubtedly more emollient than mineral oils, or because of availability of a vegetable oil as a local source of supply. Oils used for this purpose are soya oil (or soya bean oil), sunflower oil, sesame oil (sometimes referred to as (teel oil), maize oil (corn oil), and arachis oil (or nut oil). These materials are often preferred to the oils of almond and olive mainly since they are available at a lower cost. Known as fixed oils they consist chiefly of fatty acid esters of glycerol which may be triglycerides or mixed triglycerides when more than one acid is present. They are all liable to become rancid on storage due to hydrolysis or oxidation reactions and any formulation containing any of these oils should always include an antioxidant. Use the preservatives indicated for use in cold creams and cleansing creams and include: butylated hydroxyanisole 0·002 per cent as an antioxidant.

Vegetable cleansing creams can be made by replacing mineral oil with a vegetable oil in the formulae already given or as follows:

No. 2402

Vegetable oil	550
Beeswax	145
Borax	10
Water (softened or distilled)	295
	1000
Perfume	0·5 per cent

No. 2403

Vegetable oil	600
Spermaceti	20
Beeswax	150
Borax	5
Water (softened or distilled)	225
	1000
Perfume	0·5 per cent

Zinc oxide can be included in this type of cream to give a white appearance. In such cases it must be milled to a smooth cream with part of the oil before mixing with the bulk.

No. 2404

Vegetable oil	560
Beeswax	180
Lanolin anhydrous	20
Borax	10
Zinc oxide	20
Water (softened or distilled)	210
	1000
Perfume	0·5 per cent

Cleansing lotions

Cleansing milks or lotions, as their names imply, are used as an alternative cleansing agent to cleansing creams. Their popularity has increased in recent years probably because they are convenient to use, particularly during daytime when complete removal of make-up and skin soil is not required. Normally the milks are applied with a tissue or with cotton wool; hence they are easier to use than creams, and the quantity can be controlled more readily. The products can be considered as diluted forms of oil-in-water type cleansing creams varying in consistency from thin pourable milks to thick viscous lotions. They generally contain less oily material than cleansing creams and consequently do not leave a thick film of oil or grease on the surface of the skin after use. The residual film on the skin is indeed often sufficient to act as the base for make-up and with this use in mind the products are also referred to as complexion milks or beauty milks.

For cleansing or beauty milks use:

Methyl parahydroxybenzoate	0·15 per cent
Propyl parahydroxybenzoate	0·02 per cent

Formulae are as follows:

No. 2405

Mineral oil (cosmetic quality)	250
iso-Propyl myristate	150
Diethylene glycol monostearate	36
Non-ionic emulsifier	24
Cetyl alcohol	10
Water (softened or distilled)	530
	1000
Perfume	0·3–0·5 per cent

No. 2406

Mineral oil (cosmetic quality)	400
Abracol L.D.S.[1]	50
Cetyl alcohol	10
Water (softened or distilled)	540
	1000
Perfume	0·3–0·5 per cent

[1] Bush Boake Allen.

No. 2407

Mineral oil (cosmetic quality)	100
Diethylene glycol monostearate	47·5
Non-ionic emulsifier[1]	32
Spermaceti	50

[1] Oxyethylene ether of cetyl alcohol type.

No. 2407 *(continued)*

Water (softened or distilled)	770·5
	1000·0
Perfume	0·3–0·5 per cent

This formula gives a thin milk product which is an effective cleanser, although it only contains 10 per cent of mineral oil. The stability of this emulsion can be improved by homogenization—although this is not essential.

No. 2408

iso-Propyl palmitate	75
Propylene glycol monostearate (self-emulsifying type)	60
Polyethylene glycol 200 monostearate	50
Triethanolamine	5
Water (softened or distilled)	810
	1000
Perfume	0·3–0·5 per cent

No. 2409

iso-Propyl palmitate	50
Mineral oil (cosmetic quality)	100
Non-ionic emulsifier[1]	30
Glycerol	50
Water (softened or distilled)	770
	1000
Perfume	0·3–0·5 per cent

[1] Emulsene 1220 type—Bush Boake Allen.

Foundation creams

Foundation creams are applied to the skin to provide a smooth emollient base before the application of face powder and other make-up preparations. They help the powder to adhere to the skin and also act as a skin protective to prevent damaging effects caused by environmental factors such as sun or wind. It follows that they should be so designed to leave a semi-occlusive residual film on the skin which is neither too greasy nor too drying.

Modifications can also be made to formulate creams specially suitable for dry or greasy skin conditions and because a healthy condition of the skin depends to some extent upon the water or moisture content, glycols are included as humectants in moisturizing creams.

The original foundation creams were known as vanishing cream–so called because they disappear when rubbed into the skin. They are based on stearic acid which is partially saponified with an alkali, when the bulk of the acid is emulsified with the soap thus formed. The main constituent is, of course, water. Made in this fashion the cream leaves a dry but tacky residual film which also has a drying effect on the skin. It is for this reason that creams based on stearic acid soaps still find favour for use with greasy skin conditions and particular in hot climates which cause perspiration on the face and where more emollient creams are not entirely suitable.

Only the finest quantity triple-pressed stearic acid of a melting point about 56°C should be used for these products. Creams made by saponification of stearic acid vary according to the proportion of acid used. The total proportion should not exceed 25 per cent, and best results are obtained using from 16 to 20 per cent. The consistency and texture of the cream also depends upon the amount of acid saponified by alkali and also on the nature of the alkali used. For example, creams made with soda are harder than those made with potash, that is assuming the ratio between their molecular weights are taken into account, i.e.

$$10 \text{ of NaOH} = 14 \text{ of KOH}.$$

Again the different percentages used of the same alkali in two creams will give different results. This is well illustrated by reducing the quantity of KOH from 14 grams to 10 grams in the first formula given below. Instead of getting a satisfactory product a very hard cream will be produced.

In order to arrive at the approximate quantity of alkali that will be required, the following points must be observed. Stearic acid, $C_{17}H_{35}COOH$, has a molecular weight of 284. (Commercial products are never absolutely pure and the presence of other fatty acids gives a more accurate working figure of about 270.) Supposing 1 kilo of cream is being prepared containing 200 grams of fatty acid, and it is desired to saponify 28 per cent of it–i.e. 56 grams–the quantity of alkali required will be calculated thus:

$$\frac{\text{(molecular weight of alkali)} \times 56 \times 100}{284 \times \text{(no. of molecules as per equation)} \times \text{(\% purity of alkali)}}$$

To take a concrete example—caustic potash reacts with stearic acid as follows:

$$C_{17}H_{35}COOH = KOH = C_{17}H_{35}COOK + H_2O.$$

The molecular weight of KOH is 56, and commercial samples average about 80 per cent strength.

The above calculation will therefore read:

$$\frac{56 \times 56 \times 100}{284 \times 1 \times 80} = 14 \text{ grams nearly of commercial KOH}$$

In order to assist the operator in arriving at his experimental quantity of alkali the following table is appended. The figures are based on the assumption that 1 kilo of cream will contain 200 grams of stearic acid, and no other fatty substances. These bodies influence the consistency of the product, and some adjustment is usually necessary. Furthermore, an important part is played by manipulation which is referred to later:

Commercial alkali	*Average percentage strength*	*Formula*	*Molecular weight*	*Approx. weight required (grams)*
Potassium hydroxide	80	KOH	56	14
Potassium carbonate	81	K_2CO_3	138	16
Sodium hydroxide	90	NaoH	40	8
Sodium carbonate crystals	98	$Na_2CO_310.H_2O$	286	28
Borax crystals	98	$Na_2B_4O_710.H_2O$	382	37
Liquid ammonia 0·880	32 (NH_3) or 66(NH_4OH)	NH_4OH	35	10
Triethanolamine	77 of Tri- 18 of Di- 5 of Mono-	$N(C_2H_4OH)_3$	Apx. 132	20

Properties of alkalis

It is always better to use a hydroxide than a carbonate, since the gas (CO_2) liberated when the latter is added to stearic acid will not all escape. Assuming that the preparation is a thin liquid, the gas would readily come away and a perfectly clear solution would result, but since the soap and excess of fatty acid produce a viscous liquid, even while hot, it is practically impossible for the whole of the CO_2 to escape. Continuous trituration will not remove it, and after the cream has thickened and set, it will be found to be impregnated with numerous bubbles of air. In time these will rise to the top, and the cream will sink in consequence–a carbonate, therefore, possesses certain disadvantages.

Ammonia solution has a tendency to discolour creams made with it. This is only observed after a time, and it is therefore not ideal.

Borax is useful since it will produce a very white cream; the only disadvantage it possesses is that the product has a distinct tendency to grain.

Sodium and potassium hydroxide are both good, and the advantage is with the use of the latter. Manipulation is facilitated and pearliness follows very closely in its train.

Triethanolamine is excellent. Production of creams is easy and capable of great variation, according to balance of this and fatty acid. Pearliness and stability are all that may be desired, and other emollients are added with facility.

Glycerin is a constituent of many soap creams, and when used it should not exceed 10 per cent. The only real objection to glycerin is that, being hygroscopic, it has a tendency to absorb moisture after application to the skin. This is observed in the form of minute globules of water which appear here and there in the powdered surface. Frequent repowdering is generally necessary when stearic creams containing high proportions of glycerin are used. These objections are however overcome by the use of alternative glycols.

Manufacture

The methods adopted in making vanishing creams are responsible, to a very large extent, for the appearance and texture of the resulting product. They must be standardized, and adhered to, for every batch if consistent results are to be obtained. The most general method consists of melting the stearic acid on a water-bath to a temperature of about 75–85°C. The water (and glycerin if any) is brought to the same temperature, and the alkali is dissolved. This hot alkaline solution is gradually poured into the liquefied fat, while the whole is stirred briskly. The temperature is maintained at 75–85°C for about 10 minutes after all the hot alkali has been added in order to ensure that it has been completely neutralized by the stearic acid which is always in excess of the molecular quantities required. The container is removed from the source of heat, and stirring continued until the cream thickens and sets. This operation is still reverted to at intervals during the next 12 hours, and the temperature of the product is then allowed to fall. The perfume is then added when the temperature falls to about 30–35°C. Stirring is continued as the temperature falls. When cold the cream is transferred to the containers. Other methods are:

1. Adding small lumps of solid stearic acid to the boiling alkaline solution.
2. Placing all the ingredients, excepting perfume, together in a pan—cold, and heating them until saponification is completed.

The advantages or disadvantages of such modus operandi will be fully appreciated with practice.

Pearliness

Much has been written regarding the satiny appearance of vanishing creams, and in order to induce it many substances and methods of manipulation have been recommended. Among those materials which are said to produce it are: liquid paraffin, spermaceti, cocoa butter, starch, castor-oil, and almond-oil, but since there are so many degrees of pearliness, these materials can only truthfully be stated to yield a shine, and will not produce a true pearliness.

This peculiar effect is probably due to the crystallization of the stearic acid in the minutest laminae, from which the light is reflected at any angle, and strange as the statement may appear, it can be obtained without the use of direct alkali. The difficulty of producing this phenomenon seems to resolve itself into finding the best medium in which to emulsify the stearic acid so that it can rapidly form lustrous laminae. In the course of experiments conducted by H. G. Tribley, in collaboration with W. A. Poucher and subsequently confirmed by the writer, it was found that pure curd soap answered this purpose and produced a sheen within twenty-four hours. This was dissolved in boiling water and poured into the hot stearic acid while briskly stirring. The formula is appended:

No. 2410

Stearic acid	200
Curd soap	50
Water	800
	1050

This cream rolls very badly and has a tendency to dry in the pot.

A really satisfactory pearly cream may be produced, however, by using curd soap in conjunction with any approved formula. From 1 to 5 per cent yields an attractive sheen which increases in proportion to the quantity of soap used. The consistency of a cream is, of course, influenced by this addition and there is a tendency to softness. This is counter-balanced by increasing the fatty acid but not the alkali. It is usual to dissolve the soap separately in a small quantity of the prescribed amount of water and add this after the alkaline solution has been run into the stearic acid.

It has often been suggested that prolonged beating of a vanishing cream will induce a satiny appearance. This is true to some extent, but it must not be forgotten that a very fluffy product will result, and as it will necessarily contain an undue amount of air, there will be a tendency to sink rapidly after packing.

Stability

It is generally acknowledged that since soap creams contain a large proportion of water, they are very liable, under certain conditions, to lose some of it by evaporation. It is therefore imperative that the product should be packed in an air-tight container, preferably with a narrow neck.

Perfume

Perfume is a most important consideration in the preparation of cosmetic creams and it will often be found necessary to adjust the formula of the perfume in order to obtain the same odours for different products of a series. Reference has been made in another part of this work to the fact that certain substances have a tendency to discolour white creams, and on this account raw materials, such as indole, vanillin, eugenol, and musk ambrette, should be avoided. Excellent results may be obtained with materials such as the following: geranium, bois de rose, sandalwood, bergamot, patchouli, vetivert, ylang-ylang, and lavender oils, terpineol, linalol, geraniol, citronellol, phenylethyl alcohol, cinnamic alcohol, and coumarin. Some resinoids are also good and should not be overlooked. The question of irritation is also very important and several perfumery houses now offer perfume compounds which have been dermatologically tested. Tests are usually carried out with animals and include tests for occular irritation on the rabbit eye mucosa using a modified Draize eye test technique as prescribed by The Food and Drug Administration of the U.S.A. in The Federal Register of September 1964. Testing for primary irritation is measured by a patch test technique on the abraded and intact skin of the albino rabbit using a technique described in The Federal Register, and skin sensitization tests using a screening test for sensitization in the albino guinea-pig, is carried out by an experimental procedure based on that prescribed in the Food and Drug Administration of U.S.A. in 'Appraisal of the Safety of Chemicals in Food, Drugs and Cosmetics', 1959. Further tests can be carried out by a repeated patch test technique using human subjects and also using allergic subjects.

Formulae

The formulae which follow give a comprehensive range of cosmetic creams and lotions and complementary cosmetic products for use on the skin. The basic principles regarding selection of ingredients have been observed, but these can be adjusted to suit individual requirements. All the formulae given are workable as they stand and have been prepared and storage tested in the author's laboratory. An example of an original vanishing cream based on alkali saponification is as follows:

No. 2411

A	Stearic acid	200
B	Potassium hydroxide	14
	Glycerin	40

	Water (softened or distilled)	746
		1000

Perfume	0·5 per cent
Propyl parahydroxybenzoate	0·02 per cent
Methyl parahydroxybenzoate	0·15 per cent

Procedure: Heat A and B independently to 75°C. Add B to A slowly with continuous stirring adding perfume at about 35°C.

A softer cream can be obtained by decreasing the fatty acid and increasing the potash. This basic formulation is modified by reducing the content of fatty acid and introducing a suitable proportion of an oily material to counteract the drying effect on the skin. Cetyl alcohol is used for its emollient effect. Variations on these lines are described as vanishing creams or foundation creams.

No. 2412

A	Stearic acid	180
	Cetyl alcohol	5
	Alkyl myristate[1]	50
B	Potassium hydroxide	10
	Glycerin	80
	Water (softened or distilled)	675
		1000

Perfume	0·5 per cent
Methyl parahydroxybenzoate	0·15 per cent
Propyl parahydroxybenzoate	0·02 per cent

[1] Bush Boake Allen.

Procedure: Heat A and B independently to 75°C. Add B to A slowly with continuous stirring. Cool with stirring, adding the perfume at about 35°C.

No. 2413

A	Stearic acid	180
	Cetyl alcohol	5
	iso-Propyl myristate	20
B	Potassium hydroxide	10
	Glycerin	80

	Water (softened or distilled)	705
		1000

Perfume	0·5 per cent
Methyl parahydroxybenzoate	0·15 per cent
Propyl parahydroxybenzoate	0·002 per cent

Procedure: Heat A and B independently to 75°C and add B to A slowly with continuous stirring. Cool with stirring, adding the perfume at about 35°C.

The following formula indicates the use of triethanolamine for saponification, and this gives a slightly softer version of a vanishing cream:

No. 2414

A	*iso*-Propyl palmitate	30
	Stearic acid	230
B	Triethanolamine	14
	Glycerin	60
	Water (softened or distilled)	666
		1000

Perfume	0·5 per cent
Methyl parahydroxybenzoate	0·15 per cent
Propyl parahydroxybenzoate	0·02 per cent

Procedure: Heat A and B independently to 75°C and add B to A slowly with continuous stirring. Cool with stirring, adding the perfume at about 35°C.

The use of soap powder (pure curd soap) to obtain pearliness or sheen is shown in the following formula:

No. 2415

A	Stearic acid	180
	Cetyl alcohol	5
B	Soap powder	5
	Potassium hydroxide	10
	Glycerin	80
	Water (softened or distilled)	720
		1000

Perfume	0·5 per cent
Methyl parahydroxybenzoate	0·15 per cent
Propyl parahydroxybenzoate	0·02 per cent

Procedure: Heat A and B independently to 75°C and add B to A slowly with continuous stirring. Cool with stirring, adding the perfume at about 35°C.

The soap powder is introduced either as a component of B or can be dissolved separately in a suitable portion of the water and added to the cream immediately after saponification. Pearliness gradually develops and the maximum effect is obtained after about 10 days. The cream can be filled in the normal fashion and pearliness allowed to develop in the package.

Self-emulsifying glyceryl monostearate is used as an emulsifier in vanishing cream type formulations. It does not have as great a drying effect on the skin as creams prepared with alkali-stearates. Emollient and fatty materials are often included also to reduce drying effect and creams so prepared are in general referred to as foundation creams or day creams. Typical formulations are as follows:

No. 2416

A	Mineral oil	40
	Cetyl alcohol	30
	Glyceryl monostearate (self-emulsifying type)	120
B	Glycerin	120
	Water (softened or distilled)	690
		1000

Perfume	0·5 per cent
Methyl parahydroxybenzoate	0·15 per cent
Propyl parahydroxybenzoate	0·02 per cent

Procedure: Heat A and B independently to 75°C and add B to A slowly with continuous stirring. Cool with stirring, adding the perfume at about 35°C.

No. 2417

A	Glyceryl monostearate (self-emulsifying type)	120
	Stearic acid	50
	iso-Propyl myristate	20
B	Glycerin	80
	Water (softened or distilled)	730
		1000

Perfume	0·5 per cent
Methyl parahydroxybenzoate	0·15 per cent
Propyl parahydroxybenzoate	0·02 per cent

Procedure: Heat A and B independently to 75°C and add B to A slowly with continuous stirring. Cool with stirring, adding the perfume at about 35°C.

The degree of 'oiliness' can be varied to suitable requirements by increasing the content of *iso*-Propyl myristate up to 5 per cent and/or adding about 2–5 per cent of petroleum jelly. Reduce the water content to 710 and include 20 parts of lanolin if required.

Use of lanolin, lanolin derivatives or other special oily materials for skin care is shown in the following formula. *iso*-Propyl linoleate is considered of especial value for its healing effect and treatment of dry, flaky skin.

No. 2418

A	Glyceryl monostearate (self-emulsifying type)	125
	Mineral oil (cosmetic quality)	50
	Cetyl alcohol	20
	iso-Propyl linoleate[1] (or lanolin derivative)	15
B	Glycerin	80
	Water (softened or distilled)	710
		1000

Perfume	0·5 per cent
Methyl parahydroxybenzoate	0·15 per cent
Propyl parahydroxybenzoate	0·02 per cent

[1] Bush Boake Allen.

Procedure: Heat A and B independently to 75°C and add B to A slowly with continuous stirring. Cool with stirring, adding the perfume at about 35°C.

In the following formula glyceryl monostearate is used with triethanolamine stearate. Lanolin and mineral oil are used to give a non-drying emollient film on the skin.

No. 2419

A	Lanolin	10
	Mineral oil	100
	Stearic acid	25
	Glyceryl monostearate (self-emulsifying type)	30
B	Glycerin	50
	Triethanolamine	5

Water (softened or distilled)	780
	1000

Perfume	0·5 per cent
Methyl parahydroxybenzoate	0·15 per cent
Propyl parahydroxybenzoate	0·02 per cent

Procedure: Heat A and B independently to 75°C and add B to A slowly with continuous stirring. Cool with stirring, adding the perfume at about 35°C.

Coloured foundation products

Coloured foundation products are prepared in several ways. They can be based on foundation creams of the type already given to which is added from 2–5 per cent of titanium dioxide together with suitable colourants. Non-aqueous systems are used to cover the skin with a heavier type of make-up. These are referred to as cream-cake foundation. These are prepared as solid creams packed in godets similar to a pressed powder, but are thixotropic and are formulated to liquefy readily when rubbed with the finger so they can be transferred to the skin. The required thixotropy is obtained by using *iso*-Propyl myristate in a mixture of waxes used to give a suitable melting point.

No. 2420
(Cream foundation)

Petroleum jelly	800
Lanolin	75
iso-Propyl myristate	165
	1040

Butylated hydroxyanisole	0·002 per cent
Propyl parahydroxybenzoate	0·1 per cent
Perfume	0·5 per cent
Pigments Titanium dioxide	a sufficient quantity

Procedure: Melt the ingredients together and allow to cool. Gradually incorporate the titanium dioxide and pigments and finally mill. Fill warm to obtain smooth surface.

No. 2421
(Cream foundation type)

A	Olive oil	250
	Mineral oil	220
	Carnauba wax	27·5
	Ozokerite	5
	Beeswax	40
	iso-Propyl palmitate	30
B	Kaolin	220
	Titanium dioxide	205·5
	Pigments	60
		1058·0

Perfume	0·5 per cent
Propyl parahydroxybenzoate	0·15 per cent
Butylated hydroxyanisole	0·002 per cent

Procedure: Melt the constituents of A together and allow to cool. Gradually incorporate the dry powders B and mill with the melted base. Fill warm into godets.

By suitably adjusting and controlling the proportion of high melting point waxes in this type of cream formulation, the resulting product can be prepared in stick form as follows:

No. 2422

A	Mineral oil	180
	iso-Propyl myristate	150
	Petroleum jelly	20
	Kaolin	200
	Titanium dioxide	100
B	Beeswax	75
	Ozokerite	50
		775

Perfume	0·5 per cent
Propyl parahydroxybenzoate	0·10 per cent
Pigments	(about) 5·0 per cent

A good stick type foundation can also be made with the following formula:

No. 2423
(Stick type)

The oil-wax base is prepared as follows:

iso-Propyl myristate	75

Petroleum jelly	100
Castor-oil	300
Ozokerite	20
Candelilla wax	40
	535

Include 100 parts of titanium dioxide in the base to provide adequate covering properties. This should be sieved before use and first mixed to a smooth paste with iso-propyl myristate before adding the remaining ingredients. About 5 parts of pigments are required depending upon the shade required and these can be conveniently used in paste form, or the dry pigment can be used and pre-milled to form a paste with part of the castor-oil or *iso*-Propyl myristate before mixing with the remainder of the base. Use 0·1 per cent of propyl parahydroxybenzoate as a preservative and 0·002 per cent of butylated hydroxyanisole as an antioxidant. The heavy type of make-up obtained by using the stick or cream cake has now been replaced by a lighter type of coloured foundation which gives a more natural effect. These are prepared in the form of liquid creams for packing in wide-mouthed jars. The creams vary in viscosity from thin pourable lotions to thicker creams which have to be shaken out of the container. Generally the thicker product contains more powders as fillers and give more coverage to the skin.

Formulae are given:

No. 2424

(Light cream type)

A	Stearic acid	25
	Propylene glycol monostearate	60
	Mineral oil	150
B	Water (softened or distilled)	532
	Triethanolamine	13
C	Sodium lauryl sulphate	11
	Bentonite	47
	Kaolin	52
	Dry powders (titanium dioxide, pigments, talc)	110
		1000

Perfume	0·5 per cent
Methyl parahydroxybenzoate	0·1 per cent
Propyl parahydroxybenzoate	0·05 per cent

Procedure: Heat A and B independently to 75°C, then add B to A with continuous stirring. Add C at about 55°C and continue stirring. Add perfume at about 35°C. When cold, mill, and allow to stand for several hours before filling. The dry powders of section C are responsible for the final colour of the product. Thus the proportions of pigment and titanium dioxide determine the colour and covering properties, and talc powder is added to a total of 110 parts per 1000 parts of base.

No. 2425
(Liquid cream type)

A	Cetomacrogol emulsifying wax	27
	Mineral oil	50
	Glycerin	80
	Water (softened or distilled)	340
B	Titanium dioxide	100
	Pigments	10
	Water (softened or distilled)	300
		907

Perfume	0·5 per cent
Methyl parahydroxybenzoate	0·15 per cent
Propyl parahydroxybenzoate	0·05 per cent

Procedure: Prepare the emulsion part A, and allow to cool and thicken, dissolving the preservatives in the glycerine to form a solution. Sieve the colours and titanium dioxide, then mix to a smooth paste with the glycerine solution and part of the water. Pass through a triple roll mill and add this to the emulsion. Add the remainder of the water. Incorporate the perfume when making the emulsion. Milling is important to remove entrapped air which very often occurs when making this type of product.

A sophisticated liquid foundation make-up is required to be non-drying and also act as an emollient and skin protective and the following formula was devised to suit these requirements. Lanolin is included as an emollient and 10 per cent of glycols are used for their moisturizing effect. A blend of propylene glycol and glycerin gives much better results than glycerin alone, which tends to make the product too sticky on the skin.

No. 2426
(Moisturizing foundation)

A	*iso*-Propyl myristate	20
	Lanolin	35
	Stearic acid	20

	Non-ionic emulsifier[1]	20
	Glyceryl monostearate (self-emulsifying type)	15
	Cetyl alcohol	10
B	Triethanolamine	10
	Propylene glycol	60
	Glycerin	40
	Water (softened or distilled)	685
		915
	Pigments, Titanium dioxide, Talc	100
		1015

Perfume	0·3 per cent
Methyl parahydroxybenzoate	0·15 per cent
Propyl parahydroxybenzoate	0·05 per cent
Butylated hydroxyanisole	0·002 per cent

1 Polyoxethylene ether of cetyl alcohol.

Procedure: First melt together the ingredients of A and bring to a temperature of about 75°C. Use part of the glycerine and propylene glycol to mix a smooth paste with the titanium dioxide and reserve part of the glycerine and/or propylene glycerine for mixing with the pigments. Heat together the triethanolamine and water to a temperature of 75–80°C and add to A with slow continuous stirring. Continue stirring and allow the cream to cool. Add the perfume whilst mixing when the temperature falls to about 35°C. Gradually add the cream with stirring to the titanium and pigment paste already prepared.

Details of pigments and additives for individual shades to be added to 915 parts of cream base (see formula)—calculated on a weight/weight basis, are as follows:

Natural:	
Cosmetic red oxide[1]	1·43
Cosmetic yellow oxide[2]	3·33
Talc	47·62
Titanium dioxide	47·62
	100·00

1 Red oxide 11554-D. F. Anstead Ltd.
2 Yellow oxide 11556–D. F. Anstead Ltd.

Beige:

Cosmetic red oxide[1]	1·35
Cosmetic yellow oxide[2]	6·75
Cosmetic umber[3]	1·80
Talc	45·05
Titanium oxide	45·05
	100·00

Deep Beige:

Cosmetic red oxide[1]	1·31
Cosmetic yellow oxide[2]	7·87
Cosmetic umber[3]	3·06
Talc	43·88
Titanium dioxide	43·88
	100·00

Tan:

Cosmetic red oxide[1]	3·24
Cosmetic yellow oxide[2]	7·56
Cosmetic umber[3]	3·02
Talc	43·00
Titanium dioxide	43·00
	100·00

[1] Red oxide 11554–D. F. Anstead Ltd.
[2] Yellow oxide 11556–D. F. Anstead Ltd.
[3] Umber 13018–D. F. Anstead Ltd.

Highlight sticks and creams

Pearlescent pigments based on crystalline bismuth oxychloride, and lustre pigments based on titanium coated mica are used to prepare transparent or translucent make-up. The make-up contains low proportions of colour and is used either as a foundation or over a tinted foundation to give highlights and contour effects. Creams and lotions are made by adding pearlescent and lustre pigments to a suitable base. This type of preparation is also made as a translucent or highlight stick as follows:

No. 2427

Mineral oil (cosmetic grade)	300
iso-Propyl myristate	320
Petroleum jelly (white)	100
Lanolin alcohols extract[1]	20
Lanolin	40
Ozokerite	200
Carnauba wax	20
	1000

Methyl parahydroxybenzoate	0·10 per cent
Propyl parahydroxybenzoate	0·05 per cent
Butylated hydroxyanisole	0·002 per cent
Perfume	0·5 per cent

[1] Amerchol L-101 type—American Cholesterol Products, Inc.

Procedure: Melt together the petroleum jelly, lanolin alcohols, lanolin, *iso*-Propyl myristate and mineral oil. Add the ozokerite and carnauba wax. Finally add the pigments (see below) and perfume, mix and pour into containers.

A blend of pearlescent and lustre pigments is used to combine the opacity, gloss and slip properties of bismuth oxychloride with the sparkling brilliance of a titanium coated mica. To 1000 parts of base add the following:

Pearlescent pigment[1]	150
Lustre pigment[2]	100
	250

1 Bi-Lite 20—Mallinckrodt Chemical Works.

2 Timica Pearl White—The Mearl Corporation; or Timiron MP 101—Rona Pearl.

The appearance of the stick is improved by the addition of a very small proportion of a red or brown pigment which is convenient to add in the form of a cosmetic oxide paste. Cosmetics in swivel stick form in a plastic pack are attractive and imaginative variants of jar packaging, spillproof and convenient for use. The sticks are made by using waxes of suitable melting points in an oil blend containing a fatty acid and ester to form a thixotropic base which spreads readily on the skin.

A cleansing stick is prepared as follows:

No. 2428

Mineral oil	600
iso-Propyl myristate	100
iso-Propyl lanolate	20
Lanolin	40
Ozokerite wax	200
Carnauba wax	20
Titanium dioxide paste[1]	60
	1040

Methyl parahydroxybenzoate	0·10 per cent
Propyl parahydroxybenzoate	0·05 per cent
Butylated hydroxyanisole	0·002 per cent
Perfume	0·3–0·5 per cent

[1] 25 per cent dispersion in castor-oil.

Procedure: Melt together the lanolin, mineral oil, *iso*-Propyl myristate and lanolate. Add the ozokerite and carnauba waxes and melt. Finally add the dispersion of titanium dioxide and the perfume and mix well. Pour into containers. The final appearance of the all moulded stick products is improved if they are given a rapid initial cooling treatment immediately after pouring.

A stick product suitable for use at night as a non-sticky night cream or moisturizing treatment preparation is made as follows:

No. 2429

Mineral oil	360
iso-Propyl myristate	100
Petroleum jelly (white)	200
Lanolin alcohols extract[1]	20
Lanolin	60
Ozokerite	200
Carnauba wax	20
Titanium dioxide paste[2]	40
	1000

Methyl parahydroxybenzoate	0·10 per cent
Propyl parahydroxybenzoate	0·05 per cent
Butylated hydroxyanisole	0·002 per cent
Perfume	0·3–0·5 per cent

[1] Amerchol L. 101 type–American Cholesterol Products, Inc.

[2] 25 per cent dispersion in castor-oil.

The appearance of the stick is improved by including a small proportion of a suitable oil-soluble red dyestuff.

Procedure: See previous formula for cleansing sticks.

Highlight creams are a variation of the tinted foundation cream and are applied over a foundation or directly on the skin to give a pearlized or iridescent effect and highlight areas such as the forehead or cheekbones. They are prepared by adding a blend of pearlescent and lustre pigments to a suitable base and contain a small proportion of pigment. A satisfactory product can be made as follows:

No. 2430

(Cream base)

A	Propylene glycol monostearate (self-emulsifying type)	50
	Mineral oil	20
B	Glycerin	40
	Hydroxyethyl cellulose[1]	0·5
	Water (softened or distilled)	889·5
		1000·0

Perfume	0·3 per cent
Methyl parahydroxybenzoate	0·2 per cent
Propyl parahydroxybenzoate	0·02 per cent

[1] Natrosol Type 250 HHR–Hercules N. V.

Procedure: Heat ingredients of A to 75°C. In a separate vessel dissolve the preservatives in the glycerine and add the water and heat to 75°C. Add the hydroxyethyl cellulose and stir until dissolved B. Add B to A slowly and with continuous stirring. Continue stirring and allow to cool adding the perfume when the temperature falls to about 35°C. To 1000 parts of cream base add 1 part of mixed dry pigments or 5 parts of a 20 per cent dispersed pigment blend prepared as follows:

Cosmetic yellow oxide	50
Cosmetic red oxide	20
Cosmetic red oxide	30
	100

Disperse the pigments in the cream base using a Silverson type mixer. Finally add pearlescent and lustre pigments and stir slowly until the mixture is homogeneous.

No. 2431
(Highlight cream)

Cream base with pigment	1001
Pearlescent pigment[1]	150
Lustre pigment[2]	50
	1201

[1] Bi-Lite 20—Mallinckrodt Chemical Works; Shinju TM White—The Mearl Corporation; Mibiron—Rona Pearl Company.
[2] Timiron MP 101—Rona Pearl Company.

Moisturizing creams

It is known that the water (or moisture) content of the skin is responsible for keeping it soft and flexible, and a dry skin condition is not necessarily corrected by the use of fatty creams alone. A fatty cream helps to retain moisture in the skin by preventing its loss by evaporation and although this type of treatment is used at night it is not practical for daytime use. Moisturizing creams are used for this purpose. They are intended to help preserve the skin in a soft, supple condition and allay formation of lines and wrinkles. Consequently they are used on the skin before applying make-up, even before the use of a coloured foundation. They provide an ideal covering for facial skin at times when a more complete make-up is not required, and they can be used as a foundation cream base before powdering by ladies who do not normally use or require coloured foundation preparations.

A moisturizing cream purports to maintain the moisture content of the skin and prevent a dry skin condition by absorption of moisture. Since most of the water contained in a cream is lost by evaporation almost immediately during and after application, the product must provide a non-volatile residual film capable of retaining a high proportion of normal skin moisture and part of the moisture which is applied directly by the product. This is achieved by including humectants in the water phase of the emulsion, and best results are obtained by using a blend of the following materials which are sometimes referred to as moisture conditioners:

Glycerin
Propyl glycol
Sorbitol (70 per cent solution in water)

Use of glycerin or sorbitol solution on their own tend to make the product too sticky in use. The ingredients of the oil phase of the emulsion must be selected so that the product is emollient without being greasy or sticky.

Products for moisturizing are sold in either cream or lotion form and suitable formulae are as follows:

No. 2432
(Cream)

A	Lanolin	5
	Acetylated lanolin[1]	5
	Mineral oil	80
	Stearic acid	30
	Glycerol monostearate (self-emulsifying type)	30
	iso-Propyl palmitate	20
B	Glycerin	35
	Propylene glycol	25
	Triethanolamine	2
	Water (softened or distilled)	768
		1000

Perfume	0·5 per cent
Methyl parahydroxybenzoate	0·1 per cent
Propyl parahydroxybenzoate	0·05 per cent

[1] Modulan type–American Cholesterol Products, Inc.

Procedure: Heat A and B independently to 75°C and add B to A slowly with continuous stirring. Cool with stirring, adding the perfume at about 35°C.

No. 2433
(Cream)

A	Cetyl alcohol	20
	Glyceryl monostearate (self-emulsifying type)	75
	Mineral oil	75
	iso-Propyl palmitate	10
	Polyethylene glycol 400 monostearate	125

No. 2433 (*continued*)

B	Sodium lauryl sulphate	2
	Glycerin	50
	Propylene glycol	30
	Water (softened or distilled)	613
		1000

Perfume	0·5 per cent
Methyl parahydroxybenzoate	0·15 per cent
Propyl parahydroxybenzoate	0·05 per cent

Procedure: Heat A and B independently to 75°C and add B to A slowly with continuous stirring. Cool with stirring, adding the perfume at about 35°C.

The cream can be prepared slightly softer by replacing iso-propyl palmitate with *iso*-Propyl myristate. Emolliency can be increased by adding 1 or 2 per cent of lanolin.

A suitable modification follows:

No. 2434
(Cream)

A	Cetyl alcohol	20
	Glyceryl monostearate (self-emulsifying type)	75
	Mineral oil	75
	iso-Propyl myristate	10
	Polyethylene glycol 400 monostearate	110
	Lanolin	15
B	Sodium Lauryl sulphate	2
	Glycerin	25
	Propylene glycol	50
	Sorbitol (70 per cent solution)	15
	Water (softened or distilled)	603
		1000

Perfume	0·5 per cent
Methyl parahydroxybenzoate	0·15 per cent
Propyl parahydroxybenzoate	0·05 per cent

Procedure: Heat A and B independently to 75°C and add B to A slowly with continuous stirring. Cool with stirring, adding the perfume at about 35°C.

A good product with an attractive glossy appearance is obtained with the following somewhat simpler formula:

No. 2435
(Cream)

A	Lanolin	10
	Mineral oil	100
	Stearic acid	35
	Glyceryl monostearate (self-emulsifying type)	30
B	Glycerin	50
	Propylene glycol	30
	Sorbitol (70 per cent solution)	20
	Triethanolamine	5
	Water (softened or distilled)	720
		1000

Perfume	0·5 per cent
Methyl parahydroxybenzoate	0·15 per cent
Propyl parahydroxybenzoate	0·25 per cent

Procedure: Heat A and B independently to 75°C. Add B to A slowly with continuous stirring. Cool with stirring, adding the perfume at about 35°C.

A moisturizing lotion is used to give a thin protective layer of humectants and fatty materials on the face before applying make-up in the same way as a moisturizing cream. Lotions are also suitable to use as a non-greasy emollient for the hands after washing, and because the term moisturizing now implies skin protection and preservation against dryness, similar products are applied to the body after bathing. The formulations are very similar and the product can be sold as:

moisturizing lotion for the face and hands,
hand and body lotion,
skin and body lotion,

and so on. The main requirement is that they should have a softening effect without being oily or greasy. Suitable formulae are given:

No. 2436
(Lotion)

A	*iso*-Propyl palmitate	75
	Propylene glycol monostearate (self-emulsifying type)	60
	Polyethylene glycol 400 monostearate	50

No. 2436 (*continued*)

B	Triethanolamine	5
	Glycerin	25
	Propylene glycol	25
	Water (softened or distilled)	760
		1000

Perfume	0·3 per cent
Methyl parahydroxybenzoate	0·15 per cent
Propyl parahydroxybenzoate	0·05 per cent

Procedure: Heat A and B independently to 75°C and add B to A slowly with continuous stirring. Cool with stirring, adding the perfume at about 35°C.

No. 2437

(Lotion for hands, face, or body)

A	Mineral oil	100·0
	Wool wax alcohols	2·0
	Cetyl alcohol	2·0
	Stearic acid	20·0
	Sorbitan sesquioleate	2·0
B	Glycerin	7·5
	Sorbitol (70 per cent solution)	2·5
	Propylene glycol	5·0
	Triethanolamine	5·0
	Water (softened or distilled)	854·0
		1000·0

Perfume	0·3 per cent
Methyl parahydroxybenzoate	0·15 per cent
Propyl parahydroxybenzoate	0·05 per cent

Procedure: Heat A and B independently to 75°C and add B to A slowly with continuous stirring. Cool with stirring, adding the perfume at about 35°C.

No. 2438

(Hand and body lotion)

A	Mineral oil	50
	Cetostearyl alcohol	25
	Cetyl alcohol	15
	Spermaceti	5
B	Glycerin	25
	Sorbitol (70 per cent solution)	25
	Water (softened or distilled)	855
		1000

	Perfume	0·3 per cent
	Methyl parahydroxybenzoate	0·15 per cent
	Propyl parahydroxybenzoate	0·05 per cent

Procedure: Heat A and B independently to 75°C and add B to A slowly with continuous stirring. Cool with stirring, adding the perfume at about 35°C.

No. 2439
(Also for general use)

A	Glyceryl monostearate (self-emulsifying type)	25
	Mineral oil	10
	Stearic acid	25
	Lanolin anhydrous	10
B	Sorbitol (70 per cent solution)	15
	Glycerin	25
	Titanium dioxide	2
	Water (softened or distilled)	888
		1000
	Perfume	0·3 per cent
	Methyl parahydroxybenzoate	0·15 per cent
	Propyl parahydroxybenzoate	0·2 per cent

Procedure: Heat A and B independently to 75°C and add B to A slowly with continuous stirring. Cool with stirring, adding the perfume at about 35°C.

Skin foods

Skin foods are used to preserve the skin and as a treatment for dry skin conditions. They contain high proportions of oily and fatty materials together with emollients such as lanolin or lanolin derivatives, and are prepared either as oil-in-water, or water-in-oil, or in some cases as mixed emulsion systems. Many of these creams tend to be sticky and greasy, but they can be prepared without these undesirable characteristics by modifying the oil/wax phase with materials such as fatty acid esters and acetylated glycerides or other oils which have good spreading properties. Materials which increase the spreading properties help to disperse the oil-fat phase and reduce stickiness and greasiness. Thus the objective of using a skin food is to cover the skin with a well dispersed, and consequently thin residual occlusive film which is not greasy or

Plate 4 Gravfil fully automatic vacuum filling machine, capable of filling containers at up to 120 per minute–Avon Cosmetics Ltd.

sticky. When this is allowed to remain on the skin the loss of moisture is slowed down to a degree comparable with that which takes place when the skin is covered with a thick viscous film of materials such as petroleum jelly or lanolin. This type of cream is usually used at night (the period normally assigned to skin preservation and 'feeding'), but a well formulated product applied sparingly can often be used during the day as a base for face powder, particularly if the skin is dry or flaky.

Their behaviour, therefore, is in some ways comparable to a moisturizing cream, but it is likely that the higher proportion of oils and fatty materials used provides a more lasting effect which makes the product more suitable for use at night. The products are sold under various names such as tissue, night, or nourishing creams. Typical formula are as follows:

No. 2440

A	Mineral oil	380
	Petroleum jelly	80
	White beeswax	150
	Paraffin wax	10
	Lanolin	20
B	Borax	10
	Water (softened or distilled)	350
		1000

Perfume	0·5 per cent
Methyl parahydroxybenzoate	0·1 per cent
Propyl parahydroxybenzoate	0·05 per cent

Procedure: Heat A and B independently to 75°C and add B to A slowly with continuous stirring. Cool with stirring, adding the perfume at about 35°C.

No. 2441

A	Beeswax	80
	Mineral oil	150
	Lanolin derivative[1]	75
	iso-Propyl palmitate	100
	Acetylated glyceride[2] (liquid type)	20
	Lanolin	20

[1] Solulan 98 type–American Cholesterol Products Inc.
[2] Bush Boake Allen.

No. 2441 (*continued*)

B	Borax	5
	Water (softened or distilled)	550
		1000

Perfume	0·5 per cent
Methyl parahydroxybenzoate	0·1 per cent
Propyl parahydroxybenzoate	0·05 per cent

Procedure: Heat A and B independently to 75°C and add B to A slowly with continuous stirring. Cool with stirring, adding the perfume at about 35°C.

Use of lanolin derivatives is indicated in the following formula:

No. 2442

A	Lanolin absorption base[1]	80
	iso-Propyl lanolate[2]	20
	Spermaceti	40
	Stearic acid	50
	Cetyl alcohol	20
	Silicone fluid 200–350 ccs	10
	Glyceryl monostearate (self-emulsifying type)	50
	iso-Propyl myristate	100
B	Triethanolamine	10
	Propylene glycol	40
	Water (softened or distilled)	580
		1000

Perfume	0·3 per cent
Methyl parahydroxybenzoate	0·15 per cent
Propyl parahydroxybenzoate	0·02 per cent

[1] Amerchol BL type—American Cholesterol Products Inc.

[2] Amerlate P type—American Cholesterol Products Inc.

Procedure: Heat A and B independently to 75°C and add B to A slowly with continuous stirring. Cool with stirring, adding the perfume at about 35°C.

A good skin food based on lanolin derivatives can be prepared to the following formula:

No. 2443

A	Mineral oil	325
	iso-Propyl myristate	25
	Magnesium stearate	30

	Liquid lanolin fraction[1]	25
	Acetylated lanolin alcohols[2]	45
	Acetylated lanolin[3]	50
	Beeswax	50
	Sorbitan sesquioleate[4]	45
B	Water (softened or distilled)	405
		1000

Perfume	0·3 per cent
Methyl parahydroxybenzoate	0·15 per cent
Propyl parahydroxybenzoate	0·02 per cent
Butylated hydroxyanisole	0·002 per cent

[1] Amerchol L101 type–American Cholesterol Products Inc.
[2] Acetulan type–American Cholesterol Products Inc.
[3] Modulan type–American Cholesterol Products Inc.
[4] Arlacel 83 type–Honeywell and Stein Ltd.

Procedure: Melt together ingredients of A to 75°C. Heat the water to 75°C, and add slowly to A with slow stirring. Allow to cool with continuous stirring, and add the perfume when the mix has cooled to about 35°C.

The following skin food is of the oil-in-water type. It is of a lighter texture but effective in use:

No. 2444

A	Sorbitan monostearate[1]	6
	Polyoxyethylene sorbitan monostearate[2]	44
	Cetyl alcohol	25
	Stearic acid	90
	Mineral oil	75
	iso-Propyl palmitate	65
	Acetylated lanolin[3]	50
	Petroleum jelly	25
	Diethylene glycol monostearate	20
B	Sorbitol (70 per cent solution)	60
	Water (softened or distilled)	540
		1000

Perfume	0·3 per cent
Methyl parahydroxybenzoate	0·15 per cent
Propyl parahydroxybenzoate	0·02 per cent

[1] Arlacel 60 type–Honeywell and Stein Ltd.
[2] Tween 60 type–Honeywell and Stein Ltd.
[3] Modulan type–American Cholesterol Products Inc.

Procedure: Heat A and B independently to 75°C, and add B to A slowly with continuous stirring. Cool with stirring, adding the perfume at about 35°C.

Vitamin creams

Vitamin creams are also generally applied at night as a skin food and contain the oil-soluble vitamins A and D. Any suitable source of the vitamins can be used which may be a blend of fish-liver oils or a solution of the vitamins in a vegetable oil such as arachis oil. This is available in a conventional form as concentrated solution of vitamin A and D (Liq. Vitamin A and D Conc.) of The British Pharmacopoeia containing about 50,000 Units of vitamin A activity and 5000 Units of antirachitic activity (vitamin D). The solution is included in the oil phase of the emulsion, or it can be added to the warm cream after emulsification. A suitable dosage is 1 gram of concentrated solution per 100 grams of cream. Vitamin creams are useful for the treatment of dry and scaly skin conditions and also as a soothing application following depilation. A very effective product can be made to the following formula:

No. 2445

A	Mineral oil	400
	Yellow beeswax	150
	Lanolin	5
	Alkyl myristate[1]	50
	Acetylated lanolin[2]	5
	Concentrated solution of Vitamins A and D	10
B	Borax	10
	Water (softened or distilled)	370
		1000

Perfume	0·2 per cent
Methyl parahydroxybenzoate	0·1 per cent
Propyl parahydroxybenzoate	0·02 per cent

[1] Bush Boake Allen.
[2] Modulan type–American Cholesterol Products Inc.

Procedure: Heat A and B independently to 75°C and add B to A with continuous stirring. Add perfume at about 35°C and continue stirring. Fill warm.

The name Vitamin F has been given to a mixture of unsaturated fatty acids, including linoleic, linolenic, and arachidonic acids. These acids occur in large amounts in certain oils and fats of which linseed oil is a typical example. They are considered to have healing properties and to be of value for dermatitis and allergic skin conditions. They can be used in night creams either with or without vitamins A and D and also in hand creams. A convenient form of including unsaturated fatty acids (vitamin F) in cosmetic creams is to use an ester of the mixed fatty acids. A suitable material referred to as *iso*-Propyl linoleate is included in the following formula:

No. 2446

A	*iso*-Propyl linoleate[1]	10
	Mineral oil	310
	iso-Propyl palmitate	80
	Lanolin	25
	Beeswax	110
	Microcrystalline wax	50
B	Borax	8
	Water (softened or distilled)	407
		1000

Perfume	0·3 per cent
Methyl parahydroxybenzoate	0·15 per cent
Propyl parahydroxybenzoate	0·02 per cent

[1] Bush Boake Allen.

Procedure: Heat A and B independently to 75°C and add B to A slowly with continuous stirring. Cool with stirring, adding the perfume at about 35°C.

When using unsaturated esters, include butylated hydroxy-anisole 0·002 per cent as an antioxidant. The use of *iso*-Propyl palmitate is recommended as this material gives additional 'body' to the cream and in this type of product it is preferred to *iso*-Propyl myristate.

All purpose creams

Of recent years there has been a remarkable increase in the sales of preparations which are normally known as 'all purpose creams'. The products, formerly known as sports creams, are water-in-oil emulsions, generally based on wool alcohols. They are somewhat oily but non-greasy in use, and, because of their water-in-oil character spread readily on the skin to give a protective film. When applied liberally they function as a skin food or night cream or as a protective application for prevention or alleviation of sunburn, or for the treatment of roughened areas of the skin, such as occurs on the elbows or knees. Applied sparingly they function as hand creams and foundation creams.

Wool alcohols consist of the alcoholic fraction obtained by saponification of the grease of the wool of sheep and contain not less than 28 per cent of cholesterol. Its value as a water-in-oil emulsifier is due to the property of absorbing water, but it can deteriorate during storage due to oxidation and increase in acid value. This appears to affect the water holding capacity and is often the reason why water seepage occurs from emulsions made with this material. Fresh material should always therefore be used, and preferably large pieces of the wax, since small pieces oxidize more quickly in relation to the area exposed to air. Deterioration can be prevented to some extent by storing in sealed containers protected from light. When using material which has been stored for a period of time it is wise to scrape off the oxidized surface film. For the same reason an antioxidant should be introduced in all emulsions based on wool alcohols. Use of wool alcohols is given in The British Pharmacopoeia to prepare wool alcohols ointment as follows:

No. 2447

Wool alcohols	60
Hard paraffin	240
White soft paraffin	100
Liquid paraffin	600
	1000

The above ingredients are melted together and stirred until cold. An equal proportion of water can be added to this base to form the cream known as hydrous ointment.

Cosmetic creams of this type are of a softer consistency to enable them to spread more readily and the formula is modified by reducing the proportion of hard paraffin or by using microcrystalline wax as shown in the formulae which follow.

No. 2448

A	Wool alcohols[1]	25
	Microcrystalline wax (m.p. 140°/145°F)	60
	Mineral oil	210
	Petroleum jelly	50
B	Glycerin	50

[1] Hartolan—Croda Ltd.

	Magnesium sulphate	7
	Water (softened or distilled)	598
		1000

Perfume	0·3 per cent
Methyl parahydroxybenzoate	0·1 per cent
Propyl parahydroxybenzoate	0·02 per cent
Butylated hydroxyanisole	0·002 per cent

Procedure: Heat A and B independently to 75°C and add B to A slowly with continuous stirring. Homogenize at approximately 45°C. After homogenizing add the perfume and stir slowly until cool.

The stability of the cream is increased by the addition of magnesium ion to the water phase, and also by homogenization, the success of this depending on the temperature at which homogenization takes place, the critical temperature range being between 43° and 47°C.

No. 2449

A	Wool alcohols[1]	10
	Cholesterol	15
	Petroleum jelly	75
	Paraffin wax	50
	Mineral oil	200
	Sorbitan sesquioleate	10
B	Glycerin	50
	Magnesium sulphate	5
	Water (softened or distilled)	585
		1000

Perfume	0·5 per cent
Methyl parahydroxybenzoate	0·1 per cent
Propyl parahydroxybenzoate	0·02 per cent
Butylated hydroxyanisole	0·002 per cent

[1] Hartolan–Croda Ltd.

Procedure: Heat A and B independently to 75°C, and add B to A slowly with stirring. When the cream has cooled to a temperature of about 45°C pass through a homogenizer. After homogenization continue slow paddle stirring until cold. Add the perfume when the temperature has cooled to about 35°C.

Aerosol creams

A foundation cream suitable for use in an aerosol can be made to the following formula:

No. 2450
(Foundation cream)

	Cream base:	
A	Propylene glycol monostearate (self-emulsifying type)	80
	Cetyl oleyl ethoxylate (10 mols)[1]	20
	Mineral oil (cosmetic quality)	10
B	Water (softened or distilled)	870
	Glycerin	20
		1000

Perfume	0·2 to 0·3 per cent
Methyl parahydroxybenzoate	0·15 per cent

[1] Empilan KL.10 type—Albright & Wilson Ltd., Marchon Division.

Pigments: The proportions of pigments and titanium dioxide used for individual shades follow.

Procedure: Heat A and B independently to 75°C. Add B to A stirring. Cool with stirring, adding the perfume at 35°C. Sieve the pigments and titanium dioxide and add to the base. Mix well. Disperse thoroughly using a mixer of the Silverson type.

Details of the pigments required to obtain a range of shades are given below. Add the proportions given to 1000 parts of cream base.

Natural:	
Cosmetic yellow oxide[1]	2
Cosmetic red oxide[2]	1
Titanium dioxide	50
	53

Beige:	
Cosmetic yellow oxide[1]	5·0
Cosmetic red oxide[2]	2·5
Cosmetic umber[3]	3·0
Titanium dioxide	50·0
	60·5

Deep beige:	
Cosmetic yellow oxide[1]	6·00
Cosmetic red oxide[2]	2·25
Cosmetic umber[3]	3·50
Titanium dioxide	50·00
	61·75

Tan:

Cosmetic yellow oxide[1]	5·0
Cosmetic red oxide[2]	2·5
Cosmetic umber[3]	3·0
Titanium dioxide	25·0
	35·5

Deep tan:

Cosmetic yellow oxide[1]	7·0
Cosmetic red oxide[2]	6·5
Cosmetic umber[3]	4·0
Titanium dioxide	25·0
	42·5

[1] Yellow oxide 11556–D. F. Anstead Ltd.
[2] Red oxide 1154–D. F. Anstead Ltd.
[3] Umber 13018–D. F. Anstead Ltd.

A translucent foundation cream is made by including a pearlescent pigment. In this case the proportion of titanium dioxide is adjusted to a level which gives sufficient covering to the skin without affecting the translucent effect of the pearlescent pigment. Details of the proportion of pigments and the maximum amount of titanium dioxide required for selected shades are given below. The proportions given are to be added to 1000 parts of cream base (W/W).

Procedure: Sieve the pigments (and titanium dioxide if used). Add to the base and mix well. Disperse thoroughly using a mixer of the Silverson type. Finally add the pearlescent pigment and mix using a mechanical stirrer.

Natural:

Cosmetic yellow oxide	2·0
Cosmetic red oxide	1·0
Titanium dioxide	(max) 12·5
Pearlescent pigment[1]	200·0
	215·5

Beige:

Cosmetic yellow oxide	5·0
Cosmetic red oxide	2·5
Cosmetic umber	3·0
Titanium dioxide	(max) 12·5
Pearlescent pigment[1]	200·0
	223·0

Deep beige:

Cosmetic yellow oxide		6·00
Cosmetic red oxide		2·25
Cosmetic umber		3·35
Titanium dioxide	(max)	12·50
Pearlescent pigment[1]		200·00
		224·10

Tan:

Cosmetic yellow oxide		5·0
Cosmetic red oxide		2·5
Cosmetic umber		3·0
Titanium dioxide	(max)	12·5
Pearlescent pigment[1]		200·0
		223·0

Deep tan:

Cosmetic yellow oxide		7·0
Cosmetic red oxide		6·5
Cosmetic umber		4·0
Titanium dioxide	(max)	12·5
Pearlescent pigment[1]		200·0
		230·0

[1] Bi-Lite 20—Mallinckodt Chemical Works.

A translucent blusher with good pearlescent properties is made from the same base by adding lustre and pearlescent pigments and a suitable proportion of a bromo acid. Add the following pigments to 1000 parts of cream base (W/W).

Cosmetic brown oxide[1]	2·0
25 per cent dispersion of bromo acid in propylene glycol[2]	2·5
Pearlescent pigment[3]	200·0
Lustre pigment[4]	50·0
	254·5

[1] Brown oxide 13051—D. F. Anstead Ltd.
[2] D and C Red 21 Eosin—1731—D. F. Anstead Ltd.
[3] Bi-Lite 20—Mallinckodt Chemical Works.
[4] Timica Pearl White—The Mearl Corporation.

Procedure: Mix the brown oxide with the bromo acid dispersion in the base and disperse thoroughly using a mixer of the Silverson type. Add the pearlescent and lustre pigments and mix using a mechanical stirrer.

Container charge: for tinted foundation, translucent foundation and blusher.

Product:	90
Propellent: 12/114 (10 : 90)	10
	100

Container: plastic coated glass or internally lacquered aluminium monobloc with a 20 mm neck. Valve: tilt action fitted with a very small foam button.

Aerosol cleansing cream

No. 2451

Cream Base:		
A	Mineral oil (cosmetic quality)	150
	iso-Propyl palmitate	50
	Non-ionic emulsifier[1]	50
	Lanolin	10
B	Glycerin	50
	Water (softened or distilled)	690
		1000

Perfume	0·3 to 0·5 per cent
Methyl parahydroxybenzoate	0·15 per cent

1 Emulsene 1220 type–Bush Boake Allen.

Procedure: Heat A and B independently to 75°C. Add B to A with stirring. Cool with stirring, adding the perfume at 35°C.

Container charge:	
Cream base	90
Propellent–12	10
	100

Container: internally lacquered aluminium. Valve: tilt action fitted with a foam button.

Aerosol moisturizing Face, Hands and Body Lotion.

No. 2452

Lotion base:		
A	*iso*-Propyl palmitate	20
	Mineral oil (cosmetic quality)	10
	Non-ionic emulsifier[1]	50
	Lanolin	10
B	Propylene glycol	30

[1] Emulsene 1220 type–Bush Boake Allen.

No. 2452 (*continued*)

Glycerin	30
Water (softened or distilled)	850
	1000

Perfume	0·3 to 0·5 per cent
Methyl parahydroxybenzoate	0·15 per cent
Propyl parahydroxybenzoate	0·02 per cent

Procedure: heat A and B independently to 75°C. Add B to A with stirring. Cool with stirring, adding the perfume at 35°C.

Container charge:	
Lotion base	91
Propellent–12	9
	100

Container: internally lacquered aluminium. Valve: standard as supplied by manufacturer fitted with a foam button.

An aerosol non-greasy emollient cream suitable for use on the face and neck or as a hand and body cream is prepared as follows:

No. 2453

Cream base:

A	Glyceryl monostearate (self-emulsifying type)	50
	Lanolin	48
	Stearic acid	25
	Mineral oil	120
	iso-Propyl palmitate	25
B	Triethanolamine	10
	Propylene glycol	25
	Glycerin	20
	Water (softened or distilled)	677
		1000

Perfume	0·2 to 0·3 per cent
Methyl parahydroxybenzoate	0·15 per cent

Procedure: Heat A and B independently to 75°C. Add B to A with stirring. Cool with stirring adding the perfume at 35°C.

Container charge:	
Cream base	90
Propellent–12/114 (50 : 50)	10
	100

Container: internally lacquered aluminium or internally lacquered tin plate. Valve: standard fitted with a foam button or tilt action fitted with a foam button.

Aerosol hand cream

No. 2454

Cream base:		
A	Non-ionic emulsifier[1]	50
	iso-Propyl myristate	20
	Lanolin	10
B	Glycerin	15
	Water (softened or distilled)	905
		1000

Perfume	0·2 to 0·3 per cent
Methyl parahydroxybenzoate	0·15 per cent

[1] Emulsene 1220 type–Bush Boake Allen.

Procedure: Heat A and B independently to 75°C. Add B to A with stirring. Cool with stirring, adding the perfume at 35°C.

Container charge:	
Cream base	91
Propellent–12/114 (10 : 90)	9
	100

Container: plastic coated glass internally lacquered, tinplate or internally lacquered aluminium. Valve: standard as supplied by manufacturer fitted with a foam button.

Skin tonics

Skin tonics or toning lotions are used after cleansing, whether the cleansing process has been by the use of a cosmetic cleansing aid or by soap and water. They are used as toners to remove traces of grease or soap and as skin fresheners to impart a clean and fresh feeling before applying make-up. The usual skin tonic described for use on normal skins contains some alcohol and has a slight astringent effect, but the alcohol content should not be present in sufficient quantity to cause stinging.

Astringent lotions have a stronger effect than skin tonic lotions and are designed to treat conditions when the skin is greasy. This is achieved by including an astringent material such as potassium alum, or aluminium chlorhydroxide. The degree of astringency depends upon the proportion of additive used and the solutions can function either to remove grease or with a stronger astringent effect as a treatment for so-called enlarged or open pores. This effect is similar to that of a mild antiperspirant. The lotions should only be used when the skin is normally greasy due to activity of the sebaceous glands. They are not intended to replace the skin tonic or toning lotion used for a normally dry skin.

An excellent skin tonic for general use as a toner and freshener before make-up can be prepared as follows:

No. 2455
(Skin tonic)

Alcohol	200
Propylene glycol	50
Polyoxyethylene oleyl ether[1] (as perfume solubilizer)	5
Perfume	2
Water (softened or distilled)	743
	1000

Methyl parahydroxybenzoate 0·15 per cent

[1] Volpo N. 15—Croda Ltd.

Procedure: Disperse the perfume in the oleyl ether ethoxylate with the aid of heat and add the water. Dissolve the preservative in the propylene glycol and add to the mixture. Mix well and add the alcohol. A suitable proportion of a water-soluble dyestuff can be added if required.

Witch hazel has a mild astringent and cooling effect and is used in the following formula:

No. 2456
(Skin tonic)

Alcohol	150
Distilled Witch hazel	300
Glycerin	30
Rose water	520
	1000

Methyl parahydroxybenzoate 0·15 per cent

Procedure: Dissolve the preservative in the glycerine. Add the alcohol and remaining ingredients and mix. A rose perfume can be used to replace the rose water in which case it is dispersed in the solution by using a solubilizer as described in the previous formula. Alternatively a water-soluble perfume can be used. Water-soluble herbal extracts can be included in this formula to prepare a herbal skin tonic.

A lotion with mild astringent properties is as follows:

No. 2457
(Skin freshener)

Alcohol	200
Aluminium chlorhydroxide (50 per cent solution)	35
Polyoxyethylene oleyl ether[1]	30
Perfume	20
Distilled witch hazel	100
Glycerin	30
Water (softened or distilled)	585
	1000
Methyl parahydroxybenzoate	0·15 per cent

[1] Volpo N. 15–Croda Ltd.

Procedure: Disperse the perfume with the aid of heat in the oleyl ether ethoxylate and add the water. Add the witch hazel and aluminium chlorhydroxide solution. Dissolve the preservative in the glycerol and add to the solution. Finally add the alcohol and mix.

When glycerin is used in concentrations higher than 3·0 per cent the lotions tend to be sticky in use. Use of propylene glycol is preferred by the author and can be used in concentrations up to 5 per cent without having a sticky effect.

The following lotion contains a cationic surface-active agent. This has the added property of cleansing action and bactericidal activity.

No. 2458
(Skin freshener)

Alkyl trimethyl ammonium bromide[1]	2
Propylene glycol	30
Alcohol	150
Distilled witch hazel	100
Polyoxyethylene sorbitan monolaurate[2]	2
Perfume	2

[1] Cetrimide B.P.
[2] Tween 20–Honeywell & Stein Ltd.

No. 2458 (*continued*)

Water (or Rose water)	714
	1000

Methyl parahydroxybenzoate 0·15 per cent

Procedure: Disperse the perfume in the ethoxylated sorbitan monooleate and add part of the water. Dissolve the alkyl trimethyl ammonium bromide in the remaining water. Dissolve the preservative in the propylene glycol and add the alcohol. Mix the aqueous solutions and add the alcoholic lotion.

An interesting variation is a lotion in gel form used for freshening in the same manner as a skin tonic.

No. 2459
(Tonic gel)

Carbopol 940	10
Triethanolamine	13·5
Glycerin	60
Alcohol	150
Water (softened or distilled)	766·5
	1000·0

Perfume	0·2 per cent
Methyl parahydroxybenzoate	0·15 per cent

Procedure: Dissolve the preservative in the glycerin. Disperse the carbopol in the triethanolamine and glycerin solution. Thin gradually by adding the water. Finally add the alcohol and perfume.

No. 2460
(Astringent lotion)

Aluminium chlorhydrate (50 per cent solution)	160
Alcohol	200
Glycerin	50
Perfume	2
Polyoxyethylene sorbitan monooleate[1]	2
Water (softened or distilled)	586
	1000

Methyl parahydroxybenzoate 0·15 per cent

[1] Tween 20—Honeywell & Stein Ltd.

Procedure: Disperse the perfume in the ethoxylated sorbitan mono-oleate and add the water. Dissolve the preservative in the glycerine and add to the mix. Add the aluminium solution and finally the alcohol.

Note: 5·0 per cent of glycerin is included in this formula to reduce the drying effect of the aluminium salt on the skin.

No. 2461
(Astringent lotion)

Potassium alum	12
Zinc sulphate	2
Alcohol	200
Menthol	1
Camphor	1
Propylene glycol (or glycerin)	50
Distilled witch hazel	250
Perfume	2
Polyoxyethylene sorbitan monooleate[1]	2
Water (or Rose water)	480
	1000

Methyl parahydroxybenzoate 0·15 per cent

[1] Tween 20–Honeywell & Stein Ltd.

Procedure: Disperse the perfume in the polyoxyethylene sorbitan monooleate and add part of the water. Dissolve the potassium alum and zinc sulphate in the remaining water. Dissolve the preservative in the propylene glycol and add to the mix. Prepare a solution of menthol and camphor in the alcohol and add to the mix.

A perfume has been included in the above formula together with a specified surface-active material to act as a perfume solubilizer. The procedure for solubilizing the perfume in the aqueous alcoholic lotion has also been described. As an alternative a water-soluble perfume can be obtained from one of the perfumery houses and added to the aqueous portion of the lotion.

Special lotions and creams are also used as treatment preparations. A soothing lotion is used for skin blemishes, redness and inflammation due to infection or as a result of sunburn. A formula is as follows:

No. 2462
(Soothing lotion)

Calamine powder	150
Zinc oxide	50
Glycerin	150
Bentonite	20
Biosulphur fluid[1]	50

[1] Chemisches Laboratorium, 1 Berlin 41.

No. 2462 (*continued*)

Resorcin	20
Cetrimide (B.P.)	5
Alcohol	50
Water (softened or distilled)	505
	1000

Methyl parahydroxybenzoate	0·02 per cent

Procedure: First make sure that the powders bentonite, calamine and zinc are thoroughly sieved. It may even be necessary to purchase specially ground material. The powders are then mixed by trituration to a smooth paste with some of the glycerin. The remainder of the glycerin is used to dissolve the cetrimide and preservative. This mixture is then mixed with a little of the water. The sulphur and resorcin are mixed with the alcohol and added to the mixture. Add a suitable proportion of a water-soluble red dyestuff.

A soothing lotion in cream form is prepared as follows:

No. 2463
(Soothing cream)

A	Propylene glycol monostearate	40
	Stearic acid	15
	Mineral oil	100
B	Triethanolamine	5
	Glycerin	50
	Cetrimide (B.P.)	5
	Water (softened or distilled)	785
		1000

Perfume	0·2 per cent
Methyl parahydroxybenzoate	0·12 per cent
Propyl parahydroxybenzoate	0·05 per cent

Procedure: Melt together ingredients of A and B independently to a temperature of 75°C. Add B to A slowly with continuous stirring. Cool with stirring, adding the perfume during stirring at about 35°C.

Lotions containing antiseptic materials are also used for treatment of acne. Formulae are as follows:

No. 2464
(Antiseptic lotion)

Sodium bicarbonate	10
Borax	20
Menthol	0·2
Thymol	0·2

Plate 5 Cream filling line comprising Elgin Quad filler, automatic skimmer and Resina fully automatic capping machine–Avon Cosmetics Ltd.

No. 2464 (*continued*)

Methyl salicylate	0·2
Alcohol	25
Glycerin	100
Water (softened or distilled)	844·4
	1000·0

Methyl parahydroxybenzoate	0·12 per cent
Propyl parahydroxybenzoate	0·2 per cent

Procedure: Dissolve the salts in the water and add the glycerin. Dissolve the menthol, thymol, menthyl salicylate, and preservative, in the alcohol. Add purified talc or kiesulguhr. Mix well and gradually add to the aqueous solution. Filter. Add colouring material and sufficient water to produce the required volume. A suitable proportion of a water-soluble dyestuff can be added.

A cream for treatment of acne is as follows:

No. 2465
(Acne cream)

A	Mineral oil	50
	Lanolin anhydrous	20
	Abracol LDS[1]	5
	Petroleum jelly	25
B	Propylene glycol	80
	Chlorhexidine diacetate	5
	Biosulphur fluid[2]	10
	Water (softened or distilled)	805
		1000

Methyl parahydroxybenzoate	0·15 per cent

[1] Bush Boake Allen.
[2] Chemisches Laboratorium, 1 Berlin 41.

Procedure: Heat together ingredients of A and B independently to a temperature of 75°C. Dissolve the preservative in B. Add B to A slowly with continuous stirring, and continue stirring until cool.

Bleaching applications

Bleaching applications are intended to lighten the natural skin and are sometimes used to lighten areas of sunburned skin, or for treatment of freckles. They are made both as lotions and creams and contain either chemical bleaching agents or a material such as ammoniated mercury, which inactivates the enzyme responsible for producing the skin pigment melanin. Inactivation of the enzyme over a period of time results in progressive lightening.

No. 2466
(Lotion)

Hydrogen peroxide solution[1]	100
Ethylene diamine tetra-acetic acid (EDTA)	5
Hydroxyquinoline	1
Alcohol	30
Propylene glycol	50
Water (softened or distilled)	814
	1000

Methyl parahydroxybenzoate 0·02 per cent

[1] Strong hydrogen peroxide solution 30% v/v (27% w/w) of H_2O_2.

Procedure: Dissolve the preservative and hydroxyquinoline in the propylene glycol. Add the alcohol, water and EDTA. Finally add the hydrogen peroxide solution and mix well.

A bleaching cream containing 2·5 per cent of ammoniated mercury is prepared to the following formula. Higher proportions of the mercuric salt are used (up to 5 per cent), but since this is a poisonous material, manufacturers are advised to check the Pharmaceutical Poisons regulations when using the higher percentage, to conform to labelling requirements. Ammoniated mercury is also absorbed by the skin and the lower dosage is recommended to avoid the risk of sensitization or extensive absorption which can occur through prolonged usage.

No. 2467
(Bleaching cream)

Petroleum jelly	625
Ozokerite	20
Beeswax	50
Paraffin wax	20
iso-Propyl myristate	100
Ammoniated mercury (mercuric ammonium chloride)	25

No. 2467 (*continued*)

Bismuth subnitrate	80
Zinc oxide	80
	1000

Procedure: Add the ammoniated mercury to the *iso*-Propyl myristate, and mix to a smooth paste. Add to the melted waxes (70–75°C). Finally add the Bismuth subnitrate, followed by the zinc oxide. Mix to a smooth paste, and mill.

The following formulae based on the bleaching action of hydroquinone are considered to be more effective than mercury based products:

No. 2468
(Bleaching cream)

A	Cetyl alcohol	100
	Mineral oil	30
B	Sodium lauryl sulphate	15
	Hydroquinone	20
	Potassium metabisulphite	1
	Glycerin	50
	Ascorbic acid	5
	Water (softened or distilled)	779
		1000
	Perfume	0·2 per cent
	Methyl parahydroxybenzoate	0·1 per cent

Procedure: Dissolve together the hydroquinone, potassium metabisulphite, and sodium lauryl sulphate in the water. Add the glycerin and ascorbic acid, and heat gently to 70°C. Add with gentle stirring to the ingredients of A previously heated to 75°C. Continue stirring and allow to cool, adding the perfume at a temperature of 35°C.

No. 2469
(Bleaching cream)

A	Glyceryl monostearate (acid stable)	150
	Glyceryl monostearate (self-emulsifying type)	60
	Silicone fluid[1]	10
B	Propylene glycol	50
	Hydroquinone	20
	Sodium bisulphite	0·5

[1] Type MS. 200/500—Midland Silicones Ltd.

Water (softened or distilled)	709·5
	1000·0

Perfume	0·25 per cent
Propyl parahydroxybenzoate	0·02 per cent
Methyl parahydroxybenzoate	0·1 per cent

Procedure: Heat A and B independently to 75–80°C. Add B to A slowly with stirring and allow to cool to 40°C. Perfume is added at about 35–40°C.

The following cream has some bleaching effect and also has bactericidal properties. It is used to improve the conditions of the skin to treat minor skin blemishes and when pustules are present.

No. 2470

A	Non-ionic emulsifier[1]	50
	Mineral oil	50
	Petroleum jelly	50
B	Water (softened or distilled)	795
C	Benzoyle peroxide	25
	Potassium hydroxyquinoline sulphate (B.P.C.)	5
	Glycerin	25
		1000

[1] Abracol LDS type–Bush Boake Allen.

Procedure: Heat A and B independently to 75°C. Add B to A slowly with stirring. Stir until cool. Add C when the emulsion has cooled and mix well.

The active ingredients give a slight burning effect on severely scarred tissue. This is of no danger, and the burning sensation rapidly disappears after the initial application. In effect, the action is to destroy the damaged tissue and allow the new tissue to grow in a similar fashion to an abrasion technique.

Face masks

Face masks are intended to improve the condition of the skin by removing skin surface debris, and improving the circulation. This is achieved by a slight astringent action, as a corrective to make the skin firmer, and use of an absorbent agent to 'mop up' impurities. Masks should be easy to apply and remove, and should achieve maximum results in a reasonably short space of time.

Non-irritant properties are most desirable, particularly as they tend to have a drying effect by removing any surface oils present in the skin. Their value lies in the fact that while drying on the face the blood circulates freely below the skin and so leaves the patient with a fresh complexion. Special ingredients are frequently included for control of greasy conditions or infections. A typical mask for general purpose usage is made as follows:

No. 2471
(General purpose)

A	Kaolin	100
	Bentonite	100
	Zinc oxide	100
	Precipitated sulphur	10
	Bithional (or Cetrimide B.P.)	2
B	Glyceryl monostearate (self-emulsifying type)	50
	Gum acacia	5
	Glycerin	30
	Water (softened or distilled)	502
C	Menthol	1
	Alcohol	100
		1000

Perfume—non-irritant type	0·2 per cent
Alcohol	0·2 per cent

Procedure: Prepare a cream from B by dispersing the glyceryl monostearate and acacia in the water and glycerin at 70°C, using the glycerin to dissolve the preservative. Mix the powders (A) and sieve and stir into the cream base. Mix until smooth. Finally stir in the solution of menthol and perfume in alcohol.

The mask is spread evenly over the face and neck, and left undisturbed until evaporation is complete. The content of alcohol increases the rate of evaporation, and as this takes place the mask contracts and hardens. Acacia gum is used to prevent the residual film of powders from flaking away from the skin, and glycerin is used to prevent the mask drying out completely and also to assist in easy removal.

The absorbent powders remove the surface skin debris and leave the face clean whilst precipitated sulphur and biothional or cetrimide have bacteriostatic properties and assist in the treatment of acne or skin blemishes. The product may be coloured using cosmetic oxides, if required, these being mixed together with the powders of A.

A face mask for use on greasy or oily skins is prepared as follows:

No. 2472
(For greasy skin)

A	Kaolin	260	
	Bentonite	200	
	Light magnesium carbonate	140	
	Zinc oxide	20	
		620	620
B	Emulsifying wax[1]	60	
	Cetyl alcohol	15	
	Mineral oil	40	
		115	115
C	Glycerin	100	
	Sodium lauryl sulphate (needle form)	10	
	Water (softened or distilled)	775	
		885	885
D	Alcohol	300	
	Camphor	4	
	Menthol	1	
		305	305
			1925
	Perfume–non-irritant type	0·2 per cent	
	Methyl parahydroxybenzoate	0·2 per cent	

[1] Cetomacrogol 1000 (B.P.C.) type.

Procedure: Prepare as for previous formula by first preparing cream base from B and C, and add sieved powders (A). Finally add alcohol solution (C).

A modified formula is required for use on dry skins as follows:

No. 2473
(For dry skin)

A	Kaolin	280	
	Bentonite	200	
	Light magnesium carbonate	160	
	Titanium dioxide	20	
		660	660

No. 2473 (*continued*)

B	Emulsifying wax[1]	60	
	Cetyl alcohol	15	
	Mineral oil	100	
	iso-Propyl palmitate	50	
C	Glycerin	100	
	Parachlormetaxylenol	2	
	Water (softened or distilled)	673	
		1000	1000
D	Alcohol	300	300
			1960
	Perfume–non-irritant type	0·2 per cent	
	Methyl parahydroxybenzoate	0·15 per cent	

[1] Cetomacrogol 100 (B.P.C.) type.

Procedure: Prepare as for previous formulae, adding sieved powders (A) to cream base prepared from (B) and (C). Finally add the alcohol (C).

Barrier creams

Barrier creams are essentially protective products, and like hand creams their purpose is to form an occlusive and protective film on the skin, but they differ inasmuch as this film is to stop permeation to the skin from external sources. The formulation of barrier creams is governed largely by the nature of the external influences which may damage the skin. These vary from highly dermatitic defatting solvents to water and everyday dirt and soil. For protecting skin which is in prolonged contact with water, a paste or ointment type of preparation is more effective. This can be based on petroleum jelly modified by inclusion of lanolin and paraffin wax or glyceryl monostearate, as follows:

No. 2474

Paraffin wax	50
Glyceryl monostearate (non self-emulsifying type)	120
Lanolin (anhydrous)	50
Petroleum jelly	780
	1000
Perfume	0·3–0·5 per cent

Procedure: Heat all the ingredients together to 70°C until a clear solution is obtained. Mix and allow to cool.

This paste leaves a greasy but effective protective barrier covering the skin.

A less greasy film is obtained by preparing an emulsion as follows:

No. 2475

A	Glyceryl monostearate (self-emulsifying type)	120
	Paraffin wax	120
	Lanolin	60
B	Water (softened or distilled)	700
		1000

Perfume	0·3–0·5 per cent
Methyl parahydroxybenzoate	0·15 per cent

Procedure: Heat A and B independently to 70°C and add B to A with stirring. Stir until cool.

This cream gives good protection against soil and mineral oils and greases. It is readily removed after working due to the presence of glyceryl monostearate. Inert materials such as kaolin and kieselguhr help as skin protectives in barrier creams intended for use with both water and greases. Typical formulae are given:

No. 2476

A	Mineral oil	100
	Glyceryl monostearate (self-emulsifying type)	50
	Lanolin	10
B	Glycerin	100
	Water (softened or distilled)	390
C	Kaolin	350
		1000

Perfume	0·3–0·5 per cent
Methyl parahydroxybenzoate	0·15 per cent

Procedure: Heat A and B independently to 75°C and add B to A slowly with continuous stirring. Gradually add the kaolin (C) and work to a smooth paste. A suitable proportion of a water-soluble colour may be added to make the cream more attractive in appearance.

No. 2477

A	Mineral oil	100
	Glyceryl monostearate (non self-emulsifying type)	50
	Stearic acid	5
	Silicone fluid[1]	50
B	Glycerin	50
	Water (softened or distilled)	385
C	Kaolin	360
		1000

Perfume	0·3–0·5 per cent
Methyl parahydroxybenzoate	0·15 per cent

[1] MS 200/10 type—Midland Silicones Ltd.

Procedure: Melt together the ingredients of A to a temperature of 70°C. Dissolve the preservative in the glycerol and add the water and heat to 75°C. Add B to A with constant stirring and gradually add the kaolin and continue mixing until a smooth consistency is obtained. This preparation is also improved by adding a suitable proportion of a water-soluble dyestuff.

No. 2478

A	Lanolin	40
	Aluminium stearate	40
	Mineral oil	870
	Cetyl alcohol	50
B	Kaolin	1500
		2500

Perfume	0·3–0·5 per cent
Propyl parahydroxybenzoate	0·02 per cent

Procedure: Mix the aluminium stearate into a smooth paste with part of the mineral oil. Add the remainder of the mineral oil and the lanolin. Continue stirring and heat to 120°–150°C, adding the cetyl alcohol at 120°C. Add the kaolin to the completed gel, mixing until a smooth consistency is obtained.

Barrier creams prepared specifically for use when handling oily materials are based on one or more of a non oil-soluble film former such as acacia, karaya, tragacanth gum, or sodium alginate. Glycerin can be used as a plasticizer for the gum film and to increase the oil repellency of the product. A formula of this type is as follows:

No. 2479

Gum acacia	50
Gum tragacanth	50
Glycerin	20
Water (softened or distilled)	880
	1000

Perfume	0·3–0·5 per cent
Methyl parahydroxybenzoate	0·2 per cent

Procedure: Dissolve the acacia in the water. Disperse the gum tragacanth in the glycerin and add the water-acacia solution with stirring. Stir gently until homogenous.

A vanishing cream type is intended for use where the object is to prevent the hands becoming ingrained with dust and dirt. The effect is achieved by rubbing the soap into the pores which ultimately helps in the removal of soil when the hands are washed.

No. 2480
(Vanishing cream type)

A	Stearic acid	100
	Beeswax	20
	Petroleum jelly	45
B	Triethanolamine	15
	Glycerin	80
	Water (softened or distilled)	540
C	Magnesium stearate	200
		1000

Perfume	0·3–0·5 per cent
Methyl parahydroxybenzoate	0·15 per cent

Procedure: Heat A and B independently to 75°C and add B to A stirring slowly. Add the magnesium stearate (C) and stir slowly until cold.

Perfume and colour can be added to this cream.

Hand preparations

Hand preparations are either in the form of creams or lotions, having largely replaced the glycerin and rose-water types of product which were popular about 30 years ago.

The main function of a hand cream or lotion is to replace or reduce water which is lost from the surface of the skin by evaporation and by constant washing or immersion in detergent solutions. This can be achieved either by using a humectant or by preventing evaporation from the skin by means of a semi-occlusive film of water immiscible oily or fatty material. The most commonly used humectants are propylene glycol, glycerin, and sorbitol. The effects of these materials are similar and the choice is partly dependent on the viscosity required in the finished product. In oil-in-water creams, glycerin tends to produce a firmer consistency than sorbitol, and propylene glycol gives an even softer consistency. Materials used to form occlusive films may be either natural and synthetic film formers or oily substances.

Natural gums, such as karaya, acacia, tragacanth, agar agar, and synthetic film forming agents such as carboxy celluloses and polyvinyl alcohols, all help to prevent evaporation of water from the surface of the skin. The occlusive action of oily substances is similar in action inasmuch as they leave a thin occlusive film on the skin surface. Mineral oil and lanolin are used, and can be modified with alkyl fatty acid esters to prevent drag and excessive oiliness. The alkyl fatty acid esters make the oil phase less viscous and so enable the skin to be covered with a thinner oil film than would otherwise be possible.

Hand creams and lotions sometimes include a healing ingredient such as allantoin to enhance granulation of the skin. A bactericide can also be included. Alkyl esters of polyunsaturated C_{18} fatty acids—linoleic and linolenic—help to prevent scaling of the skin surface and are useful additions to the modern type of hand cream or lotion. Generally both hand creams and lotions are oil-in-water emulsions which on evaporation leave the skin feeling relatively non-oily and not sticky. Creams and lotions based on either natural or synthetic film formers are not so widely popular because as the water evaporates from the product the high concentration of gum makes the cream feel sticky, although this can be overcome to some extent by the use of humectants. Formulae are given for several types of hand creams and lotions:

No. 2481

(Lotion type)

A	*iso*-Propyl myristate	20
	iso-Propyl linoleate[1]	20

[1] Bush Boake Allen.

	Mineral oil	20
	Stearic acid	30
	Emulsifying wax[2]	2·5
B	Glycerin	30
	Water (softened or distilled)	877·5
		1000·0

Perfume	0·25 per cent
Methyl parahydroxybenzoate	0·15 per cent

[2] Emulsifying wax B.P.

Procedure: Heat A and B independently to 75°C and add B to A slowly with continuous stirring. Cool with stirring, adding the perfume at about 35°C.

No. 2482
(Lotion type)

A	*iso*-Propyl myristate	40
	Mineral oil	20
	Stearic acid	30
	Emulsifying wax[1]	2·75
	Lanolin	25
B	Glycerin	30
	Triethanolamine	10
	Water (softened or distilled)	842·25
		1000·00

Perfume	0·25 per cent
Methyl parahydroxybenzoate	0·15 per cent
Propyl parahydroxybenzoate	0·05 per cent

[1] Emulsifying wax B.P.

Procedure: Heat A and B independently to 75°C and add B to A slowly with continuous stirring. Cool with stirring, adding the perfume at about 35°C.

No. 2483
(Lotion type)

A	Lanolin	10
	Emulsifying wax[1]	40
	Mineral oil	10
B	Cetrimide	10
	Glycerin	20
	Water (softened or distilled)	910
		1000

[1] Cetomacrogol 1000 (B.P.C.) type.

Perfume	0·25 per cent
Methyl parahydroxybenzoate	0·15 per cent
Propyl parahydroxybenzoate	0·05 per cent
Water-soluble dyestuff	*q.s.*

Procedure: Heat A and B independently to 75°C. Add B to A slowly with continuous stirring, adding the perfume at about 35°C.

The lotion develops a slight pearlescence on standing.

No. 2484
(Lotion type)

A	Mineral oil	50
	Stearic acid	25
	Glyceryl monostearate (self-emulsifying type)	25
	Lanolin	10
B	Glycerin	20
	Triethanolamine	10
	Water (softened or distilled)	860
		1000

Perfume	0·25 per cent
Methyl parahydroxybenzoate	0·15 per cent
Propyl parahydroxybenzoate	0·05 per cent

Procedure: Heat A and B independently to 75°C and add B to A slowly with continuous stirring. Cool with stirring, adding the perfume at about 35°C.

No. 2485
(Cream)

A	Fatty alcohol ethylene oxide condensate[1]	20
	Stearic acid	250
	Mineral oil	50
	Cetyl alcohol	50
B	Glycerin	30
	Sodium hydroxide (pellets)	2
	Water (softened or distilled)	598
		1000

Perfume	0·25 per cent
Methyl parahydroxybenzoate	0·15 per cent
Propyl parahydroxybenzoate	0·05 per cent

[1] Lubrol type—I.C.I. Ltd.

Procedure: Heat A and B independently to 75°C and add B to A slowly with continuous stirring. Cool with stirring, adding the perfume at about 35°C.

The following cream containing a silicone, is excellent both for treatment of roughened hands, and also as a protective. It can be packed either in a jar or flat tin for convenience.

No. 2486
(Cream)

A	Cetyl oleyl ethoxylate[1] (10 mols)	90
	Polyethylene glycol 400 monostearate	140
	Mineral oil	120
	Silicone oil[2]	10
	Paraffin wax	80
	Petroleum jelly	40
	iso-Propyl myristate	80
B	Water (softened or distilled)	460
		1020

Perfume	0·25 per cent
Methyl parahydroxybenzoate	0·15 per cent
Propyl parahydroxybenzoate	0·05 per cent

[1] Empilan KL10 type–Albright & Wilson Ltd., Marchon Division.
[2] Silicone MS 200 type–Midland Silicones Ltd.

Procedure: Heat A and B independently to 75°C. Add B to A with continuous stirring. Cool with stirring, adding the perfume at about 35°C.

The following formula gives a pourable product which does not whiten on the hands and leaves the hands feeling soft and velvety. If a slightly thicker cream is required, this can be obtained by increasing the proportion of cetyl alcohol, and experiments using, for example, 0·25, 0·5, 0·75 and 1·0 per cent of cetyl alcohol can be carried out to determine the viscosity which is most suitable. It should be noted that the final viscosity of the product should be assessed after the cream has been allowed to stand overnight.

No. 2487

A	Fatty alcohol ethylene oxide type[1]	10
	iso-Propyl myristate	50
	Lanolin	50
	Mineral oil	50
	Abracol LDS	100
	Spermaceti	25
	Acetoglyceride Liquid type	50

[1] Lubrol type–I.C.I. Ltd.

No. 2487 (*continued*)

	Petroleum jelly	35
B	Glycerin	50
	Water (softened or distilled)	580
		1000

Perfume	0·3 per cent
Methyl parahydroxybenzoate	0·15 per cent
Propyl parahydroxybenzoate	0·02 per cent

Procedure: Heat A and B independently to 75°C and add B to A slowly with continuous stirring. Cool with stirring, adding the perfume at about 35°C.

No. 2488
(Liquid)

A	*iso*-Propyl myristate	40
	Mineral oil	20
	Stearic acid	30
	Emulsifying wax[1]	2·75
	Lanolin	25
B	Glycerin	30
	Triethanolamine	10
	Water (softened or distilled)	842·25
		1000·00

Perfume	0·3 per cent
Methyl parahydroxybenzoate	0·15 per cent
Propyl parahydroxybenzoate	0·02 per cent

[1] Emulsifying wax B.P.

Procedure: Heat A and B independently to 75°C and add B to A slowly with continuous stirring. Cool with stirring, adding the perfume at about 35°C.

Honey and almond cream is used for the hands. It is a pourable lotion and a small proportion of powdered soap is included in the formula, which gives the lotion an attractive pearly appearance. Pearling takes place more rapidly if the soap powder is dissolved in a small proportion of the available water and added to the lotion after emulsification. Almond oil is given in the formula which follows, but this can be reduced and partly replaced by iso-propyl palmitate, if required.

No. 2489

A	Almond oil	50
	Beeswax	10
	Spermaceti	20

B	Borax	3
	Soap powder	25
	Glycerin	50
	Water (softened or distilled)	650
		808

Perfume (almond type)	0·25 per cent
Methyl parahydroxybenzoate	0·15 per cent
Propyl parahydroxybenzoate	0·02 per cent
Butylated hydroxyanisole	0·002 per cent

Procedure: Dissolve the preservative in the glycerin, then heat A and B independently to 70°–75°C and add B to A with continuous stirring. Allow to cool with stirring, adding the perfume at about 35°C.

Glycerine jellies are prepared as a mucilage using tragacanth, or methylcellulose. Sodium alginate can also be used. A typical formula using tragacanth is as follows:

No. 2490

Tragacanth	10
Boric acid	15
Glycerin	120
Alcohol	120
Water (softened or distilled)	735
	1000

Perfume	0·25 per cent
Methyl parahydroxybenzoate	0·15 per cent

Procedure: The preservative is first dissolved in the glycerin using gentle heat. Mix the tragacanth with the alcohol and add the glycerin and perfume. Dissolve the boric acid in the water, and add the solution in small quantities at a time, to the tragacanth paste, stirring gently after each addition to avoid entrapment of air.

The consistency of the jelly can be varied by controlling the proportion of tragacanth. To some extent this depends upon the quality and gelling strength of the tragacanth used. This can be determined by experiment.

Cellulose ethers are used either as partial or total replacement materials for natural gums to prepare mucilages suitable for glycerin hand jellies. They are available under several trade names ('Cellulos', 'Edifas', and 'Methofas'–I.C.I. Ltd.) as a methyl ethyl cellulose derivative or sodium carboxymethyl cellulose and each product is available in several viscosity grades. Methyl ethyl cellulose is slowly soluble in cold water, is insoluble in hot water but dissolves on cooling. Solutions are best prepared by adding the materials slowly to about half the required amount of water at a temperature of about 80°C with thorough stirring, and allowing the mix to soak for about 1 hour before adding the remaining water (cold) continuously until solution is effected. In addition to glycerin, propylene glycol or sorbitol solutions are equally suitable materials to use to prepare either hand or lubricating jellies.

Methyl ethyl cellulose forms colourless, odourless neutral mucillages which are stable to acids and alkalis, but precipitation occurs by addition of 7 to 10 per cent of soluble organic salts. Sodium carboxymethyl cellulose dissolves in cold and hot water, and solutions are easily prepared by adding the material to cold or warm water with constant stirring until solution is effected. The solutions are stable to alkalis and are compatible with other water-soluble polymers such as tragacanth or polyvinyl alcohol. The viscosity is, however, affected by acids and metallic ions. Mucillages prepared with cellulose ethers do not readily support bacterial or fungus growth, but nevertheless it is advisable to add a preservative such as 0·15 per cent of methyl parahydroxybenzoate, which should be dissolved in the glycol being used in the formula and added together with the last addition of water.

Carboxy vinyl polymers (Carbopol Resins—B. F. Goodrich Chemical Company) of high molecular weight are available as dry fluffy powders in acid form, which form viscous mucillages when neutralized. Three types are available—type '934' is used for thick viscous products; type '940' probably the most versatile polymer for application in the cosmetic field forms clear sparkling gels with both aqueous and aqueous-alcoholic systems; and type '941' which is used to prepare suspensions or pourable gel-like fluids of relatively low viscosity. Mucillages or gels are prepared by first dispersing the polymer in the dispersion medium and then neutralizing the dispersion by adding a suitable neutralizing agent. Some difficulty is often experienced in obtaining complete dispersion of the polymer as it tends to form clumps which wet on the outside and prevent further wetting of the interior. The best way to avoid this lumping is to mix the polymer either with another powder or to disperse it in a viscous fluid, whichever material is being used in a formulation. The dispersion or suspension is then added very slowly and carefully to the dispersion medium whilst it is being stirred rapidly. When complete dispersion has been obtained the speed of the mixer is reduced and the neutralizing agent added when gel formation takes place.

Triethanolamine and monoethanolamine give excellent results in most aqueous and aqueous-alcoholic systems, although sodium and ammonium hydroxide also give satisfactory gels. The gels are stable in the pH range of 5·0–11·0 and have excellent storage stability surviving even freezing conditions and they do not support bacterial or fungus growth. The carboxy vinyl polymers have several interesting applications in the cosmetic field, some of which will be found in other sections of the text. A hand or lubricating jelly with glycerin is given in the next formula.

No. 2491

Carboxy vinyl polymer[1]	10
Glycerin	400
Water (softened or distilled)	590
	1000
Triethanolamine	1·35
Perfume (water-soluble)	1 per cent
Methyl parahydroxybenzoate	0·15 per cent

[1] Carbopol 940–B.F. Goodrich Chemical Company.

Procedure: Mix the glycerin with part of the water, and add the polymer resin with rapid stirring until thoroughly dispersed. Then add the triethanolamine and the remaining water and stir briskly until gelling occurs. Finally add the water-soluble perfume.

A good hand or lubricating jelly suitable for tube packing can be prepared using a mixture of tragacanth and polymer resin as follows:

No. 2492

Carboxy vinyl polymer (type 940)	20
Tragacanth	5
Glycerin	400
Water (Softened or distilled)	563
Triethanolamine	12
	1000
Perfume (water soluble)	1·0 per cent
Methyl parahydroxybenzoate	0·15 per cent

Procedure: Dissolve the preservative in the glycerin. Mix the tragacanth and add the polymer resin, glycerin and water, stirring rapidly until the mix is thoroughly dispersed. Allow to stand to permit escape of entrapped air and then add the triethanolamine and stir gently until gelling occurs.

Hand and body lotions

Both the preceding formulae can be varied by including additional humectants to give additional moisturizing properties. Comparable products in either gelled or emulsified form are used to rub on the body generally after bathing and are really an extension of a low fat hand preparation intended to lubricate the body and prevent dryness. They are sold under various names such as skin lotion, moisturizing lotion, hand and body lotion, hand and body cream, moisturizing lotion for hands and body, skin moisturizers for hands and body. Suitable formulae are as follows:

No. 2493
(Hand and body lotion)

A	Fatty alcohol ethylene oxide condensate [1]	30
	Diethyleneglycol monostearate	45
	Cetyl alcohol	10
	Mineral oil	250
	iso-Propyl myristate	50
B	Glycerin	30
	Propylene glycol	20
	Water (softened or distilled)	565
		1000

Perfume	0·5 per cent
Methyl parahydroxybenzoate	0·15 per cent
Propyl parahydroxybenzoate	0·02 per cent

[1] Lubrol type—I.C.I. Ltd.

Procedure: Heat A and B independently to 75°C. Add B to A slowly with continuous stirring. Cool with stirring, adding the perfume at about 35°C.

No. 2494
(Hand and body moisture cream)

A	Lanolin	10
	Mineral oil	100
	Stearic acid	25
	Glyceryl monostearate (self-emulsifying type)	30
B	Glycerin	30
	Sorbitol solution	20
	Triethanolamine	5

	Water (softened or distilled)	780
		1000

Perfume	0·5 per cent
Methyl parahydroxybenzoate	0·15 per cent
Propyl parahydroxybenzoate	0·02 per cent

Procedure: Heat A and B independently to 75°C. Add B to A slowly with continuous stirring. Cool with stirring, adding the perfume at about 35°C.

The viscosity of this cream can be varied by altering the proportion of self-emulsifying glyceryl monostearate. A skin and body lotion containing a higher proportion of oil and fatty material suitable for treatment of dry skin conditions can be made as follows:

No. 2495
(Skin and body lotion)

A	Petroleum jelly	50
	Mineral oil	400
	Lanolin	25
	Sorbitan sesquioleate[1]	30
	Beeswax	20
	Zinc stearate	5
B	Borax	5
	Water (softened or distilled)	465
		1000

Perfume	0·5 per cent
Methyl parahydroxybenzoate	0·15 per cent
Propyl parahydroxybenzoate	0·02 per cent

[1] Arlacel C–Honeywill-Atlas Ltd.

Procedure: Heat A and B independently to 75°C. Add B to A slowly with continuous stirring. Cool with stirring, adding the perfume at about 35°C.

Skin and body perfumes

In addition to their lubricating and moisturizing effects hand and body lotions are also a means of perfuming the skin. Good quality perfumes should always be used, since their perfuming effect may well be an alternative to the use of toilet water or cologne although they do not contain as high a proportion of perfume. Colognes and toilet waters, when applied as body lotions, have a drying effect due to the high alcohol content and used in this way are more suitable to use as a skin freshener in hot weather conditions. A body perfume can be made with a lower alcohol content by using a perfume solubilizer, and is particularly attractive when prepared as a gel. Emollient or moisturizing materials can also be included to reduce the drying effect, as shown in the following formulation:

No. 2496

(Skin and body perfume)

Carboxy vinyl polymer[1]	3
Diisopropyl adipate	20
Polyoxyethylene cholesterol[2] (24 mols)	30
Alcohol	500
Polyoxyethylene oleyl ether[3] (15 mols)	75
Perfume	8
Triethanolamine (10 per cent solution in water)	20
Water (softened or distilled)	344
	1000

Methyl parahydroxybenzoate 0·10 per cent

[1] Carbopol 941–B. F. Goodrich Chemical Company.
[2] Solulan C. 24–American Cholesterol Products Inc.
[3] Volpo N. 15–Croda Ltd.

Procedure: Disperse the polymer resin in water using a high speed stirrer and continue mixing until free from clumps. Dissolve diisopropyl adipate in alcohol and add to the solution. Add the triethanolamine solution with gentle stirring until gel is formed. Finally mix the perfume with the oxyethylene oleyl ether and add to the base and mix.

The viscosity of this lotion varies according to the composition of the perfume and its effect on the gel system, but can be controlled by adjusting the proportion of resin in the formula.

A most attractive product can be obtained by suitably colouring the solution and including up to 1·0 per cent of a pearlescent pigment (Timica Pearl White–Cornelius Chemical Co. Ltd.). This should be made into a paste with the cholesterol derivative before including in the mix.

A solid perfume or pomade is used to apply concentrated perfume without use of alcohol. It is made as follows:

No. 2497

Paraffin wax	150
Petroleum jelly	600

Mineral oil	100
iso-Propyl myristate	50
Perfume	100
	1000

Procedure: Warm together the paraffin wax, petroleum jelly, and mineral oil. Add the perfume, stir and pour into moulds or godets and allow to set.

A cream perfume or perfume sachet is an alternative method of applying perfume to the skin. Prepared as an emulsion they contain from 5 to 10 per cent of perfume compound as part of the oil phase of the emulsion. Packaging is important and adequate sealing is essential to prevent moisture loss during storage. Most perfume oils emulsify to form attractive white emulsions, but perfumes of a dark colour are not so attractive in this type of product. Tests should be carried out to determine whether the colour of the cream deteriorates on storage due to the high content of perfume compound. A suitable formula is as follows:

No. 2498

A	*iso*-Propyl myristate	25
	Glyceryl monostearate (self-emulsifying type)	25
	Stearic acid	150
	Beeswax	30
B	Glycerin	80
	Potassium hydroxide	5
	Water (softened or distilled)	605
C	Perfume	80
		1000

Methyl parahydroxybenzoate 0·15 per cent

Procedure: Heat A and B independently to 65–70°C and add B to A slowly with continuous stirring. Add the perfume oil (C) at 50–55°C and stir until cream begins to thicken. Allow to stand, then restir.

No. 2499

A	Lanolin alcohols fractionated[1]	50
	Acetylated lanolin alcohols[2]	20
	Acetylated ethoxylated ester of lanolin alcohols[3]	30
	Glyceryl monostearate (self-emulsifying type)	70

No. 2499 (*continued*)

	Stearic acid	150
	Beeswax	60
B	Propylene glycol	50
	Water (softened or distilled)	570
		1000

Methyl parahydroxybenzoate 0·15 per cent

[1] Type Amerchol L. 101–American Cholesterol Products, Inc.

[2] Type Acetulan–American Cholesterol Products, Inc.

[3] Type Solulan 98–American Cholesterol Products, Inc.

Procedure: Melt together the ingredients of A and heat to 70–75°C. In a separate vessel dissolve the preservative in the propylene glycol. Add the water and heat to 75°C. Add B to A slowly with continuous stirring. Continue stirring and allow to cool and add the perfume at a temperature of about 35°C. From 5 to 8 per cent of perfume compound can be added.

Plate 6 Yorkway vessel used for producing various products such as cream sachet–Avon Cosmetics Ltd.

CHAPTER THIRTEEN

Sunburn Preparations

The beneficial effects of sunlight on the human organism are well known. Ultra-violet light in contact with the skin yields vitamin D which acts as a tonic and stimulant, incidentally assisting the body to ward off disease. Providing sun-bathing is not indulged in to excess, the epidermis is able to tolerate the mildest erythema without discomfort. Vivid erythema is produced after approximately one hour's exposure of the skin to the sun on a cloudless summer day, and for normal untreated skins, this is about the maximum period, wisely indulged in. Beyond that painful burns result and excessive periods of exposure to the sun yield blisters.

There is no short cut to a good suntan. It is a progressive business with the light rays slowly building up and reacting with the skin's melanin pigment which becomes darker when exposed to ultra-violet rays. At the same time the epidermis becomes thicker and so the body develops its own natural protection. The same ultra-violet rays which tan cause erythema and burning, but a person living in a perpetually sunny climate will gradually develop sufficient melanin accompanied by epidermal thickening to allow of the skin's maximum tolerance to sunshine. Others who do not have the advantage of this 'natural' protection, and who only indulge in sun-bathing on holidays, must treat the skin with some protective preparation if the tan is to be pleasantly produced.

Sunburn is caused by ultra-violet light and its prevention is only possible if a large percentage of these rays responsible are not allowed to come in contact with the skin. Sunlight is a form of

radiant energy composed mainly of rays of wavelengths from 2800 Å, where Å signifies an Angstrom unit (10^{-8} cm). It includes infra-red, visible and ultra-violet radiation, the last mentioned being responsible for most of the harmful effects.

Various investigations have shown that the rays chiefly responsible for sunburn are those with wavelengths 2900 Å to 3100 Å. The maximum effect corresponding to the peak of the erythema curve occurs with ultra-violet light of a wavelength 2967 Å, and the effectiveness falls away quite sharply on either side of this figure. The wavelengths of the radiations which produce tanning have not been established with such certainty, but there is general agreement that those between 3000 Å and 3200 Å are the most effective, that is by radiations which are accompanied by erythema and sunburn. Radiations up to 4200 Å, that is without accompanying erythema and burning, are also considered to be important but tanning in this range is said to be due to darkening of preformed pigment.

Synthesis of vitamin D_2 is also promoted by radiations in the region of 3000 Å, but a series of small doses of radiation is as effective in the treatment of vitamin deficiency as a single massive one, so that the filtering out of the greater part of these radiations will in no way jeopardize vitamin D_2 formation, although it will prevent burning. Thus sunburn cannot really be prevented without affecting both tanning and vitamin D synthesis but its prevention need not interfere with these processes to a marked extent.

Sunburn preventives

Preventives owe their efficacy to the use of a sunscreening agent which absorbs all, or a majority, of the radiations of wavelengths between 2900 Å and 3100 Å. Various factors other than the duration of the exposure should also be taken into account. The background of the exposed subject is important. Snow, for example, reflects a higher proportion of ultra-violet radiation than does sand, and consequently increases the effect on the individual. Some consider that the sea must increase the effectiveness of ultra-violet radiation, whilst wetting of the skin or the application of certain oils will cause sensitization to erythemogenic light. Atmospheric conditions are also important. The light from an overcast sky contains a relatively higher percentage of ultra-violet radiation than direct sunlight. The ideal sunscreening agent should therefore:

1. Absorb light preferentially over the range 2900–3100 Å.
2. Be stable to heat, light, and perspiration.
3. Be non-toxic and non-irritant.
4. Not be rapidly absorbed.
5. Be approximately neutral.
6. Be readily soluble in suitable vehicles.

There are numerous substances which are suitable for use as sunscreens. Some of these are appended:

Para-aminobenzoic acid, its derivatives and glyceryl esters.

Anthranilates, salicylates, cinnamic acid derivatives, dihydroxy and trihydroxy cinnamic acid derivatives, certain hydrocarbons, diazoles and triazoles.

Coumarin derivatives, quinine salts, quinoline derivatives, hydroquinone.

A method of comparing the relative absorption properties of sunscreen agents is to measure the amount of light absorbed by films of standard thickness using ultra-violet light of a wavelength 2967–3000 Å, that is corresponding to the peak of what is referred to as the 'erythema curve'. Data concerning the absorption spectra is readily available from manufacturers and can be used to tabulate the materials available in the form of an index. This information provides a guide to the selection of a suitable material for use in a particular formulation. It will be found that it is necessary to use different sunscreening agents at different concentrations to provide the equivalent effect. Thus whereas one compound is effective if used at a concentration of 2·0 per cent, it may be necessary to use 8·0 per cent of another. Economic factors, therefore, must be considered when selecting a sunscreening agent. The choice of material should also be considered in relation to the end-product. If the product is to be in the form of an emulsion, for example, a sunscreen agent may be dissolved in the aqueous or oil phase according to its solubility, and may thus affect the balance of ingredients of an emulsion system.

An oily sunscreen agent used at 8·0 to 10·0 per cent in a product will be an effective portion of the oil phase, whereas an alternative material used at a concentration of 2·0 per cent will require addition of 6 to 8 parts of oil to obtain the same type of system. For this reason in the formulae which are given to illustrate the types of product which are used, the active light absorption material is referred to as a 'sunscreen agent' and the proportion as 5·0 per cent. Adjustments must be made according to the physical properties of the active material, those of an oily nature being considered as part of the oil phase of the product. If a water-soluble material is used this is dissolved in the aqueous phase and the oil phase of the system in the case of an emulsion increased by 5·0 per cent.

The end-product can be in the form of an aqueous or alcoholic lotion, a fatty cream, an oil or an emulsion and choice of vehicle or other components of the product can in some cases contribute to its effectiveness. Certain natural oils such as coconut oil, peanut oil (arachis), and olive oil have a fairly high absorption for ultra-violet light in the 'erythema range', whereas mineral oils do not have any absorption in this range. When large areas of the body are to be covered with oil it is also useful to remember that a natural oil does not give such a greasy film as a mineral oil, as it is absorbed to some extent by the skin.

Effective bases can be prepared by using mixtures of natural oils and mineral oils, or by blending these with fatty acid esters such as *iso*-Propyl myristate or *iso*-Propyl palmitate. Vehicles of this type satisfy the popular demand for sunburn oil which are not too greasy in use. When a natural oil is included in a formulation, however, an antioxidant should always be included to prevent rancidity. It is also necessary to use an oil-soluble perfume in products based on mineral oils alone, but if a fatty acid ester or a natural oil is included in the product these materials will help perfume solubility problems.

Some measure of protection from sunburn is also provided by materials such as white or yellow petroleum jelly, and lanolin which coat the skin with a protective barrier. There are also several opaque materials which provide a protective coating on the skin and act as light sensitizing agents. These include materials such as titanium dioxide, zinc oxide, talc, kaolin, stearates, and carbonates. Specialized products for use as make-up for the face or for use in cases of lip allergies can be prepared containing these protective barrier materials with recognized sunscreening agents.

In the selection of formulae which follow chemical compounds known to be effective, sunscreen agents are given in some formulae, but proprietary products are not specified. In these cases the active material referred to simply as 'sunscreen agent' should be adjusted in the manner already mentioned. The proportion of proprietary sunscreens recommended by manufacturers should be taken as a guide. It is not wise to reduce the proportion used to an absolute minimum. Consider also that a formulation for a sunburn preparation which is effective in U.K. conditions will not necessarily provide adequate protection for exposure to the more intense sunlight which can occur abroad.

No. 2500
(Oil type)

Homomenthyl salicylate	80
Mineral oil	920
	1000

For oil type preparations add a suitable quantity of an oil-soluble brown dyestuff, and an oil-soluble perfume. As a preservative include 0·1 per cent of propyl parahydroxybenzoate. Dissolve this with the aid of gentle heat and add to the remaining ingredients previously mixed.

No. 2501
(Oil type)

Homomenthyl salicylate	80
Mineral oil	800
iso-Propyl myristate[1]	120
	1000

No. 2502
(Oil type)

Homomenthyl salicylate	80
Mineral oil	500
iso-Propyl myristate[1]	420
	1000

[1] *iso*-Propyl palmitate can be used as an alternative material to *iso*-Propyl myristate in this type of application.

The three formulae given above illustrate the use of blends of mineral oil with a fatty acid ester. Application tests will clearly indicate differences in use of these three preparations, and final selection can then be made.

No. 2503
(Cream type)

A	Homomenthyl salicylate	80
	Non-ionic emulsifier[1]	75
	Mineral oil	20
	Spermaceti	50
B	Glycerin	50
	Water (softened or distilled)	725
		1000

Perfume	0·3 per cent
Methyl parahydroxybenzoate	0·15 per cent
Propyl parahydroxybenzoate	0·02 per cent

[1] Abracol L.D.S. type–Bush Boake Allen.

Procedure: Melt together the ingredients of A and heat to about 70°C. In a separate vessel dissolve the preservatives in the glycerin with the aid of heat. Add the remaining ingredients of B and heat to about 75°C. Add B to A with slow, continuous stirring. Stir until cool, adding the perfume at about 30°C.

No. 2504
(Cream type)

A	Sunscreen agent	50
	Non-ionic emulsifier	75
	Cetyl alcohol	10
	iso-Propyl myristate	150
	Mineral oil	170
B	Glycerin	10
	Water (softened or distilled)	545
		1000

Perfume	0·3 per cent
Methyl parahydroxybenzoate	0·15 per cent
Propyl parahydroxybenzoate	0·02 per cent

Procedure: Melt together the ingredients of A and heat to about 70°C. In a separate vessel dissolve the preservatives in the glycerin with the aid of heat. Add the remaining ingredients of B and heat to about 75°C. Add B to A with slow continuous stirring. Stir until cool, adding the perfume at about 30°C.

No. 2505
(Cream type)

A	Mineral oil	100
	Castor-oil	50
	iso-Propyl myristate	25
	Glyceryl monostearate (self-emulsifying type)	50
	Sunscreen agent	50
B	Glycerin	25
	Water (softened or distilled)	650
		1000

Perfume	0·3 per cent
Methyl parahydroxybenzoate	0·15 per cent
Propyl parahydroxybenzoate	0·02 per cent
Butylated hydroxyanisole	0·002 per cent

Procedure: Melt together the ingredients of A and heat to about 70°C. In a separate vessel dissolve the preservatives and antioxidant in the glycerin with the aid of heat. Add the remaining ingredients of B and heat to about 75°C. Add B to A with slow continuous stirring. Stir until cool, adding the perfume at about 30°C.

No. 2506
(Cream type)

A	*iso*-Propyl myristate	150
	Acetylated monoglyceride (liquid type)	50
	Sunscreen agent	50
	Cetyl alcohol	25
	Glyceryl monostearate (self-emulsifying type)	130
B	Propylene glycol	80
	Sodium lauryl sulphate (needles)	1
	Water (softened or distilled)	514
		1000

Perfume	0·3 per cent
Methyl parahydroxybenzoate	0·15 per cent
Propyl parahydroxybenzoate	0·02 per cent

Procedure: Melt together the ingredients of A and heat to about 70°C. In a separate vessel dissolve the preservative in the propylene glycol with the aid of heat. Add the remaining ingredients of B and heat to about 75°C. Add B to A with slow continuous stirring. Stir until cool, adding the perfume at about 30°C.

No. 2507
(Lotion type)

A	Homomenthyl salicylate	80
	Mineral oil	70
	Beeswax	20
	Stearic acid	60
B	Triethanolamine	30
	Propylene glycol	50
	Water (softened or distilled)	690
		1000

Perfume	0·3 per cent
Methyl parahydroxybenzoate	0·15 per cent
Propyl parahydroxybenzoate	0·02 per cent

Procedure: Heat ingredients of A to a temperature of about 70°C. Heat ingredients of B to a temperature of about 75°C and add B to A. Stir until cool, adding the perfume at about 35°C.

No. 2508
(Lotion type)

A	Sunscreen agent	50
	Mineral oil	100
	Stearic acid	20
	Paraffin wax	10
	Beeswax	20
	Petroleum jelly	50
	Silicone fluid[1]	80
	Polyethylene glycol monostearate	50
B	Triethanolamine	20
	Water (softened or distilled)	600
		1000

Perfume	0·3 per cent
Methyl parahydroxybenzoate	0·15 per cent
Propyl parahydroxybenzoate	0·002 per cent

[1] MS 200 type–Midland Silicones Ltd.

Procedure: Heat ingredients of A to a temperature of about 70°C. Heat ingredients of B to a temperature of about 75°C and add B to A. Stir until cool, adding the perfume at about 35°C.

No. 2509
(Aqueous type)

Sunscreen agent[1]	50
Distilled extract of witch hazel	100
Propylene glycol	100

No. 2509 (*continued*)

Water (softened or distilled)	750
	1000

Perfume (water-soluble type)	2–3 per cent
Methyl parahydroxybenzoate	0·15 per cent

Add a suitable proportion of a water-soluble brown dyestuff.

Procedure: Dissolve the preservative in the propylene glycol, and add the remaining ingredients. Mix well.

No. 2510

(Aqueous-alcoholic type)

Sunscreen agent[1]	50
Alcohol	150
Glycerin	80
Water (softened or distilled)	720
	1000

Perfume (water-soluble type)	2–3 per cent
Methyl parahydroxybenzoate	0·15 per cent

[1] Water-soluble type.

Add a suitable proportion of a water-soluble brown dyestuff.

Procedure: Dissolve the preservative in the glycerin and add remaining ingredients. Mix well.

No. 2511

(alcoholic type)

iso-Propyl myristate	400
Sunscreen agent	80–100
Alcohol	520
	1000

Perfume	0·2 per cent
Methyl parahydroxybenzoate	0·1 per cent

Procedure: Mix the ingredients together until uniformly mixed.

This type of lotion is particularly clean in application and provides an excellent non-oily non-greasy vehicle for the application of a sunscreen agent. The lotion does not leave any noticeable residue, yet prevents the skin from becoming dry and scaly from the combined efforts of sun and wind.

A sunscreen prepared in the form of a gel is useful to apply on certain occasions and as the amount of the residual film on the skin is negligible it can be used without fear of soiling clothes. A suitable formula follows:

No. 2512
(Gel type)

Carboxy vinyl polymer[1]	20
Propylene glycol	90
Triethanolamine	15
Sunscreen agent[2]	50
Distilled water	825
	1000

Perfume (water-soluble)	2·5 per cent
Methyl parahydroxybenzoate	0·15 per cent

[1] Carpobol—B. F. Goodrich Co. Ltd.
[2] A water-soluble sunscreen is essential for this particular formulation.

Procedure: Dissolve the sunscreen agent in a small portion of the water. Dissolve the carboxy vinyl polymer in the triethanolamine and gradually add the main proportion of water. Dissolve the preservative in the propylene glycol and add to the mix. Finally, add the sunscreen agent dissolved in sufficient water. When the gel is uniform, add the water-soluble perfume.

No. 2513
(Gel type)

A	Phosphated oleyl ether[1]	40
	Polyoxyethylene oleyl ether[2] (10 mols)	120
	Polyoxyethylene oleyl ether[3] (3 mols)	40
	Mineral oil	120
	Sunscreen agent	80
B	Glycerin	100
	Water (softened or distilled)	500
		1000

Perfume	0·1–0·2 per cent
Sorbic acid	0·2 per cent

[1] Crodafos—Croda Ltd.
[2] Volpo N. 10—Croda Ltd.
[3] Volpo N. 3—Croda Ltd.

Procedure: Heat A and B independently to 75°C. and add B to A slowly with continuous stirring. Cool with slow stirring, to avoid entrapment of air, adding the perfume at about 35°C.

Certain individuals experience discomfort from exposure to the sun even if this is restricted to very small periods of time. This is often due to an allergic reaction caused by some earlier exposure to the sun which has caused the condition of photosensitivity. It is not uncommon for this reaction to occur on areas of the body such as the face, lips or neck to which many cosmetic materials have been applied, any one of which may be responsible for starting the sensitivity condition. For these special cases the skin or lips must be coated before exposure to the sun with a thick layer of a suitable sunscreen cream. A suitable product which is also recommended for use on dry as well as sensitive skins can be made in the form of a skin food to the following formula:

No. 2514

A	Sunscreen agent	100
	Mineral oil	300
	Petroleum jelly	30
	Lanolin	20
	Paraffin wax	10
	Beeswax	150
	Oil solution of vitamins A and D	10
B	Borax	10
	Water (softened or distilled)	370
		1000

Methyl parahydroxybenzoate	0·12 per cent
Propyl parahydroxybenzoate	0·02 per cent

Use of perfume is not recommended for this particular application.

Procedure: Heat ingredients of A to a temperature of about 70°C. Add the preservative and dissolve in the mix. In a separate vessel heat the ingredients of B together to a temperature of about 75°C. Add B to A with slow continuous stirring, and stir until cool. Fill warm.

The addition of titanium dioxide to this base to act as a light scattering agent will give an even greater degree of protection in cases of photosensitivity. Use up to 5·0 per cent of titanium dioxide and mill this to a smooth paste with part of the mineral oil before incorporating in A of the emulsion system. A somewhat similar product containing pigments, known in Europe as a sun maquillage, is used as a tinted foundation for protection of sensitive skin conditions during summer or at other times when sensitivity arises from exposure to sun or wind. A useful formula is as follows:

No. 2515

A	Mineral oil	100
	Sunscreen agent	100
	Non-ionic emulsifier[1]	30
	Cetyl alcohol	20
B	Water (softened or distilled)	510
C	Titanium dioxide	100
	Pigments	40
	Glycerin	50
	Water (softened or distilled)	50
		1000

Methyl parahydroxybenzoate	0·15 per cent
Propyl parahydroxybenzoate	0·02 per cent

[1] Emulsene 1220–Bush Boake Allen.

If this product is intended for sale specifically for use in cases of photosensitivity, choice of perfume should be restricted to one which has been tested and found non-sensitizing. It is recommended that the proportion used in such a product should be not more than 0·2 per cent.

Procedure: Heat A and B independently to about 75°C incorporating the preservatives in A. Add B to A with slow continuous stirring, adding any perfume when the temperature has fallen to about 35°C. Mix together into a paste the ingredients of C and add to the cream base. Mill.

Insect repellents

Whether sunbathing is carried out at the seaside or in the country, use of an insect repellent is often desirable.

Dimethyl phthalate is an excellent insect repellent, effective against most flying insects, including mosquitos. It is usually applied to the skin in the form of a cream or lotion containing about 35·0 per cent of dimethyl phthalate, although the author has found concentrations of 20.0 per cent to be effective in many conditions of use. Dimethyl phthalate is a faintly yellow coloured oily liquid miscible with most of the oily materials used either as an ingredient of or as a vehicle of cosmetic preparations. It can, therefore, be easily included in a sunscreen preparation conveniently combining the functions into one application. It has no serious irritant or toxic effects, and can be applied freely to the arms, legs, and body, but very often causes smarting when applied to the face. For this reason it should not be applied to particular sensitive areas near the eyes or to mucous membrane and preparations containing dimethyl phthalate should be labelled with a caution to this effect.

Dibutyl phthalate is a similar material also used as an insect repellent, but is considered to be slightly less effective than dimethyl phthalate.

2-ethyl-hexanediol-1,3 or ethohexadiol, also known as Rutgers 612, is also used as an effective insect repellent. It is a viscous and odourless hygroscopic liquid soluble in alcohol and miscible with oils, and suitable for use in many of the formula types which follow. This material also causes slight irritation when used on the face and neck and sensitive areas such as around the eyes, and products containing it should also carry a cautionary label to this effect. Dimethyl and dibutyl phthalates are effective plasticizers and should not be packed in plastic containers or allowed to come into contact with spectacle frames made from plastic materials. Contact with fabrics made from synthetic fibres should also be avoided. A warning to this effect should be included on the label of all products. The following formulae indicate the types of preparations used as insect repellents.

No. 2516

(Alcoholic type)

Dimethyl phthalate	250–400
Alcohol	750–600
	1000
Perfume	0·2 per cent

No. 2517

(Oily type)

Dimethyl phthalate	300
Castor-oil	20
iso-Propyl myristate	180
Alcohol	500
	1000
Perfume	0·2 per cent

No. 2518
(Oily type)

Dimethyl phthalate	350
iso-Propyl myristate	250
Mineral oil	400
	1000

Perfume 0·2 per cent

Add a sufficient quantity of an oil-soluble brown dyestuff if required.

A popular method of using dimethyl phthalate in cream form is to prepare a paste by simple admixture with stearates and gel the mix by heating.

No. 2519
(Paste type)

Dimethyl phthalate	750
Magnesium stearate	150
Zinc stearate	100
	1000

Perfume 0·2 per cent

Procedure: Mix to a paste then heat and stir gently until the mixture gels to a creamy consistency.

Use of an insect repellent with a sunburn preparation is illustrated in the following formulae:

No. 2520
(Oily type)

Sunscreen agent[1]	50
Dimethyl phthalate	250
iso-Propyl myristate	250
Mineral oil	450
	1000

Perfume 0·2 per cent

[1] Oil soluble type.

Add a suitable quantity of an oil-soluble brown dyestuff if required.

No. 2521
(Cream type)

A	*iso*-Propyl palmitate	50
	Sunscreen agent	80
	Dimethyl phthalate	200

No. 2521 *(continued)*

	Polyethylene glycol monostearate	100
	Cetyl alcohol	30
	Glyceryl monostearate (self-emulsifying type)	50
B	Glycerin	20
	Water (softened or distilled)	470
		1000

Perfume	0·2 per cent
Methyl parahydroxybenzoate	0·15 per cent
Propyl parahydroxybenzoate	0·02 per cent

Procedure: Heat ingredients of A to a temperature of about 70°C. Heat ingredients of B to a temperature of 75°C in a separate vessel. Add B to A with slow continuous stirring. Stir until cool adding the perfume at a temperature of about 35°C.

It is only necessary to use small proportions of perfume in all products containing insect repellents. Usually 0·1 to 0·2 per cent will be found quite adequate and a select light type of perfume such as a rose or cologne, preferably one which is known to have non-irritant and non-sensitizing properties.

Artificial suntans

An artificial suntan is obtained by staining the skin and nowadays dihydroxyacetone, which produces a semi-permanent stain, is mainly used for this purpose Dihydroxyacetone (dihydroxy-propanone) is a white crystalline powder with a characteristic odour and a sweet taste. It is not a dye or a stain but a staining agent which produces a brown stain due to chemical reaction with certain amino acids present in skin keratin. The light brown staining which occurs from three to six hours after application is gradually increased by exposure to light. It is, however, difficult to obtain an even level of staining due no doubt to variations in thickness of the epidermal layer and distribution of keratinized tissue. Products based on dihydroxyacetone must, therefore, aim to provide uniform distribution of the active material and it is the author's opinion that this is best achieved by using an emulsified product. If the product is based on a solution the vehicle should include a glycol which helps to stabilize the solution and provide a continuous film.

The colour of the stain obtained closely resembles natural suntan, but a warning should be made that this does not give any protection against sunburn. For this reason products can also be formulated to contain both dihydroxyacetone and a sunscreen agent to give an artificial quick tanning effect and also reduce the hazard of sunburn. Dihydroxyacetone is not considered to cause any irritant or sensitizing effects. As a result of experimental work the author considers that concentrations of from 3·0 to 3·5 per cent dihydroxyacetone produce the most natural looking effect provided the vehicle used gives a continuous film on the skin. The pH value of solutions should also be adjusted to a value of 6.0–6.5.

Concentrations of dihydroxyacetone above 3·5 per cent and solutions of lower pH value cause unsightly patchy staining. It is better to achieve a gradual increase in the colour of the stain by using several applications of a weak solution applied at intervals rather than by applying a single application of a more concentrated solution.

No. 2522

Dihydroxyacetone	30
Propylene glycol	60
Alcohol	30
Water (softened or distilled)	880
	1000

Perfume (water-soluble type)	2–3 per cent
Methyl parahydroxybenzoate	0·15 per cent

Procedure: Dissolve the preservative and dihydroxyacetone in propylene glycol and part of the water. Add alcohol and perfume and remainder of water. Mix and adjust the pH value of 6·0 with the addition of lactic acid.

No. 2523

A	Non-ionic emulsifier[1]	100
	Mineral oil	100
B	Dihydroxyacetone	30
	Propylene glycol	60

[1] Abracol L.D.S. type–Bush Boake Allen.

No. 2523 (*continued*)

Water (softened or distilled)	710
	1000

Methyl parahydroxybenzoate	0·15 per cent
Propyl parahydroxybenzoate	0·02 per cent

Use 0·1 to 0·2 per cent of a perfume which is non-irritant and non-sensitizing for artificial tanning products.

Procedure: Melt together ingredients of A and heat to temperature of 75°C. In a separate vessel dissolve the dihydroxyacetone and preservative in propylene glycol. Add the water and heat to 75–80°C. Add B to A with continuous stirring. Continue stirring until cool, adding perfume at about 35°C.

Products containing the staining effect of dihydroxyacetone and the protective action of a sunscreen agent are prepared as follows:

No. 2524

(Lotion type)

Sunscreen agent (water-soluble type)	100
Dihydroxyacetone	30
Propylene glycol	50
Alcohol	250
Water (softened or distilled)	570
	1000

Perfume (water-soluble type)	1·0 per cent
Methyl parahydroxybenzoate	0·15 per cent

Procedure: Dissolve the dihydroxyacetone and preservative in the propylene glycol. Add the sunscreen agent previously dissolved in the water. Add the perfume and alcohol and mix. Finally adjust the pH value to about 6·0–6·5 by addition of a solution of lactic acid. Do not use proportions of alcohol greater than 25 per cent as the more rapid dryout of a higher alcohol content will cause streaking.

No. 2525

A	Sunscreen agent (oil-soluble type)	100
	Non-ionic emulsifier[1]	25
	Ethylene glycol monostearate	20
B	Propylene glycol	80
	Dihydroxyacetone	30

[1] Emulsene 1220—Bush Boake Allen.

Water (softened or distilled)	745
	1000

Perfume	0·1–0·2 per cent
Methyl parahydroxybenzoate	0·15 per cent
Propyl parahydroxybenzoate	0·02 per cent

Procedure: Melt together ingredients of A and heat to 75°C. Dissolve the dihydroxyacetone in B and heat to a temperature of 75–80°C. Add B to A with continuous stirring. Cool with stirring adding the perfume at about 35°C.

Sunburn correctives

Correctives are intended to allay the skin irritation resulting from sunburn and are generally familiar as calamine or sunburn lotion. They contain one or more of the following:

Zinc oxide, carbonate, hydroxide.
Calamine (a prepared basic zinc carbonate).
Glycerin.

An antiseptic such as zinc phenolsulphonate (zinc sulphocarbolate), and camphor or phenol which also acts as a mild analgesic. Rose water is recommended as a suitable and adequate type of perfume for lotions intended for treatment of an inflammed or sensitive area of skin. An effective calamine type lotion is prepared as follows:

No. 2526

Calamine	150
Zinc oxide	50
Camphor	10
Alcohol	100
Glycerin	100
Lime water (or rose water)	590
	1000

Methyl parahydroxybenzoate	0·12 per cent

Procedure: Mix the calamine and zinc oxide to a smooth paste with the glycerin. Add sufficient lime water to prepare a cream. Add the camphor dissolved in the alcohol and finally add sufficient water to volume.

Calamine is coloured with ferric oxide and the appearance of this lotion can be improved by increasing the proportion of ferric oxide to obtain a more pronounced colour, which blends better with tanned or sunburned skin.

No. 2527
(Cream type)

A	Calamine	100
	Zinc oxide	50
	Mineral oil (or vegetable oil)	250
	Emulsifying wax[1]	50
	Camphor	10
B	Glycerin	50
	Water (softened or distilled) (or rose water)	490
		1000

Methyl parahydroxybenzoate	0·12 per cent
Propyl parahydroxybenzoate	0·02 per cent

[1] Emulsifying Wax B.P.

Procedure: Melt the emulsifying wax and mineral oil, and heat to a temperature of about 70°C. Add part of the water heated to 75°C and mix to a cream consistency. Dissolve the preservative and camphor in the glycerin and mix with the calamine and zinc oxide to a smooth paste. Add water to a creamy consistency, and mix with the emulsifying wax cream mixture.

Aerosol preparations

Sunscreen and insect repellent preparations are suitable for packing in aerosols as the aerosol form provides a convenient non-spillable package and a method of application giving rapid and even coverage of the product over large areas of the body. Formulae for aerosol preparations are given corresponding to the various types already described and the proportion of sunscreen agent should be adjusted according to the type and recommendations of the manufacturer in the manner previously indicated.

No. 2528
(Alcoholic type)

Sunscreen base:	
Sunscreen agent (alcohol-soluble)	50
iso-Propyl myristate	20
Alcohol 99% v/v	930
	1000

Perfume 0·2 per cent

Procedure: Mix raw materials together. Add the perfume and mix.

Container charge:	
Sunscreen base	40
Propellant 11/12 (50 : 50)	60
	100

Container: plain tin plate, aluminium monobloc or 2 piece aluminium with internally lacquered tin plate base. Valve: standard as supplied by the manufacturer.

No. 2529
(Oil based)

Sunscreen base:	
iso-Propyl myristate	100
Sunscreen agent (oil-soluble)	50
Mineral oil (cosmetic quality)	850
	1000

Perfume 0·2 per cent

Procedure: Dissolve the perfume in the *iso*-Propyl myristate, and add the sunscreen agent and mineral oil. Add a suitable proportion of an oil-soluble dyestuff if required and mix well.

Container charge:	
Sunscreen base	40
Propellant 11/12 (50 : 50)	60
	100

Container: Plain tinplate, aluminium monobloc or 2 piece aluminium with internally lacquered tinplate base. Valve: standard as supplied by the manufacturer.

Sunscreen preparations in cream form are generally preferred as a soft cream or mousse which is first expelled on to the hand before being applied to other areas of the body. Two formulae are given:

No. 2530
(Cream)

Cream base:		
A	Non-ionic emulsifier[1]	100
	Sunscreen agent (oil-soluble)	50
	Mineral oil (cosmetic quality)	30
	Cetyl alcohol	20

[1] Abracol LDS type—Bush Boake Allen.

No. 2530 (*continued*)

B	Propylene glycol	50
	Water (softened or distilled)	750
		1000

Perfume	0·2 per cent
Methyl parahydroxybenzoate	0·15 per cent
Propyl parahydroxybenzoate	0·02 per cent

Procedure: Heat A and B independently to 75°C and add B to A slowly with continuous stirring, adding the perfume at about 35°C.

Container charge:	
Cream base	92
Propellant–12/114 (50 : 50)	8
	100

Container: internally lacquered aluminium, internally lacquered tinplate, with a side stripe over the soldered seam. Valve: standard as supplied by the manufacturers, the cup to be lacquered and fitted with a foam button.

No. 2531

(Cream)

Cream base:

A	Sunscreen agent (oil-soluble)	50
	Stearic acid	17
	iso-Propyl myristate	60
	Propylene glycol monostearate (self-emulsifying type)	35
B	Triethanolamine	8
	Water (softened or distilled)	830
		1000

Perfume	0·2 per cent
Methylparahydroxybenzoate	0·15 per cent
Propyl parahydroxybenzoate	0·02 per cent

Procedure: Heat A and B independently to 75°C and add B to A slowly with continuous stirring. Cool with stirring adding the perfume at about 35°C.

Container charge:	
Cream base	90
Propellant–12/114 (50 : 50)	10
	100

Container: internally lacquered aluminium or internally lacquered tinplate with a side stripe over the soldered seam. Valve: standard as supplied by the manufacturer, the cup to be lacquered. Fitted with a foam button.

The artificial tanning properties of dihydroxyacetone are combined with a sunscreen agent in the following formula:

No. 2532
(Artificial tan)

Artificial tan and sunscreen base:	
Sunscreen agent (alcohol-soluble)	80
Propylene glycol	50
Dihydroxyacetone	30
Alcohol 99% v/v	840
	1000
Perfume	0·2 per cent

Procedure: Mix all ingredients together and add the perfume and re-mix.

Container charge:	
Artificial tan and sunscreen base	40
Propellant–11/12 (50 : 50)	60
	100

Container: internally lacquered aluminium or internally lacquered tinplate. Valve: standard as supplied by the manufacturer. The cup to be laquered.

An aerosol sunscreen combined with insect repellent properties is prepared as follows:

No. 2533

Sunscreen-insect repellent base:	
Dimethyl phthalate	200
Sunscreen agent (oil-soluble)	100
iso-Propyl myristate	450
Mineral oil (cosmetic quality)	250
	1000
Perfume	0·2 per cent

Procedure: Dissolve the perfume in the *iso*-Propyl myristate and add the remaining ingredients. Add a suitable proportion of an oil-soluble dyestuff if required and mix well.

Container charge:	
Sunscreen-insect repellent base	45
Propellent–11/12 (50 : 50)	55
	100

Container: plain tinplate or plain aluminium. Valve: standard as supplied by the manufacturer.

No. 2534
(Insect repellent)

Insect repellent base:

Dimethyl phthalate	500
iso-Propyl myristate	500
	1000

Perfume 0·1 per cent

Procedure: Mix and add the perfume.

Container charge:

Insect repellent base	45
Propellant—11/12 (50 : 50)	55
	100

Container: plain tinplate or plain aluminium. Valve: standard as supplied by the manufacturers.

CHAPTER FOURTEEN

Toilet Powders

Face powders constitute an indispensable article of my lady's toilet. She is able to find a powder of whatever quality may suit her purse, and the first thing that attracts her is perfume. This is, however, by no means the only feature which appeals to the discerning woman; she is interested also in colour, texture and adherence to the skin.

Assuming, then, that the perfume makes the first sale of a face powder, will an intriguing odour produce repeat sales? No, not necessarily; it is the appearance and finish given to the skin by the powder which induces the majority of women to buy, repeatedly, any particular make. This characteristic may almost be described as 'texture' although generally ladies apply the term rather to the 'fineness' of the powder—that is to say, to the dimensions of the silk sieve through which the powder passes in the course of its manufacture, thus producing fineness of sub-division and even distribution of all ingredients including colour. Texture, then, is a most important characteristic of a face powder, and is influenced by the following considerations:

(1) the powder must have good covering power and so hide skin blemishes;
(2) it must adhere perfectly to the skin and not blow off easily;
(3) it must not be completely dissipated in a few minutes and so make re-powdering continually necessary;
(4) the finish given to the skin must be preferably of a matt or peach-like character;

(5) shine on or around the nose must be completely eliminated. The powder must be absorbent;
(6) there must be sufficient 'slip' to enable the powder to spread on the skin by the puff without producing a blotchy effect;
(7) the constituents of the powder must be such that a clown-like effect is impossible. The preference should be rather towards one of transparency.

Raw materials

The main substances used to impart all these good qualities are as follows: Light kaolin, zinc oxide, titanium dioxide, light calcium carbonate (precipitated calcium carbonate), purified talc (talcum), light magnesium carbonate, stearates of zinc and magnesium.

Other materials which have been commonly used as ingredients of face powders but are now only used when indicated either by reasons of ease of availability or cost include the following: Zinc carbonate, barium carbonate, kieselguhr, and the salts of bismuth. Starch is an economical material included in some powders, but is not often used in modern formulations. Barium sulphate and calcium sulphate are often used in compacts because they help to prevent the compacts breaking.

Covering power

Covering power is obtained by the use of light kaolin, zinc oxide, or titanium dioxide. These materials give degrees of covering power in the order mentioned. Kaolin is light in use and has the greatest grease-resisting properties, being excellent therefore for oily skins; zinc oxide is heavier and titanium dioxide the heaviest. Using these materials face powders can thus be prepared with differing covering properties. Titanium dioxide is, however, often the material chosen to provide the covering properties of a face powder, a suitable proportion being used in preference to the higher proportion of zinc oxide and kaolin, which are required to obtain the same effect.

Zinc oxide may be used on economical grounds and for this reason is often preferred in talcum powders where less covering power is required. Only the finest qualities of zinc oxide should be used, since lower grades contain gritty particles. Zinc oxide also has a tendency to form balled particles and it is a good plan to sieve the material separately before mixing with any of the other ingredients of a particular formula. Only the finest B.P. quality of light kaolin should also be used again to ensure that the material is free from grittiness.

Adhesiveness

Adhesiveness is obtained primarily with the stearates of zinc and magnesium; zinc stearate is commonly used. Magnesium stearate has a slightly better covering power and is considered more adhesive than the zinc salt. Magnesium stearate is thus often preferred for high quality face powders, whereas zinc stearate is used extensively in talcum powders. Both materials are obtainable from high grade stearic acids and superfine qualities now prepared specially for the cosmetic industry are available with excellent colour and texture, and a minimum of odour. Both are light and fluffy products and when used in suitable proportions give to the powder an attractive velvety softness. Moreover they are 'waterproof' and tend to maintain the complexion in inclement weather. Stearates of lithium and calcium are also available and used with similar effect to zinc and magnesium stearate. Lithium stearate in particular has good covering properties, and a fluffy texture.

Magnesium and calcium salts of myristic acid are also available as speciality materials for use in high quality products. Magnesium myristate has a better texture than magnesium stearate, and a higher specific volume. It also has a more oily or greasy feel and is correspondingly more adhesive to the skin. Because of these properties very good effects can be obtained by using only small proportions of the material.

Complete dissipation on the skin in a short time is very noticeable in those powders which are prepared with an exceptionally large proportion of talcum. Talc is used, however, as the main ingredient of many face powders. Talc of cosmetic quality is purified hydrated magnesium silicate. It is obtained from several countries, the best known available being of Italian of Indian origin. The Italian variety is by far the best quality and recommended both for face powders and talcum powders. It is much whiter in colour and finer in texture than the Indian. Italian talc is also much less liable to be contaminated with micro-organisms, an important factor when considering manufacture of talcum powder for baby use. Talc is, however, often sterilized before use which overcomes this particular problem. Talc imparts the property of silkiness and provides slip to a product and these properties can be demonstrated by rubbing between the fingers or on the back of the hand. This test also illustrates the degree of fineness and a sample should not show many particles of a crystalline character when reviewed obliquely after spreading finely in this manner. Special grades of micronized talc are available, and are of even texture and bulk, and used in compacted products. They are also essential for aerosol products.

Peach-like finish

The peach-like finish given to the skin by face powders can be achieved by use of rice starch. Other starches are used in cheaper powders, maize starch being often employed. As is well known, however, the particles of rice starch are much finer than those of any other cereal and a distinct preference is, in consequence, shown for it. Rice starch is prone to absorb moisture and should be placed in a drying oven for two to three hours before use.

Potato starch should be avoided. The question of a shiny nose, is, incidentally closely associated with starches. This particular part of the anatomy is inclined to perspire and powders which are absorbent, such as starch or kaolin, more readily rectify this blemish. Rice starch is particularly absorbent and thus favoured for this additional reason. The objection taken to its use is centred in the belief that as starch absorbs moisture so it swells and distends the pores. In modern types, therefore, it is seldom found, and the peach finish is obtained instead with light precipitated calcium carbonate.

Occasionally speciality materials such as silica or powdered silk are included in face powders. Silica is available in a very finely divided form of a very fluffy nature, and used to increase the bulk. It also has very good adhesion properties and is used to prepare powders of a light transparent type.

Powdered silk is obtained by a partial hydrolysis of silk protein and the hydrolysate is then ground to an impalpable powder, and special grades are available to the cosmetic industry which are light and bulky and of a good white colour. It should not be used in excess of 30 per cent. Light magnesium carbonate is also used for the same purpose, but at a lower concentration. Both materials are used to absorb perfume which then helps to disperse the perfume throughout the remaining ingredients of a formulation:

	(per cent)
Light kaolin	20–50
Zinc oxide or carbonate	5–30
Titanium dioxide	1–5
Rice starch	20–50
Light magnesium carbonate	5–10
Light calcium carbonate	5–30
Magnesium stearate	1–5
Zinc stearate	1–10
Talcum	50–75
Magnesium myristate	1–5

The process of sifting and mixing is important. Several raw materials, particularly starch, chalk, and talcum, are bought in sacks, and although frequently lined, it is amazing how much extraneous matter seems to find its way inside during transit. This matter is, however, readily eliminated by means of a 120-mesh sieve, and silk is unquestionably the finest material for this purpose. Substances such as zinc oxide do not sift readily; they clog the sieve, particularly if it is a metal one. Silk does not possess this disadvantage, and is almost universally employed. For small operations the raw materials may be brushed through twice, but for quantity production machines are indispensable. The powders are emptied into a hopper on the floor level and carried by cup-conveyor to be deposited on the end of an inclined revolving silk drum. This has a small 'knocker' which prevents clogging. The extraneous matter falls from the lower end of the drum, while the fine ingredients pass through in the course of its revolutions. Below is a rapidly rotating shaft having bent arms, which ensure perfect mixing.

Face powder bases

The most popular face powders consist of any of the above substances, judiciously combined, but all should contain either some zinc oxide or titanium dioxide as the material to impart a degree of covering power. The proportion of ingredients used can be altered to give slight variation in the properties of the powder, but, as a rule, powders are classified as being of a light, medium, or heavy type, although they must not be either too light or too heavy. A selection of formulae for face powder bases follow:

No. 2535
(Light type)

Zinc stearate	70
Zinc oxide	100
Light calcium carbonate	200
Talc	630
	1000

No. 2536
(Medium type)

Zinc stearate	70
Zinc oxide	100
Titanium dioxide	20
Light calcium carbonate	200
Talc	610
	1000

No. 2537
(Heavy type)

Magnesium stearate	50
Light kaolin	200
Zinc oxide	150
Light calcium carbonate	400
Talc	200
	1000

No. 2538
(Heavy type)

Titanium dioxide	50
Zinc oxide	150
Magnesium stearate	50
Light calcium carbonate	150
Light kaolin	250
Talc	350
	1000

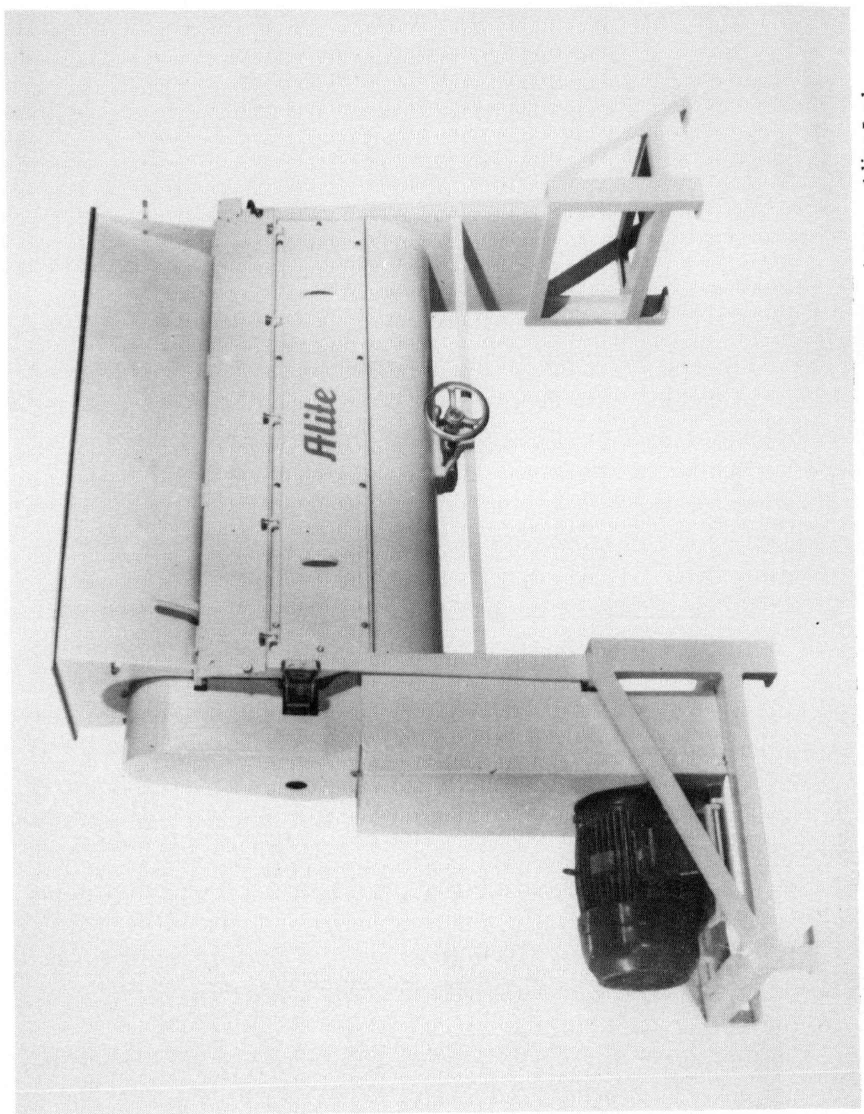

Plate 7 Sifter-mixer used for manufacture of talcum and face powders—Arenco-Alite Ltd.

No. 2539
(Medium type)

Zinc stearate	150
Zinc oxide	175
Light calcium carbonate	200
Talc	475
	1000

Colours

The colouring of a face powder is designed to produce a wide range of shades suitable to blend with and enhance the natural complexion, and to do this a long programme of experimental work must be undertaken. A suitable range is thus based on shades designed for the blonde and fair skinned, the brunette and dark skinned user, and those who require a face powder to tone down a naturally florid type complexion. The so-called 'natural' range originally consisting of pinkish shades was designed for use by the fair skinned, but these are now required to be of a more subtle character to avoid an artificial appearance. Similarly the original cream or rachel shades based on high proportions of ochres are modified to give more subtle tones. Blue, green and brown and even black pigments are used to achieve the desired effects which are essentially designed to blend with and enhance the natural complexion.

The names ascribed to modern shades are consequently more often selected for their sale and psychological appeal rather than a connection with the actual colours used. The exceptions to this fashion are shades such as tan or olive which are always acceptable for dark or suntanned skins.

Colours used for face powder must have good dispersing properties. A concentrated pigment invariably contains hard aggregates which are formed during the drying proceeds of manufacture, and these are not broken down by subsequent milling. Consequently if a concentrated pigment is used it must first be milled, not only to reduce the particle size but as a means of removing all hard aggregates which occur in concentrated pigments. Colours specially prepared for face powder are available from manufacturers. These are reduced colours prepared during the manufacture of the pigment itself when the pigment is reduced on bases used in the face powder compositions. Experimental work on blending must therefore be carried out using pigments which have either been specially milled to remove agglomerates or preferably using the specially manufactured semi-prepared colours.

For experimental work a further dilution of the pigment should be made using one part of the pigment mixed with nine parts of calcium carbonate. This mixture will not require any further milling, but should be mixed in a small sifter and mixing machine specially reserved for this purpose. A similar procedure of dilution should be carried out when the process is transferred to the manufacturing scale, that is before the addition of colours to the bulk powder used for the powder base. Careful attention to this procedure ensures even dispersion of the pigments and reduces the limit of error both in experimental work and on the manufacturing scale.

The colours for face powder, rouges and eye make-up powders are mainly based on inorganic earths, oxides, and ultramarines. All these naturally occurring materials are suitable for use in powder based products. Iron oxide pigments for example, vary in colour through red, yellow and brown, including ochres, ambers, and siennas. These materials, however, occur in various degrees of purity and generally contain high levels of lead and arsenic. They are now manufactured synthetically by a wet oxidation process and by variations in the process a complete range of colour shades is obtained. Generally known as 'synthetic inorganics' or 'cosmetic oxides' they contain low levels of impurities, are of small particle size, and high colour strength. Most important they can be manufactured to agreed colour standards and an adequate range of colours from greenish yellow through ranges to purple, reds, browns, and blacks can be obtained by judicious blending of selected batches of the manufactured synthetics. Some indication of these is given below:

Cosmetic yellow ochre
Cosmetic red oxide
Cosmetic umber
Cosmetic brown oxide
Cosmetic black oxide
Cosmetic hydrated chromium oxide (blue/green)
Cosmetic oxide (yellow/green)

Ultramarines are also available as for example:

Cosmetic ultramarine blue
Cosmetic ultramarine violet

extending similarly through a range of shades. These colours are available in a reduced form specially prepared for use in coloured powder products. This helps to obtain good consistent dispersions in the product and also reduced variations of supplies to a minimum. Nevertheless manufacturers are advised to prepare standard dispersions in a medium such as talc or light calcium carbonate in order to compare the tinctorial properties of individual batches of colour as received from the supplier. Small variations in batches of colour can cause trouble and checks against a standard avoid adjustments having to be made to the end product. Use of reduced colours also minimizes bleeding effects which can be troublesome in powder products and occurs particularly with the concentrated forms. Resistance to bleeding is one of the most important factors required for a face powder colour. This is the result of a solvent effect and is caused mainly by perfumes. Any perfumes to be used in a face powder should be checked to see that no solubilization takes place. A test should also be made to check for bleeding in the presence of water which can be present in use as skin preparation.

Pigments which bleed cause the product to change colour on the skin or powder puff and are responsible for streaky effects in use. It is for this reason that water-soluble dye-stuffs should never be used in powder used for face or eye make-up. Similarly any organic pigments which are sometimes considered for their brightening effect should be carefully screened before use.

Another type of inorganic pigment used for colouring powder products is the range of metallic colourants used to give gold, silver, and similar metallic effects. Aluminium powders are used to obtain silver effects and gold or bronze colours are obtained using copper and copper alloys which have also been reduced to fine powders by milling.

Other coloured metallic shades such as blues and greens are prepared from powdered aluminium by treating the powder with a solution of a basic dyestuff. In addition to the normal range of powder colours, special shades are prepared which are designed to give a luminous and translucent effect on the skin. The luminous or transluscent and shimmer effect is obtained by using the following materials:

(1) natural pearl essence (guanine);
(2) bismuth oxychloride;
(3) synthetic pearl pigments.

Natural pearl essence (guanine) is obtained from fish scales and is generally available as a dispersion in castor-oil. It occurs as thin narrow platelets and produces excellent shimmering silver effects as a result of light reflection from the crystals as they occur in depth through a dispersing medium. The high cost of the material does, however, restrict its use.

Bismuth oxychloride is used as a substitute for natural pearl essence and although it gives a good pearl effect does not have the brilliant lustre of the natural product. It is now prepared in special grades for cosmetic use and is also available as bismuth pearl pigments.

Synthetic pearl pigments can be obtained in various shades including white, gold, and pink. They are based on mica flakes coated with titanium dioxide. Several grades are available and these vary in their brilliance or lustre effect according to the particle size of the platelets.

To prepare face powders incorporating one or more of the pearlized colourants, reduce the content of the cosmetic earths and oxides so that the base colour is less opaque. This permits maximum effect of the pearlized material and the natural complexion is enhanced by the translucent lustre or frosted effect. Sometimes a pearlized or frosted powder is dusted over the more dense normal face powder to produce highlights. Similarly when formulating, reduce the proportion of heavy opaque materials which tend to diminish the lustre. On the other hand the proportion of talc, which itself has a faint lustre, can be increased.

A basic light type formula (No. 2535) can be modified to prepare a pearlized or frosted face powder as follows:

No. 2540

Magnesium stearate	70
Zinc oxide	80
Light calcium carbonate	100

TABLE 14.1
Rose compounds

Raw material	*Grade I*	*Grade II*	*Grade III*
Rhodinol	100	120	–
Nerol	20	–	–
Citronellol	100	40	–
Linalol	–	50	100
Geraniol 1 (Palmarosa)	120	200	–
Geraniol 2 (Java citronella)	–	70	250
Geraniol 3 (Ceylon citronella)	–	–	100
Phenylethyl alcohol	120	300	250
Phenylethyl propionate	200	–	–
Citronellyl butyrate	1	–	–
Geranyl acetate	3	60	8
Alpha ionone	120	50	–
100 per cent ionone	–	30	80
Patchouli oil	1	1	1
Vetivert oil	–	1	2
Nonyl aldehyde	1	1	1
Ethyl cinnamate	–	1	2
Rose otto	60	10	–
Rose absolute	30	–	–
Diphenyl methane	–	–	10
Sandalwood oil	–	10	70
Musk ketone	3	5	5
Vanillin	1	1	1
Benzyl benzoate	70	50	120
Musk extract, 3 per cent	20	–	–
Civet extract, 3 per cent	30	–	–
Total	1000	1000	1000

It should be noted that there are also many other raw materials which may be used, but which have not been mentioned above. Among these are the various geranium oils–particularly French geranium distilled over roses, eugenol, cinnamic alcohol, heliotropin, coriander oil, etc.

When experiments are commenced upon a bouquet compound there are no definite limits as to the possible raw materials which may be used, and in consequence the creation takes upon itself a much more difficult aspect than that of the flower compound already dealt with. In the case of a jasmin bouquet, substances outside the usual straightforward jasmin compound constituents have to be tried and blended for several weeks to gauge the odour of the finished creation. The ultimate object is to produce something having a jasmin odour note, but an underlying fragrance of an intangible nature—a background of numerous flowers as it were, with the fragrance of jasmin flowers predominating. With no price to consider, the perfumer has the whole gamut of odours to experiment with regardless of cost. In a low-priced article, however, considerable reliance has to be placed on the crystalline synthetics for the background, and oils of the jasmin type, plus

TABLE 14.2
Jasmin bouquet

Raw material	*Grade I*	*Grade II*
Jasmin compound	100	100
Bergamot oil	100	100
Ylang-ylang oil, Manilla	90	–
Ylang-ylang oil, Bourbon	–	120
Neroli oil, Bigarade	10	–
French petitgrain oil	–	30
Orange blossom absolute	15	–
Rose otto	15	–
Rose compound	–	30
Rhodinol	–	70
Concrete orris oil	25	–
Beta ionone	–	5
Methyl ionone	130	–
Alpha ionone	–	100
Carnation compound	120	–
iso-Eugenol	–	100
Acetyl *iso*-Eugenol	–	20
Amber synthetic	100	100
Heliotropin	60	60
Vanillin	10	10
Coumarin	40	40
Musk ketone	30	30
Benzoin resinoid, Siam	25	25
Peru balsam	10	10
Jasmin chassis absolute	50	–
Benzyl acetate	–	48
Amyl cinnamic aldehyde	–	2
Civet extract, 3 per cent	50	–
Ambergris extract, 3 per cent	20	–
Total	1000	1000

benzyl acetate as the principal odour note. For the purpose of these notes we will assume that the perfumer has already available in his laboratory a good-quality artificial jasmin compound made without natural jasmin; then it can be used in both the expensive and cheap perfume. Using this as a base, the fresh flowering note may be improved and modified with bergamot and ylang-ylang oils. In the latter case there are three qualities available, the Manila, Bourbon, and the commoner cananga oil. The quality of odour, as usual, bears a direct ratio to the cost price. These oils may be further modified with the neroli oil, orange blossom absolute, rose otto, concrete orris, methyl ionone and artificial carnation. Body and fixative value may be given to these oils by the use of amber synthetic, heliotropin, vanillin, coumarin, artificial musks, benzoin and Peru balsam. Finally, the jasmin flowery note is imparted by additions of natural jasmin chassis absolute. In the example below the cheaper substitutes are placed immediately below each ingredient. (See Table 14.2.)

When using a well-blended compound in either face powders or compacts, it is best to employ about 2 per cent of perfume. This is first rubbed down with an absorbent powder, such as light magnesium carbonate or calcium carbonate and added to the other powders before sifting and mixing. In the case of compacts it is well to remember that the perfume does have a slight influence on the final colour and tends to darken it. When compacts dry out after 2 or 3 days, it will be noticed that the shade of colour is slightly lighter than when first pressed. In general, about one-quarter the weight of colours is necessary to give the same shade to a finished compact as is already standardized in the existing loose face powder.

Compact powders

Compact powders have enjoyed many years of popularity chiefly on account of their convenience, for they can be carried in the handbag and deftly used without any danger of soiling clothing. They consist of an ordinary face powder mixed with a suitable blending agent and the mix is compressed either by hand operation or machine into cake form. The composition of the mix must be such that the compacted powder must come off easily when rubbed with a powder puff, and the addition of binding agent must be so controlled that the compact must not break under normal conditions of use and storage. There are three distinct processes of production, i.e. damp, wet, and dry. As a rule the bases for the first type contain a fairly large percentage of talcum, but are never overloaded with starch as this makes the compact too hard, and when rubbed with a puff it becomes shiny and 'slate-like'. On the other hand, the presence of some starch is desirable because of its binding properties which prevents the tablet disintegrating to powder. When only a small proportion of starch is included in the powder base another binding agent must be included in the formula, and materials such as light magnesium carbonate, acacia or tragacanth gums are used for this purpose. Gum resins such as benzoin or tolu and styrax are also useful. Compact powders prepared in this way are comparatively hard, although correct formulation and manufacturing technique does ensure that the powder is readily transferred to the puff. A distinction should clearly be made between the compact powder and the modern type of 'cream' compact or compressed powder which contains fatty materials as part of the binding material and produces a softer product which although when correctly made maintains its compressed form and is more readily transferred to the puff. It is probably true to say that the original type of compact powder is now only used for the manufacture of rouge and eye make-up preparations.

Cake make-up

Cake make-up is in some ways a type of powder compact, but in this case the normal constituents of a face powder are mixed with both oil and water miscible materials to a stiff paste-like consistency sometimes compared to a powder cream. This is dried, finely ground by milling and compressed into a cake. They generally contain a higher proportion of pigments than loose face powder. The cake is used by moistening with a wet application sponge which reconstitutes a power cream in situ. This is applied in wet form and allowed to dry on the skin as a fairly thick film. Consequently its use and popularity is governed to a large extent by the tastes of a market or a fashion vogue which requires a heavy type of make-up. As a heavy type make-up it is acceptable for use in artificial lighting and is also useful to hide skin blemishes. A patent issued in 1937 (British Patent No. 501732) prescribes the composition of a cake make-up as follows:

(1)	Pigments and colours	12–50 per cent
(2)	Oily and Waxy constituents	0·8–24 per cent
(3)	Water-soluble (water miscible dispersing agents)	1·0–13 per cent
(4)	Fillers	35–80 per cent

Two types of formulae are given for the powder base as follows:

No. 2541

Zinc oxide	150
Light calcium carbonate	300
Talc	430
Pigments	120
	1000

No. 2542

Light calcium carbonate	100
Calcium sulphate	120
Rice starch	40
Light kaolin	300
Talc	320
Pigments	120
	1000

Use either of the following formulae to prepare the cream base used as a binder. The second formula is more fatty and forms a thicker cream when the end product is applied to the skin.

No. 2543

A	White petroleum jelly	100
	Lanolin	50
	Glyceryl monostearate (self-emulsifying type)	100
B	Propylene glycol	100
	Water (softened or distilled)	650
		1000

Methyl parahydroxybenzoate 0·2 per cent

No. 2544

A	Petroleum jelly	100
	Lanolin	30
	Mineral oil	100
	Glyceryl monostearate (self-emulsifying type)	75
B	Glycerin	100
	Sodium lauryl sulphate (powder or needles)	20
	Water (softened or distilled)	575
		1000
	Methyl parahydroxybenzoate	0·2 per cent

First prepare the coloured powder by milling the pigments with a suitable proportion of calcium carbonate as a diluent. Disperse the perfume (2 per cent) in the remaining calcium carbonate. Mix the remaining power ingredients in a sifter mixer and add the pigment mixture and perfume mixture. Set aside until required.

To prepare the cream base melt together the ingredients of A to a temperature of about 70°C. Heat ingredients of B to about 75°C. Add the preservative and add B to A with slow continuous stirring until a smooth cream is formed. Now mix the powder base with the cream base in a dough type mixer using 2 parts of powder base to 1 part of cream base. Mixing is best carried out by placing about one third of the cream base into the mixer and adding sufficient of the powders to form a paste. Following this by adding more cream base followed by powder base and so on until the whole has been added. This process takes some time and stirring is continued until a perfectly smooth and even coloured paste is obtained. It is permissible to add a small amount of additional water if the mass does not appear sufficiently smooth. Care should be taken not to add too much water in an attempt to reduce mixing time, since this will only add to the length of drying process later.

The mass is then removed and broken into suitably sized pieces which are spread on to trays lined with greaseproof paper, ready for drying. Drying is carried out by racking the trays and placing in a current of warm (not hot) air. Fierce heat not only destroys the perfume but also causes alteration to the colour on the edges of the mass. Drying is best carried out in two stages. At the half way period the mass is removed and then crumpled to a relatively fine powder and drying is continued until the mass is dry enough to be sieved. When completely dry the powder is sieved, re-mixed, and pressed into cakes using either a hand or machine operated press.

Compressed powders

The modern type of compact powder is that known as compressed powder, also known as cream powder, or simply pressed powder, and although it cannot be said to have replaced loose face powder it has certainly reduced the demand. This, in like manner to the original compact powders is due, no doubt, to convenience in use, and also because it contains fatty materials which help adherence to the skin, and gives good covering power without the necessity of applying a foundation cream.

Both in composition and behaviour the compressed powder can be regarded as intermediate between the original compact and the cake make-up. Compressed powders are also made with suitable colours for use as eye-shadow powders. They are made by using either a wet mix and drying process in a somewhat similar fashion to a cake make-up or by a dry compression method. With either process it is essential to control the pressure of the compressing machine so that the product is neither too hard, and thus resembles a compact or cake make-up, nor is it too soft so that it crumbles to the touch. The finished product can be varied and modified to specific requirements depending upon the following factors.

1. Composition of the dry powder base
2. The composition and proportion of fatty material used.
3. Mixing and sieving techniques.
4. The pressure control of the pressing machine during compression of the prepared base. Care must be maintained to ensure that neither the pressure nor the weight of the mix used in each stamping varies, otherwise this will result in variable and faulty products.

Manufacture of compressed powder products, therefore, requires considerable technique, and experimental batches should always be made to vary the fatty content, and to judge its effect using a certain degree of compression per standard weight of powder. Then adjust the content of fatty material and again observe the effect of the same degree of compression. By following this rather laborious technique the correct conditions for a specific mix, fat content, and pressure will be established to obtain a satisfactory product. The ideal product should not be too hard, not too soft. It should rub off easily on to the puff and contain sufficient fatty material to make it adhere to the skin. Suitable powder bases can be prepared as follows:

No. 2545

Titanium dioxide	75
Kaolin	300
Zinc stearate	100
Talc	525
	1000

No. 2546

Zinc oxide	100
Titanium dioxide	50
Magnesium stearate	100
Kaolin	200
Talc	550
	1000

No. 2547

Zinc oxide	100
Titanium dioxide	50
Zinc stearate	100
Kaolin	250
Talc	500
	1000

The powder base usually contains a higher proportion of zinc and/or titanium oxides than that used in loose face powder, and consequently it is also necessary to use a higher percentage of pigments. Use about 2 per cent of perfume. From 3 to 4 per cent of fatty material is added by spraying onto the powder base. The fatty material can be lanolin (anhydrous) used alone or mixed with suitable oily materials to break up the stickiness and consequent difficulty in handling, or a lanolin derivative such as acetylated lanolin alcohols (Acetulan—American Cholesterol Products, Ltd.). Suitable blends are as follows:

No. 2548

Acetylated lanolin alcohols	900
Petroleum jelly	100
	1000

No. 2549

Lanolin anhydrous	800
iso-Propyl myristate	200
	1000

No. 2550

Lanolin	700
Cetyl alcohol	150
iso-Propyl myristate	150
	1000

No. 2551

Acetylated lanolin alcohols	700
Cetyl alcohol	150
Petroleum jelly	150
	1000

Melt the fat-wax mixture and spray onto the powder base. Adequate mixing with the base is best obtained by first making a concentrated mix with part of the powders, adding and mixing with the bulk and milling with a Kek type pin mill. Finally, pass the mix through a sifter-mixer machine and compress.

The alternative method of manufacture is to prepare the fatty material into an oil-in-water emulsion before mixing with the powder base. If this method is used about 10–15 per cent of the emulsion is mixed with the powder, depending on the fat concentration to give a concentration of 3–5 per cent fatty material in the finished dried product. The method of mixing is the same as for the non-aqueous additive. If the emulsion is used the mix must be spread out on trays for 2 or 3 days to dry out before compressing. Formulae for emulsion bases are as follows:

No. 2552

A	Lanolin	150
	iso-Propyl myristate	25
	Petroleum jelly	50
	Glyceryl monostearate (self-emulsifying type)	100
B	Water (softened or distilled)	675
		1000

Methyl parahydroxybenzoate 0·02 per cent

No. 2553

A	Lanolin	100
	Cetyl alcohol	25
	Stearic acid	200
	Mineral oil	50
B	Glycerin	80
	Borax	15
	Water (softened or distilled)	630
		1000

Methyl parahydroxybenzoate 0·02 per cent

It will be seen from these formulae that the composition of the fatty additive can be varied considerably and each formulation can easily be made in small quantities and used to check the adhesive properties by pressing between the figures or stamping in a hand operated press.

Having obtained a satisfactory powder product the adjustments to pressure must then be made with the compressing machine to obtain the correct pressure combined with the most suitable amount of take-off of the finished compressed powder.

Talcum powder

Talcum powder continues to enjoy a very wide sale. It is used to absorb moisture after bathing and especially in warmer countries to absorb perspiration. It acts as an efficient lubricant and preventive and tends to allay any irritation of the skin due to chafing. In view of the fact that this type of powder contains a very large proportion of talc, some remarks regarding standards of purity are given.

1. The primary physical characteristic is that of colour. Best quality materials are white, second quality products are greyish-white, and others which should be avoided, are distinctly grey.
2. The next important point to notice is that of 'slip'. This peculiar characteristic is best illustrated in talc of good quality. Inferior qualities do not exhibit the requisite amount of 'slip', and since the lubricating action is most essential for good quality products, only good quality talc with adequate 'slip' properties should be used.
3. The last physical characteristic worthy of mention is that of lustre. A good sample should be lustrous without exhibiting any undue amount of glitter when examined in a fine film. Unusual radiance or glitter is often due to insufficient grinding or sifting of the powdered material and such samples should not be used.

The physical characteristics of talc vary according to their country of origin and it is generally acknowledged that the finest talc of good white colour, high lustre, and adequate slip properties is that of Italian origin. Indian talc of good quality is also available. Samples of talc, do, however, vary and in addition to careful examination of physical characteristics a bacteriological examination is desirable since certain qualities have been found to be heavily contaminated with bacteria. In such cases the talc should be sterilized before use.

Talcum should also be submitted to a chemical examination. This consists of estimating the percentage of acid soluble matter. Absolutely pure talc is not obtainable since the mineral deposits contain varying amounts of impurities, consisting chiefly of iron. Any sample containing over 4 per cent of matter soluble in hydrochloric acid should be rejected. Much has been said regarding the comparative values of Italian and American talcs, but the author has examined numerous specimens and has found the former generally superior. The average sample has given the following figures:

Italian talc	1 to 2 per cent soluble in HCl
American talc	3 to 4 per cent soluble in HCl

The absence of a large quantity of impurity is desirable, largely on account of its deleterious influence on the perfume, and to some extent owing to the irritation it may cause to the skin. In the former case, it will not only tend to alter the odour, but it often affects the colour of certain perfume ingredients. From these hints, the reader will readily appreciate the importance of the high degree of chemical purity required in a good talcum.

Many of the talcum powders sold are nothing more nor less than French chalk which has been suitably perfumed. Better quality products contain a metallic stearate to give additional lubricating properties. Zinc stearate is considered to have beneficial dermatological properties being soothing and mildly antiseptic and is preferred to magnesium stearate.

Zinc oxide is also used in talcum powders. This is a mild astringent for the skin and is also soothing and has some healing effect to skin chafing or minor abrasions. It is also useful to relieve the prickling and irritation which occurs in prickly heat.

Light magnesium carbonate or light calcium carbonate are included to increase fluffiness and also for their value as absorbing materials to aid the addition of perfume during the manufacturing procedure, as described under face powders. The following are typical formulae for talcum powders:

No. 2554

Zinc stearate	50
Light calcium carbonate	250
Talc	700
	1000
Perfume	*q.s.*

For all talcum powders mix the perfume with the light calcium carbonate and allow to stand, preferably overnight, before mixing with the other ingredients. A concentration of 0·5 per cent of perfume is generally adequate. Concentrations higher than this are often overpowering in their effect.

No. 2555

Zinc stearate	50
Zinc oxide	50
Light magnesium carbonate	150
Talc	750
	1000

After-shave talcum is smoothed into the skin after drying the face to give a clean matt appearance. Its alternative name is invisible talc and it is made by adding a light suntan pigment to an ordinary talcum formulation. The content of talc should be high, however, as the powder is not required to have noticeable covering properties.

No. 2556

Zinc oxide	40
Light calcium carbonate	80
Talc	880
	1000

Perfume	0·2 per cent
Cosmetic oxide pigment	Sufficient

Deodorant powders

To prepare a deodorant powder include zinc oxide and zinc stearate for their mild astringent effect, and a suitable bactericide, such as chlorhexidine diacetate or bithional, to suppress proliferation of the micro-organisms responsible for development of perspiration odour. Suitable formulae are as follows:

No. 2557

Zinc oxide	50
Zinc stearate	50
Chlorhexidine diacetate	3
Light calcium carbonate	100
Talc	797
	1000

Perfume	0·2–0·5 per cent

No. 2558

Zinc oxide	100
Zinc stearate	50
Bithional	5
Light calcium carbonate	50
Talc	795
	1000

Perfume	0·2–0·5 per cent

Baby powders

Baby powders closely resemble talcum powders, having similar composition but they usually contain an antiseptic material. Boric acid has been used for this purpose for many years. It is known, however, that boric acid accumulates in the body when administered orally and similarly when it is applied as a powder it can be absorbed by broken skin tissue or mucous membrane. Because infants and young children are particularly susceptible to boric acid poisoning it is only recommended to be used in powders at concentrations up to 5 per cent, and even so, such powders should not be applied freely to any broken areas of skin. Formulae for zinc, starch, and boric powders used unperfumed for their absorbent and antiseptic properties are prepared as follows:

No. 2559

Powdered boric acid	30
Zinc oxide	100
Rice starch	870
	1000

Alternative antiseptic materials are now preferred to boric acid and modern baby powers are, therefore, often similar in composition to deodorant powders. Zinc or magnesium stearate are used as adhesives and give a soft velvety feel to the finished product. Starch powder is used as a moisture absorbent. A suitable absorbent dusting powder or baby powder is made simply using zinc oxide, starch, and talc, as follows:

No. 2560

Zinc oxide	150
Rice starch	200
Talc	650
	1000

Alternative formulae are given:

No. 2561

Chlorhexidine diacetate	3
Zinc oxide	100

No. 2561 (*continued*)

Light magnesium carbonate	120
Talc	777
	1000

No. 2562

Chlorhexidine diacetate	3
Zinc stearate	50
Zinc oxide	100
Light magnesium carbonate	150
Talc	697
	1000

No. 2563

Cetrimide (B.P.)	2
Magnesium stearate	70
Light magnesium carbonate	120
Zinc oxide	150
Talc	658
	1000

For baby powder use a rose or light floral bouquet perfume at a concentration of from 0·2 to 0·3 per cent. Preferable the perfume should be one which has been tested and shown to be non-irritant and non-sensitizing. Perfumes used for baby powders or deodorant powders should also be tested for discolouration by exposing the final product to sunlight. This is because antiseptic materials of a phenolic nature react with certain perfumery materials to form colour complexes.

Talcum powders containing antiseptic materials are also used for prickly heat and fungus infections of the skin. Undecylenic acid and/or zinc undecylenate are also used specifically for treatment of athlete's foot and other similar infections. A suitable powder is prepared as follows:

No. 2564

Zinc stearate	20
Calcium carbonate	50
Menthol	2·5
Undecylenic acid	20
Zinc undecylenate	100
Talc	807·5
	1000·0

Procedure: Mix the menthol by grinding with part of the calcium carbonate and mix with the remaining ingredients, and finally sieve and mix.

To prepare an aerosol talc use a grade of micronized talc which will pass through a 300 or 325 mesh sieve or preferably one with a maximum particle size of 2 microns. Fine quality grades of metallic stearates are also required and a small proportion of a low viscosity fatty acid ester is included as a lubricant which helps to prevent the powder from clogging the value. A suitable formula is given:

No. 2565

Talc base:		
A	Micronized talc	910
	Magnesium myristate	60
B	*iso*-Propyl myristate	30
		1000

Perfume 1·0–2·0 per cent

Procedure: Mix the micronized talc and magnesium myristate (A) and prepare separately a mixture of perfume and *iso*-Propyl myristate, (B). Weigh A and B separately into the aerosol container. Container charge:

Container charge:	
Talc base	13
Propellant 12	87
	100

A high pressure propellant is used to reduce the cold wet effect experienced by using low pressure propellents.

Container: plain tinplate or lacquered or plain aluminium 2 piece or monobloc. Valve: special powder valve which allows solid particles to pass without clogging.

A deodorant talc suitable for use on the body or as a foot spray is made on similar lines by including a suitable bactericide.

No. 2566

Deodorant base:	
Micronized talc	905
Magnesium myristate	60

No. 2566 *(continued)*

Chlorhexidene diacetate	5
iso-Propyl myristate	30
	1000

Perfume	1·0–2·0 per cent

Procedure: Prepare a trituration of Chlorhexidene diacetate with part of the talc and add to the remaining powders and mix well. Proceed as for the previous formula.

Container: internally lacquered aluminium. Valve: special powder valve which allows solid particles to pass without clogging.

APPENDIX

TABLE I

Specific gravity of mixtures of alcohol and water

Specific gravity at 60°F (15·5°C)	*Absolute alcohol*		*Per-centage of proof spirit*	*Specific gravity at 60°F (15·5°C)*	*Absolute alcohol*		*Per-centage of proof spirit*
	By volume	*By weight*			*By volume*	*By weight*	
1000	0·00	0·00	0·00	965	30·34	24·97	53·04
999	0·66	0·53	1·16	964	31·18	25·68	54·51
998	1·34	1·07	2·33	963	31·99	26·37	55·93
997	2·02	1·61	3·52	962	32·79	27·06	57·33
996	2·72	2·17	4·73	961	33·56	27·73	58·68
995	3·42	2·73	5·98	960	34·33	28·39	60·03
994	4·14	3·31	7·24	959	35·06	29·03	61·32
993	4·88	3·90	8·51	958	35·79	29·66	62·60
992	5·63	4·51	9·82	957	36·50	30·28	63·85
991	6·40	5·13	11·16	956	37·20	30·90	65·09
990	7·18	5·76	12·53	955	37·89	31·50	66·29
989	7·98	6·41	13·94	954	38·57	32·09	67·48
988	8·80	7·08	15·38	953	39·22	32·67	68·62
987	9·65	7·76	16·85	952	39·85	33·25	69·76
986	10·51	8·46	18·34	951	40·50	33·81	70·87
985	11·40	9·18	19·87	950	41·13	34·37	71·98
984	12·29	9·91	21·44	949	41·74	34·92	73·05
983	13·20	10·65	23·02	948	42·35	35·46	74·12
982	14·13	11·42	24·66	947	42·95	36·00	75·17
981	15·08	12·20	26·32	946	43·54	36·54	76·21
980	16·04	12·99	27·00	945	44·13	37·07	77·24
979	17·02	13·80	29·70	944	44·71	37·60	78·26
978	18·00	14·61	31·42	943	45·28	38·12	79·26
977	18·99	15·43	33·15	942	45·85	38·64	80·26
976	19·98	16·25	34·87	941	46·40	39·15	81·23
975	20·97	17·08	36·61	940	46·95	39·65	82·19
974	21·96	17·90	38·35	939	47·50	40·15	83·15
973	22·94	18·72	40·06	938	48·04	40·65	84·10
972	23·91	19·53	41·77	937	48·57	41·15	85·04
971	24·85	20·34	43·47	936	49·10	41·64	85·97
970	25·83	21·14	45·14	935	49·63	42·13	86·89
969	26·77	21·93	46·77	934	50·15	42·62	87·81
968	27·69	22·71	48·38	933	50·67	43·11	88·71
967	28·69	23·48	49·98	932	51·18	43·59	89·61
966	29·48	24·23	51·53	931	51·68	44·06	90·49

TABLE I—*continued*

Specific gravity of mixtures of alcohol and water

Specific gravity at 60°F (15·5°C)	*Absolute alcohol*		*Per-centage of proof spirit*	*Specific gravity at 60°F (15·5°C)*	*Absolute alcohol*		*Per-centage of proof spirit*
	By volume	*By weight*			*By volume*	*By weight*	
930	52·18	44·53	91·36	897	67·08	59·37	117·54
929	52·67	45 00	92·93	896	67·50	59·80	118·26
928	53·16	45·47	93·09	895	67·92	60·23	118·98
927	53·65	45·94	93·95	894	68·33	60·66	119·70
926	54·14	46·40	94·80	893	68·74	61·09	120·42
925	54·62	46·87	95·65	892	69·14	61·52	121·14
924	55·10	47·33	96·49	891	69·55	61·95	121·85
923	55·58	47·79	97·33	890	69·95	62·38	122·56
922	56·05	48·25	98·16	889	70·35	62·81	123·27
921	56·52	48·71	98·98	888	70·75	63·24	123·97
920	56·99	49·17	99·80	887	71·15	63·67	124·06
91976	57·10	49·28	100·00	886	71·55	64·10	125·37
				885	71·90	64·53	126·07
919	57·46	49·63	100·62	884	72·34	64·96	126·77
918	57·92	50·08	101·43	883	72·24	65·39	127·46
917	58·38	50·53	102·24	882	73·13	65·81	128·14
916	58·83	50·98	103·05	881	73·52	66·24	128·82
915	59·29	51·43	103·84	880	73·91	66·66	128·50
914	59·74	51·88	104·63	879	74·30	67·09	130·18
913	60·19	52·33	105·42	878	74·68	67·51	130·86
912	60·63	52·77	106·20	877	75·06	67·93	131·53
911	61·07	53·21	106·97	876	75·44	68·35	132·19
910	61·51	53·65	107·74	875	75·82	68·77	132·86
909	61·95	54·10	108·52	874	76·19	69·19	133·53
908	62·39	54·54	109·29	873	76·57	69·62	134·19
907	62·83	54·98	110·06	872	76·94	70·04	134·84
906	63·26	55·42	110·82	871	77·32	70·46	135·50
905	63·70	55·87	111·59	870	77·69	70·88	136·16
904	64·13	56·31	112·35	869	78·06	71·30	136·81
903	64·56	56·75	113·10	868	78·43	71·72	137·46
902	64·98	57·18	113·84	867	78·80	72·14	138·10
901	65·41	57·62	114·59	866	79·17	72·55	138·74
900	65·83	58·06	115·33	865	79·53	72·97	139·38
899	66·25	58·50	116·07	864	79·89	73·39	140·02
898	66·67	58·93	116·81	863	80·25	73·81	140·65

TABLE I—*continued*

Specific gravity of mixtures of alcohol and water

Specific gravity at 60°F (15·5°C)	*Absolute alcohol*		*Percentage of proof spirit*	*Specific gravity at 60°F (15·5°C)*	*Absolute alcohol*		*Percentage of proof spirit*
	By volume	*By weight*			*By volume*	*By weight*	
862	80·61	74·22	141·28	827	91·98	88·27	161·26
861	80·97	74·64	141·91	826	92·26	88·65	161·76
860	81·32	75·05	142·54	825	92·55	89·03	162·26
859	81·68	75·47	143·16	824	92·83	89·41	162·75
858	82·03	75·88	143·78	823	93·11	89·79	162·24
857	83·38	76·30	144·40	822	93·38	90·16	163·72
856	83·73	76·71	145·01	821	93·65	90·53	164·20
855	83·08	77·12	145·62	820	93·92	99·90	164·67
854	83·42	77·53	146·23	819	94·19	91·27	165·14
853	83·77	77·94	146·83	818	94·45	91·63	165·60
852	84·11	78·35	147·43	817	94·71	92·00	166·06
851	84·44	78·76	148·03	816	94·97	92·36	166·51
850	84·78	79·17	148·62	815	95·22	92·72	166·96
849	85·12	79·58	149·21	814	95·47	93·08	167·41
848	85·46	79·98	149·80	813	95·72	93·44	167·86
847	85·80	80·39	150·39	812	95·97	93·80	168·28
846	86·12	80·79	150·97	811	96·21	94·15	168·71
845	86·44	81·20	151·55	810	96·45	94·50	169·13
844	86·77	81·60	152·12	809	96·69	94·85	169·55
843	87·09	82·00	152·68	808	96·93	95·20	169·96
842	87·42	82·40	153·25	807	97·16	92·55	170·37
841	87·74	82·80	153·81	806	97·39	95·89	170·77
840	88·06	83·20	154·37	805	97·62	96·23	171·17
839	88·37	83·60	154·92	804	97·84	96·57	171·56
838	88·68	83·99	155·47	803	98·06	96·91	171·95
837	88·99	84·39	156·02	802	98·28	97·25	172·23
836	89·30	84·78	156·56	801	98·49	97·59	172·71
835	89·61	85·17	157·10	800	98·70	97·91	173·07
834	89·91	85·56	157·63	799	98·91	98·24	173·44
833	90·22	85·95	158·16	798	99·12	98·57	173·80
832	90·52	86·34	158·68	797	99·32	98·90	174·16
831	90·82	86·73	159·21	796	99·52	99·22	174·52
830	91·11	87·11	159·73	795	99·72	99·55	174·87
829	91·40	87·50	160·24	794	99·92	99·87	175·21
828	91·69	87·88	160·75	79359	100·00	100·00	175·35

TABLE II

Dilution of alcohol by volume with distilled water

Percentage strength of alcohol required by volume	*Add to 1000 of alcohol at*								
	90	*85*	*80*	*75*	*70*	*65*	*60*	*55*	*50*
	Per cent by volume								
85	66								
80	138	69							
75	219	145	72						
70	311	231	153	77					
65	414	330	247	164	82				
60	537	445	354	265	176	88			
55	679	579	481	383	286	190	95		
50	847	739	630	524	417	313	205	104	
45	1053	933	814	695	578	461	345	229	114
40	1308	1173	1040	907	776	645	514	385	256
35	1633	1480	1329	1178	1029	880	700	583	436
30	2062	1886	1711	1535	1363	1189	1017	845	675
25	2661	2452	2243	2036	1828	1622	1417	1212	1007
20	3558	3298	3040	2783	2526	2270	2014	1760	1506
15	5053	4710	4369	4028	3689	3349	3011	2673	2336
10	8045	7537	7029	6522	6016	5511	5005	4502	2999

Examples: To convert 90 per cent to 45 per cent add to 1 litre 1053 cc Aqua Destil.
To convert 75 per cent to 20 per cent add to 1 litre 2783 cc Aqua Destil.

TABLE III

Dilution of alcohol by weight with distilled water

Percentage strength of alcohol used	*To produce 1000 of alcohol at*				
	50	60	70	80	90
	Per cent by weight				
96	453	555	665	783	913
95	460	564	676	796	927
94	467	573	686	808	942
93	474	582	697	820	956
92	481	590	707	832	970
91	489	599	718	845	985
90	496	609	728	858	
89	504	618	740	871	
88	511	627	752	884	
87	519	637	763	898	
86	527	646	774	912	
85	535	656	786	926	
84	543	667	798	940	
83	552	677	811	955	
82	560	687	823	969	823
81	569	698	836	984	
80	578	709	849		
79	587	720	863		
78	597	732	877		
77	606	744	891		
76	616	756	905		
75	626	768	920		
74	636	781	935		
73	647	794	951		
72	658	807	967		
71	669	821	983		
70	681	835			
69	692	849			
68	705	864			
67	717	880			
66	730	896			
65	743	911			

TABLE III—*continued*

Dilution of alcohol by weight with distilled water

Percentage strength of alcohol used	To produce 1000 of alcohol at				
	50	60	70	80	90
	Per cent by weight				
64	756	928			
63	770	946			
62	785	963			
61	800	981			
60	815				
59	831				
58	847				
57	864				
56	881				
55	901				
54	918				
53	938				
52	958				

Examples: To make a kilo of 90 per cent by weight take

(1) 913 grams of 96 per cent alcohol and add distilled water to make 1000 grams, or

(2) 956 grams of 93 per cent.

Alcohol table

(to required weight and volume)

Showing the strength of alcohol (percentage absolute) with its specific gravity, also the weight or volume required to make one kilogram or one litre of another strength by the addition of distilled water at 15·5°C or 60°F

			Quantity of alcohol of known strength required to make 1000									
	Specific gravity by		*50 per cent by*		*60 per cent by*		*70 per cent by*		*80 per cent by*		*90 per cent by*	
Strength of absolute alcohol	*Weight*	*Volume*	*Weight grams*	*Volume cc*	*Weight grams*	*Volume cc*	*Weight grams*	*Volume cc*	*Weight grams*	*Volume cc*	*Weight grams*	*Volume cc*
per cent												
96	0·8065	0·8125	453	520	555	625	665	730	783	833	915	833
93	0·8149	0·8237	474	537	582	645	697	752	820	860	956	967
90	0·8229	8·8339	496	555	609	666	728	777	858	888		
85	0·8359	0·8496	535	588	656	705	786	823	926	941		
80	0·8484	0·8639	578	602	709	750	849	785				
75	0·8605	0·8773	626	666	678	800	920	933				
70	0·8724	0·8900	681	701	835	852						
65	0·8842	0·9021	743	769	911	923						
60	0·8958	0·9134	815	833								
55	0·9072	0·9242	901	981								

Index